Springer Proceedings in Mathematics & Statistics

Volume 192

Springer Proceedings in Mathematics & Statistics

This book series features volumes composed of selected contributions from workshops and conferences in all areas of current research in mathematics and statistics, including operation research and optimization. In addition to an overall evaluation of the interest, scientific quality, and timeliness of each proposal at the hands of the publisher, individual contributions are all refereed to the high quality standards of leading journals in the field. Thus, this series provides the research community with well-edited, authoritative reports on developments in the most exciting areas of mathematical and statistical research today.

More information about this series at http://www.springer.com/series/10533

Natália Bebiano
Editor

Applied and Computational Matrix Analysis

MAT-TRIAD, Coimbra, Portugal, September 2015
Selected, Revised Contributions

Springer

Editor
Natália Bebiano
Department of Mathematics
University of Coimbra
Coimbra
Portugal

ISSN 2194-1009 ISSN 2194-1017 (electronic)
Springer Proceedings in Mathematics & Statistics
ISBN 978-3-319-49982-6 ISBN 978-3-319-49984-0 (eBook)
DOI 10.1007/978-3-319-49984-0

Library of Congress Control Number: 2016957843

Mathematics Subject Classification (2010): 15A86, 15B05, 05C50, 18F25, 16G99, 47A56, 47A12, 94B10, 62J12, 15A22, 93B40, 47A75

Printed on acid-free paper

This Springer imprint is published by Springer Nature
The registered company is Springer International Publishing AG
The registered company address is: Gewerbestrasse 11, 6330 Cham, Switzerland

Preface

The MAT-TRIAD 2015, sixth in the series of international conferences on matrix analysis and its applications, was held at the Department of Mathematics, University of Coimbra, Portugal, during 7–11 September. Following the tradition of its predecessors, this meeting gathered researchers around topics in matrix theory and its role in theoretical and numerical linear algebra, numerical and functional analysis, graph theory and combinatorics, coding theory and statistical models with matrix structure. A total of 170 participants from 39 countries, from Europe, North and South America, Africa and Asia, have attended the conference in the University of Coimbra, UNESCO World Cultural Heritage since 2013. The audience was multidisciplinary allowing the participants to exchange diversified ideas and to show the wide applicability of different methods. There were two kinds of lectures: invited talks of one hour presented by distinguished experts and half an hour contributions. The winners of the Young Scientists Award of MAT-TRIAD 2013 presented invited talks. The conference included two lectures specially dedicated to young participants.

MAT-TRIAD 2015 was sponsored by the International Linear Algebra Society (ILAS), Department of Mathematics, University of Coimbra (DMUC), Center of Mathematics, University of Coimbra (CMUC), Center for R&D in Mathematics (IDMA), Center for Mathematical Analysis, Geometry and Dynamical Systems (CAMGSD), Center for Functional Analysis, Linear Structures and Applications (CEAFEL), Polytechnic Institute of Tomar (IPT), Fundação para a Ciência e Tecnologia (FCT), Programa Operacional Factores de Competitividade (COMPETE), Quadro de Referência Estratégica Regional (QREN), Fundo Europeu de Desenvolvimento Regional—União Europeia.

The Conference Scientific Committee consisted of Tomasz Szulk (Poland)—Chair, Natália Bebiano (Portugal), Ljiljana Cvetkovicć (Serbia), Heike Faßbender (Germany) and Simo Putanen (Finland). The Organizing Committee was constituted by Natália Bebiano—Chair, Francisco Carvalho, Susana Furtado, Celeste Gouveia, Rute Lemos and Ana Nata, all from Portugal.

We would like to publicly acknowledge the financial support of the sponsors, as well as the hospitality of the Department of Mathematics of the University of

Coimbra, and the strong encouragement of its Center of Mathematics. We are also very grateful for the secretarial help of Dra. Rute Andrade.

Selected papers of MAT-TRIAD 2015 are presented in the volume *Applied and Computational Matrix Analysis* in the series Proceedings of Mathematics & Statistics published by Springer Verlag. With the publication of these proceedings, we hope that a wider mathematical audience will benefit from the conference research achievements and new contributions to the field of matrix theory and its applications.

More details of the program and the book of abstracts can be found at http://www.mattriad.ipt.pt.

Coimbra, Portugal
August 2016

Natália Bebiano

Acknowledgements

It is a pleasure to thank Doctor Ana Nata for her valuable support in the preparation of this book, done with competence and with good cheer. Without her help, this project might have been postponed indefinitely.

Contents

Contributors

P. Almeida Department of Mathematics, CIDMA - Center for Research and Development in Mathematics and Applications, University of Aveiro, Aveiro, Portugal

Amir Amiraslani STEM Department, University of Hawaii-Maui College, Kahului, HI, USA

Prashant Batra Institute for Reliable Computing, Hamburg University of Technology, Hamburg, Germany

Natália Bebiano Department of Mathematics, CMUC, University of Coimbra, Coimbra, Portugal

M. Bendaoud Moulay Ismail University, ENSAM, Meknès, Al Mansour, Morocco

Durmuş Bozkurt Science Faculty Department of Mathematic, Selcuk University, Konya, Turkey

Domingos Moreira Cardoso Center for Research and Development in Mathematics and Applications, Department of Mathematics, University of Aveiro, Aveiro, Portugal

Carlos A. Coelho Centro de Matemática e Aplicações (CMA/FCT-UNL), Caparica, Portugal; Departamento de Matemática (DM/FCT-UNL), Faculdade de Ciências e Tecnologia, Universidade Nova de Lisboa, Caparica, Portugal

Mohammed El Oued High Institute of Maths and Computer Science of Monastir, Monastir, Tunisia

Fatemeh Esmaeili Department of Mathematics, CMUC, University of Coimbra, Coimbra, Portugal

Heike Faßbender Institut Computational Mathematics, AG Numerik, Technische Universität Braunschweig, Braunschweig, Germany

M. Fonseca UNL, Lisbon, Portugal

Carlos Gamas Department of Mathematics, University of Coimbra, Coimbra, Portugal

Luís M. Grilo Unidade Departamental de Matemática, Instituto Politécnico de Tomar, Tomar, Portugal; Centro de Matemática e Aplicações (CMA/FCT-UNL), Caparica, Portugal

Milan Hladík Faculty of Mathematics and Physics, Department of Applied Mathematics, Charles University, Prague, Czech Republic

Jaroslav Horáček Faculty of Mathematics and Physics, Department of Applied Mathematics, Charles University, Prague, Czech Republic

M. Jabbar Moulay Ismail University, ENSAM, Meknès, Al Mansour, Morocco

Charles R. Johnson Department of Mathematics, College of William and Mary, Williamsburg, VA, USA

Radosław Kala Department of Mathematical and Statistical Methods, Poznań University of Life Sciences, Poznań, Poland

Alexander Kovačec Department of Mathematics, University of Coimbra, Coimbra, Portugal

Hatice Kübra Duru Science Faculty Department of Mathematic, Selcuk University, Konya, Turkey

Ricardo Mamede Department of Mathematics, CMUC, University of Coimbra, Coimbra, Portugal

Augustyn Markiewicz Department of Mathematical and Statistical Methods, Poznań University of Life Sciences, Poznań, Poland

Sérgio Mendes ISCTE - Lisbon University Institute, Lisbon, Portugal

J. Mexia UNL, Lisbon, Portugal

Diego Napp Department of Mathematics, CIDMA - Center for Research and Development in Mathematics and Applications, University of Aveiro, Aveiro, Portugal

Ana Nata Department of Mathematics, CMUC, Polytechnic Institute of Tomar, Tomar, Portugal

Jolanta Pielaszkiewicz Linköping University, Linköping, Sweden; Linnaeus University, Växjö, Sweden

Sofia J. Pinheiro Center for Research and Development in Mathematics and Applications, Department of Mathematics, University of Aveiro, Aveiro, Portugal

Raquel Pinto Department of Mathematics, CIDMA - Center for Research and Development in Mathematics and Applications, University of Aveiro, Aveiro, Portugal

João P. da Providência Department of Physics, University of Beira Interior, Covilhã, Portugal

João da Providência Department of Physics, CFisUC, University of Coimbra, Coimbra, Portugal

Simo Puntanen School of Information Sciences, University of Tampere, Tampere, Finland

Dietrich von Rosen Linköping University, Linköping, Sweden; Swedish University of Agricultural Sciences, Uppsala, Sweden

Manideepa Saha Department of Mathematics, National Institute of Technology Meghalaya, Shillong, Meghalaya, India

M. Sarih Faculty of Sciences, Meknès, Morocco

Nikta Shayanfar Institut Computational Mathematics, AG Numerik, Technische Universität Braunschweig, Braunschweig, Germany

A. Silva UniCV, Praia, Cabo Verde; UNL, Lisbon, Portugal

Martin Singull Linköping University, Linköping, Sweden

Marisa Toste CIDMA - Center for Research and Development in Mathematics and Applications, Superior School of Technologies and Management of Oliveira Do Hospital, Polytechnic Institute of Coimbra, Coimbra, Portugal

Kenji Toyonaga Department of Integrated Arts and Science, Kitakyushu National College of Technology, Kitakyushu, Japan

Richard Uhrig Department of Mathematics, College of William and Mary, Williamsburg, VA, USA

Michal Černý Faculty of Computer Science and Statistics, University of Economics, Prague, Czech Republic

Birkhoff Polynomial Basis

Amir Amiraslani, Heike Faßbender and Nikta Shayanfar

Abstract The Birkhoff interpolation problem is an extension of the well-known Lagrange and Hermite interpolation problems. We propose a new set of basis polynomials for representing the Birkhoff interpolation polynomial. The proposed basis extends the definition of the Newton basis for non-distinct interpolation nodes. This approach allows to determine the Birkhoff interpolation polynomial via a special linear system of equations. When applied to the special cases of Taylor, Lagrange and Hermite interpolations, this approach reduces to the well-known solutions of these problems expressed in the Newton basis. A number of examples are studied.

Keywords Polynomial bases · Polynomial interpolation · Differentiation matrix · Birkhoff matrix

1 Introduction

The following general interpolation problem [18, 24], known as the Birkhoff interpolation problem, will be considered:

> Let $\{z_i\}_{i=0}^{k}$ be a set of distinct interpolation nodes and $\{f_{i,j}\}$ be a set of $n+1$ data values where $n \geqslant k$ and $f_{i,j}$ is seen as the jth derivative of a function f at node z_i, that is, $f_{i,j} = f^{(j)}(z_i)$.
>
> $$\text{Find} \quad P \in \mathbb{P}_n \quad \text{such that} \quad P^{(j)}(z_i) = f_{i,j},$$
>
> where $\mathbb{P}_n$ is the set of complex polynomials of degree at most n.

A. Amiraslani
STEM Department, University of Hawaii-Maui College, Kahului, HI 96732, USA
e-mail: aamirasl@hawaii.edu

H. Faßbender · N. Shayanfar (✉)
Institut Computational Mathematics, AG Numerik, Technische Universität Braunschweig, 38092 Braunschweig, Germany
e-mail: n.shayanfar@tu-braunschweig.de; nikta.shayanfar@gmail.com

H. Faßbender
e-mail: h.fassbender@tu-braunschweig.de

N. Bebiano (ed.), *Applied and Computational Matrix Analysis*,
Springer Proceedings in Mathematics & Statistics 192,
DOI 10.1007/978-3-319-49984-0_1

Note that it is not required that at each node z_i a complete sequence of derivatives $f_{i,j} = f^{(j)}(z_i)$ for $j = 0, 1, \ldots, t_j$ for some $t_j \in \mathbb{N}_0$ is given. It is sufficient that some of this information is given; that is, derivatives of f at z_i may be given without specifying all lower order derivatives (or $f(z_i)$ itself). The total number of derivatives given at a node z_i is referred to as the confluency s_i of the node z_i, $i = 0, \ldots, k$, using the standard notation $f^{(0)}(z) = f(z)$. However, for each z_i at least one $f_{i,j}$ must be given. Hence, more precisely, for the nodes z_i, $i = 0, \ldots, k$, with the confluency s_i, we are looking for a polynomial of degree $n = s_0 + \cdots + s_k - 1$, such that it satisfies the interpolation conditions $P^{(j)}(z_i) = f_{i,j}$.

Example 1 Suppose that the following information about the function $f(x)$ at the distinct nodes z_0, z_1, z_2 is given:

$$f(z_0) = f_{0,0}, \quad f''(z_0) = f_{0,2}, \quad f'(z_1) = f_{1,1}, \quad f(z_2) = f_{2,0}.$$

The corresponding sequence of s_i is as follows:

$$s_0 = 2, \quad s_1 = 1, \quad s_2 = 1.$$

We seek the polynomial $P(x) \in \mathbb{P}_3$ satisfying the interpolation data:

$$P(z_0) = f_{0,0}, \quad P''(z_0) = f_{0,2}, \quad P'(z_1) = f_{1,1}, \quad P(z_2) = f_{2,0}.$$

The more well-known Lagrange and Hermite interpolation problems are special cases of the Birkhoff interpolation problem. Results on the existence and uniqueness of a solution of these interpolation problems are given in [7] which is an easily readable account of several different interpolation schemes. For ease of further reference, we briefly review both interpolation problems.

Definition 1 (*Lagrange interpolation problem*)
Given $n + 1$ distinct nodes $z_i, i = 0, \ldots, n$ and the associated functional values $f_i, i = 0, \ldots, n$ of the function $f(x)$ at these points, we seek a polynomial $P(x) \in \mathbb{P}_n$ satisfying

$$P(z_i) = f_i, \qquad i = 0, \ldots, n.$$

It is immediate that $s_i = 1$ for $i = 0, \ldots, n$.

The Hermite interpolation matches an unknown function not only at observed values (z_i, f_i), but also at observed values of consecutive sequences of derivatives at z_i. That is, at a node z_i not only f_i, but also the sequential derivatives of up to order $s_i - 1$, that is $f^{(j)}(z_i)$, $j = 0, \ldots, s_i - 1$, are given. Our definition of the Hermite interpolation problem makes use of repeated nodes (as needed, e.g., for determining the interpolation polynomial via divided differences).

Definition 2 (*Hermite interpolation problem*)
Consider $n + 1$ interpolation nodes $\underbrace{z_0, \ldots, z_0}_{s_0 \ times}, \ldots, \underbrace{z_k, \ldots, z_k}_{s_k \ times}$, where

$$\sum_{i=0}^{k} s_i = n + 1.$$

Let us assume that we are given $n + 1$ specified values $f^{(j)}(z_i) := f_{i,j}$ for some function $f(x)$ where $i = 0, \ldots, k, \ \ j = 0, \ldots, s_i - 1$. The Hermite interpolation problem is to find a polynomial $P(x) \in \mathbb{P}_n$, that satisfies

$$P^{(j)}(z_i) = f_{i,j}, \qquad i = 0, \ldots, k, \quad j = 0, \ldots, s_i - 1. \tag{1}$$

Clearly, $s_i, i = 0, \ldots, k$, gives the total number of the derivatives given at the node $z_i, i = 0, \ldots, k$.

Note that the orders of the derivatives in the Hermite interpolation form an unbroken sequence, and if some (or all) of the sequences are broken, we have the Birkhoff interpolation. In fact, the Birkhoff interpolation generalizes the Hermite one, in the following sense: In the Hermite interpolation problem, for each node $z_i, i = 0, \ldots, k$, all the functional values for $f^{(0)}(z_i), f^{(1)}(z_i), \ldots, f^{(s_i-1)}(z_i)$ have to be given. The Birkhoff interpolation problem does not require all derivatives to be given. It is possible to consider derivatives without specifying (all) lower derivatives. However, we still denote the number of the given derivatives at node z_i, by $s_i, i = 0, \ldots, k$.

A special case of the Hermite interpolation problem is the Taylor interpolation problem in which just one node z_0 and an unbroken sequence of derivatives at that node is given.

Definition 3 (*Taylor interpolation problem*)
Consider one interpolation node z_0. Assume that $n + 1$ specified values $f^{(j)}(z_0) := f_{0,j}, j = 0, \ldots, n$ for some function $f(x)$ are given. The Taylor interpolation problem is to find a polynomial $P(x) \in \mathbb{P}_n$, that satisfies

$$P^{(j)}(z_0) = f_{0,j}, \qquad j = 0, \ldots, n.$$

Obviously, $s_0 = n + 1$ and the usual Taylor expansion polynomial

$$P(x) = \sum_{j=0}^{n} \frac{f^{(j)}(z_0)}{j!}(x - z_0)^j, \tag{2}$$

solves the Taylor interpolation problem.

As the Lagrange and Taylor interpolation problems are special cases of the Hermite one, it suffices to state that there exists a unique solution to the Hermite interpolation problem.

Theorem 1 ([7, P. 24]) *Consider the Hermite interpolation problem defined in Definition 2. There exists a unique polynomial $P(x) \in \mathbb{P}_n$ such that the interpolation conditions* (1) *are held.*

In contrast to the Taylor, Lagrange and Hermite interpolation problems, the additional freedom in the Birkhoff interpolation problem implies that the interpolation problem not necessarily have a solution for every choice of data values. In this paper, we assume that the Birkhoff interpolation problem considered does have a solution, see e.g. [3, 17, 18, 24] for a discussion of this important aspect. The purpose of this paper is to introduce a new approach for solving the Birkhoff interpolation problem. We observe that the interpolating polynomial can be represented essentially via the well-known Newton basis. The Newton polynomials are usually defined for distinct nodes. Here we will consider this set of polynomials for non-distinct nodes and call the so obtained set of polynomials $\{\mathscr{B}_k(x)\}_{k=0}^n$ the Birkhoff polynomials; they form a polynomial basis of the space $\mathbb{P}_n$ of complex polynomials of degree at most n. Our main goal is to show that, in the presence of confluent nodes, the solution of the Birkhoff interpolation problem can be computed from an easy to set up linear system. The resulting interpolating polynomial is expressed in the Birkhoff basis. When applied to the special cases of Taylor, Lagrange and Hermite interpolations, this approach reduces to the well-known solutions of these problems expressed in Newton bases.

The Birkhoff interpolation problem has numerous applications. An equivalence between the Birkhoff interpolation problem and a sequence of problems from linear optimal control is studied in [29]. The Birkhoff interpolant may be useful in the development of numerical solutions of ordinary differential equations with defect control [16]. Moreover, they may arise when using collocation to solve two-point boundary value problems [14]. Another problem that has been studied and was shown to be related to Birkhoff interpolation is the study of optimal digital to analog conversion using linear system theory [29].

In 1906, George David Birkhoff introduced the Birkhoff interpolation problem [4], that has been studied in the literature since then. Later in 1931, the problem was restated by Polya [20], as a differential equation in which a combination of initial and terminal values suffice to construct a unique solution. In [26, 27], 15 open questions on Hermite-Birkhoff interpolation problems were stated. Over 20 years later some of these questions have been answered in [24]. A great deal of research focuses on the Birkhoff interpolation problem for special nodes or uniform interpolation conditions, see [8–12, 21] among others. A solution to the Birkhoff interpolation problem in a barycentric form via a contour integral formula has been obtained in [5]. Particularly, the specific case of prescribed function values and only first derivative values is discussed in [13], while applying quantifier elimination to the Birkhoff interpolation problem is presented in [15].

The organization of the paper is as follows: In Sect. 2, we recall the precise statement of the problem via an incidence matrix as in the classical theory. Next, in Sect. 3, we introduce the Birkhoff matrix which will replace the usual incidence matrix in the statement of the Birkhoff interpolation problem in our further discussion. Moreover, the notion of a differentiation matrix will be reviewed. Section 4 presents our new approach for solving the Birkhoff interpolation problem via a linear system of equations. The resulting interpolating polynomial is expressed in the generalized Newton basis, called Birkhoff basis. In Sect. 5, we discuss the Taylor, Lagrange, and Hermite interpolation problems in the context of our new approach. Some illustrative examples are provided in Sect. 6.

2 Statement of the Birkhoff Interpolation Problem via an Incidence Matrix

The Birkhoff interpolation problem can be characterized with the help of incidence matrices. In general, a $(k+1) \times (t+1)$ matrix $\mathbf{J} = [J_{i,j}]_{i=0,j=0}^{k,t}$ is an *incidence matrix* if its entries are either 0 or 1. Here we let t be the highest order of the given derivatives in the interpolation problem. Obviously, $n+1 \leq (t+1)(k+1)$, as for fixed k and fixed t at each of the $k+1$ nodes at most $t+1$ functional values can be given, while k and n are as given in our initial problem statement.

Definition 4 (*Birkhoff incidence matrix*)
A $(k+1) \times (t+1)$ matrix $\mathbf{J} = [J_{i,j}]_{i=0,j=0}^{k,t}$ is called a Birkhoff incidence matrix for a specific interpolation problem if $J_{i,j} = 1$ in case $f_{i,j}$ is specified and $J_{i,j} = 0$ otherwise.

Note that the indices of the Birkhoff incidence matrix begin with 0, since the nodes and derivatives, z_i, $f_{i,j}$, start with the index 0.

Example 2 Consider the Birkhoff interpolation problem given in Example 1. The associate Birkhoff incidence matrix is given by

$$\mathbf{J} = \begin{pmatrix} 1 & 0 & 1 \\ 0 & 1 & 0 \\ 1 & 0 & 0 \end{pmatrix}.$$

In the Birkhoff incidence matrix $\mathbf{J}$, the confluency s_i, $i = 0, \ldots, n$ is the sum of the elements in the specific row of $\mathbf{J}$ corresponding to z_i, and the sum of all elements of $\mathbf{J}$ equals to $n+1$.

In the literature, often the Birkhoff incidence matrix $\mathbf{J}$ is defined as a $(k+1) \times (n+1)$ matrix which has exactly $n+1$ ones, while the Birkhoff incidence matrix in Definition 4 is of size $(k+1) \times (t+1)$. Following [19], $t \leqslant n$, one could easily extend our Birkhoff incidence matrix to one of size $(k+1) \times (n+1)$ by adding sufficient columns with all zero entries. In [29], the Birkhoff incidence matrix needs to be square which is not required here. To sum up, our Birkhoff incidence matrix is of the smallest possible size for the information encoded.

Now we rewrite the initial statement of the problem using the Birkhoff incidence matrix:

Definition 5 (*Birkhoff interpolation problem*)
Let $\{z_i\}_{i=0}^k$ be a set of distinct interpolation nodes, $\{f_{i,j}\}$ be a set of $n+1$ data values and $n \geqslant k$. Let t be the highest order of the given derivatives and $\mathbf{J}$ be the corresponding $(k+1) \times (t+1)$ Birkhoff incidence matrix. The Birkhoff interpolation problem is about finding a polynomial $P(x) \in \mathbb{P}_n$ which satisfies the conditions

$$P^{(j)}(z_i) = f_{i,j}, \quad \text{if} \quad J_{i,j} = 1 \quad \text{for} \quad i = 0, \ldots, k, \ j = 0, \ldots, t. \tag{3}$$

Special cases of the Birkhoff interpolation problem can be identified from the associated Birkhoff incidence matrix $\mathbf{J}$ as follows:

- **Lagrange interpolation**: The Lagrange interpolation problem is given if $t = 0$, $n = k$, and $\mathbf{J}$ is a $(k+1) \times 1$ matrix in which $J_{i,0} = 1$ for every i, $i = 0, \ldots, k$; that is, $\mathbf{J} = (1, \ldots, 1)^T \in \mathbb{R}^{k+1}$.
- **Taylor interpolation**: The Taylor interpolation problem is given if $k = 0$, $t = n = s_0 - 1$, and $\mathbf{J}$ is a $1 \times (t+1)$ matrix in which $J_{0,j} = 1$ for every j, $j = 0, \ldots, t$; that is, $\mathbf{J} = (1, \ldots, 1) \in \mathbb{R}^{1\times(t+1)}$.
- **Hermite interpolation**: The Hermite interpolation problem is given if $t - 1 = \max_{i=0,\ldots,k} s_i$, $\mathbf{J}$ is a $(k+1) \times (t+1)$ Birkhoff incidence matrix in which each row starts with a one in the first column and there does not exist any zero in the sequence of consecutive ones in each row. Simply put, $J_{i,0} = 1$ and for $j = 1, \ldots, t$, $J_{i,j} = 1$ implies that $J_{i,k} = 1$, for every $k \leq j$.

3 Two Important Matrices

In this section, we introduce two important types of matrices, the Birkhoff and the differentiation matrices, which will be of use in order to state our main result.

3.1 *Birkhoff Matrix*

Here, we define a new matrix called Birkhoff matrix which gives similar information as the more compressed Birkhoff incidence matrix. Recall that we have $k+1$ distinct nodes z_i, each with s_i functional values $f^{(j)}(z_i) = f_{i,j}$, $j = 0, \ldots, t$, where $\mathbf{J}_{i,j} = 1$. Moreover, $n = s_0 + \cdots s_k - 1$ and t is the highest order of all given derivatives.

Definition 6 (*Birkhoff matrix*)
Define the $(n+1) \times (t+1)(k+1)$ block diagonal matrix

$$\mathbf{B} = \begin{pmatrix} \mathbf{B}_0 & & & \\ & \mathbf{B}_1 & & \\ & & \ddots & \\ & & & \mathbf{B}_k \end{pmatrix},$$

in which the block $\mathbf{B}_i, i = 0, \ldots, k$ of size $s_i \times (t+1)$ is associated with the node z_i. Let $f_{i,j_1}, \ldots, f_{i,j_{s_i}}$ be the given data for the node z_i where $j_k < j_\ell$ for $k < \ell$. Then $\mathbf{B}_i$ is given by

$$\begin{aligned} e_1^T \mathbf{B}_i &= e_{j_1+1}^T, \\ e_2^T \mathbf{B}_i &= e_{j_2+1}^T, \\ &\vdots \\ e_{s_i}^T \mathbf{B}_i &= e_{j_{s_i}+1}^T, \end{aligned}$$

where e_p^T is the pth row of the identity matrix $\mathbf{I}_{t+1}$ of size $t+1$.

A more intuitive way on how to construct $\mathbf{B}_i, i = 0, \ldots, k$ is as follows: we start with $\mathbf{I}_{t+1}$. The r-th row $r = 1, \ldots, t+1$ of the identity matrix is associated with the $(r-1)$st derivative at $z_i, i = 0, \ldots, k$. Hence, the rth row e_r^T appears in $\mathbf{B}_i$ for every $f_{i,r-1}, r = 1, \ldots, t+1$ with $\mathbf{J}_{i,r-1} = 1$, in other words, it is given as the interpolation condition (3). We simply eliminate the rows of the identity matrix where no information is given at z_i. The Maple code in Table 1 describes how to obtain the Birkhoff matrix $\mathbf{B}$ from the Birkhoff incidence matrix $\mathbf{J}$.

Table 1 Construction of Birkhoff matrix from Birkhoff incidence matrix

```
 for i from 0 to k do
   B[i]:=IdentityMatrix(t+1):
   u:=0:
   for j from 0 to t do
      if J[i+1,j+1]=0 then
B[i]:=DeleteRow(B[i],j-u+1):
       u:=u+1:
    else B[i]:=B[i]
    end if
  end do:
end do:
i:='i': BB:=B[0]:
for i from 1 to k do
    BB:=DiagonalMatrix([BB,B[i]]):
end do:
B:=convert(BB, Matrix);
```

Comparing the Birkhoff incidence and the Birkhoff matrices, we can see that while each row $i, i = 0, \ldots, k$ of the Birkhoff incidence matrix $\mathbf{J}$ contains exactly s_i entries 1, each row of the Birkhoff matrix $\mathbf{B}$ has exactly one 1 and each block $\mathbf{B}_i, i = 0, \ldots, k$ has s_i entries 1. In fact, each interpolation condition (3) generates one 1 in the Birkhoff incidence matrix and one row in the Birkhoff matrix. More precisely, the ith row, $i = 0, \ldots, k$ of the Birkhoff incidence matrix $\mathbf{J}$ corresponds to block $\mathbf{B}_i$ of the Birkhoff matrix.

Proposition 1 *There exists an element* 1 *in the* $(j + 1)$*th column of* $\mathbf{B}_i$, *if and only if* $J_{i,j} = 1, \; i = 0, \ldots, k, \; j = 0, \ldots, t.$

Example 3 For the problem in Example 1, we have $k = 2, n = 3$ and $t = 2$. The corresponding Birkhoff matrix corresponding is the 4×9 block matrix $\mathbf{B} = Diag$ $[\mathbf{B}_0, \mathbf{B}_1, \mathbf{B}_2]$ with the blocks

$$\begin{aligned}
\mathbf{B}_0 &= \begin{pmatrix} 1 & 0 & 0 \\ 0 & 0 & 1 \end{pmatrix}, \\
\mathbf{B}_1 &= \begin{pmatrix} 0 & 1 & 0 \end{pmatrix}, \\
\mathbf{B}_2 &= \begin{pmatrix} 1 & 0 & 0 \end{pmatrix}.
\end{aligned}$$

Let us consider again the three special cases of the Birkhoff interpolation problem reviewed in the introduction and identify their associated Birkhoff matrix $\mathbf{B}$.

- **Lagrange interpolation** The Lagrange interpolation problem is given if $t = 0, n = k$, and each block $\mathbf{B}_i$ is the 1×1 scalar 1. Hence, $\mathbf{B}$ is an $(n + 1) \times (n + 1)$ identity matrix.
- **Taylor interpolation** The Taylor interpolation problem is given if $k = 0, t = n = s_0 - 1$, and $\mathbf{B}$ is the $(n + 1) \times (n + 1)$ identity matrix.
- **Hermite interpolation** The Hermite interpolation problem is given if $\mathbf{B}$ is an $(n + 1) \times (t + 1)(k + 1)$ Birkhoff matrix in which each $s_i \times (t + 1)$ diagonal block $\mathbf{B}_i$ contains the first s_i rows of the $(t + 1) \times (t + 1)$ identity matrix, that is $\mathbf{B}_i = [\mathbf{I}_{s_i} \quad \mathbf{0}_{s_i \times (t+1-s_i)}]$.

3.2 Differentiation Matrix

The term differentiation matrix was used by E. Tadmor in his review on spectral methods [28], and denotes the transformation between grid point values of a function and its approximate derivative. In order to introduce the differentiation matrix, we first need the notation of degree-graded polynomials.

Definition 7 Any sequence of polynomials $\{p_j(x)\}_{j=0}^{\infty}$ with p_j of degree j is called degree graded.

Degree-graded polynomials satisfy the following interesting property:

Lemma 1 ([2]) *Any sequence of degree-graded polynomials forms a linearly independent set. These polynomials satisfy the following recurrence relation:*

$$xp_j(x) = \alpha_j p_{j+1}(x) + \beta_j p_j(x) + \gamma_j p_{j-1}(x), \qquad j = 0, 1, \ldots, \tag{4}$$

where $\alpha_j, \beta_j, \gamma_j$ *are complex and* $p_{-1}(x) := 0$, $p_0(x) := 1$ *and if* κ_j *is the leading coefficient of* $p_j(x)$, *then*

$$0 \neq \alpha_j = \frac{\kappa_j}{\kappa_{j+1}}, \qquad j = 0, 1, \ldots.$$

Moreover, for a finite family of degree-graded polynomials, we have the following useful result:

Lemma 2 *For the degree-graded family* $\{p_j(x)\}_{j=0}^n$ *let*

$$\Pi(x) := \begin{pmatrix} p_0(x) \\ p_1(x) \\ \vdots \\ p_n(x) \end{pmatrix}.$$

Then there exists a nilpotent matrix $\mathbf{D}$ *of degree* $n+1$, *called differentiation matrix,*

$$\mathbf{D} = \begin{pmatrix} 0 & 0 & \cdots & 0 \\ & & & \vdots \\ & \mathbf{Q} & & \\ & & & 0 \end{pmatrix}, \tag{5}$$

where $\mathbf{Q}$ *is an* $n \times n$ *lower triangular matrix defined according to the basis of* $\mathbb{P}_n$, *such that the* q*th derivative of the vector* $\Pi(x)$ *can be computed via:*

$$\Pi^{(q)}(x) = \mathbf{D}^q \Pi(x), \qquad q \geqslant 0.$$

The lower triangular matrix $\mathbf{Q}$ does depend on the basis used to represent the polynomials in $\mathbb{P}_n$. The differentiation matrix has been obtained for different bases, especially Chebyshev and Jacobi polynomials [25], Jacobi and Bernstein basis [22], Hermite basis [6], etc.

Here we will consider the Newton basis. The differentiation matrix has been obtained in [1].

Definition 8 (*Newton basis*)
Given $n+1$ distinct nodes τ_i, $i = 0, \ldots, n$, the set of $n+1$ Newton polynomials $\mathcal{N}_i(x)$, $i = 0, \ldots, n$ with

$$\mathcal{N}_i(x) = \prod_{j=0}^{i-1}(x - \tau_j), \qquad i = 0, \ldots, n, \tag{6}$$

is called the Newton basis of $\mathbb{P}_n$. By standard convention, $\mathcal{N}_0(x) = 1$.

The Newton polynomials $\mathcal{N}_i, i = 0, \ldots, n$ form a degree-graded sequence of polynomials; thus according to Lemma 1 they are linearly independent, and they satisfy the general recurrence relation of degree-graded polynomials (4) with $\alpha_j = 1$, $\beta_j = \tau_j$ and $\gamma_j = 0$, as

$$\mathcal{N}_0(x) = 1, \qquad \mathcal{N}_{j+1}(x) = (x - \tau_j)\mathcal{N}_j(x), \quad j = 0, \ldots, n-1.$$

Lemma 3 ([1]) *The qth order derivative of*

$$\Pi(x) := \begin{pmatrix} \mathcal{N}_0(x) \\ \mathcal{N}_1(x) \\ \vdots \\ \mathcal{N}_n(x) \end{pmatrix}$$

is given by

$$\Pi^{(q)}(x) = \mathbf{D}^q \begin{pmatrix} \mathcal{N}_0(x) \\ \mathcal{N}_1(x) \\ \vdots \\ \mathcal{N}_n(x) \end{pmatrix},$$

where **D** *is as in* (5) *with* **Q** *such that*

$$q_{i,j} = \begin{cases} i, & i = j, \\ (\tau_{j-1} - \tau_{i-1})q_{i-1,j} + q_{i-1,j-1}, & i > j, \end{cases} \qquad i = 1, \ldots, n, \tag{7}$$

where $q_{0,j} := 0$, $q_{i,0} := 0$.

Example 4 For $n = 3$, the differentiation matrix **D** for the Newton basis has the following form:

$$\mathbf{D} = \begin{pmatrix} 0 & 0 & 0 & 0 \\ 1 & 0 & 0 & 0 \\ \tau_0 - \tau_1 & 2 & 0 & 0 \\ (\tau_0 - \tau_2)(\tau_0 - \tau_1) & -2\tau_2 + \tau_1 + \tau_0 & 3 & 0 \end{pmatrix}. \tag{8}$$

4 New Approach to the Birkhoff Interpolant

The aim of this section is to develop a new approach for computing the Birkhoff interpolant assuming the solvability of the problem [3, 17, 18, 24].

Consider the Birkhoff interpolation problem for the interpolation nodes $z_i, i = 0, \ldots, k$ with confluency $s_i, i = 0, \ldots, k$. For each node $z_i, i = 0, \ldots, k$, a total number of s_i derivatives of $f(x)$ are given. As in the Hermite interpolation problem (see Definition 2), let us introduce repeated nodes such that we have s_i nodes $z_i, i = 0, \ldots, k$:

$$\begin{aligned} \tau_0 = \tau_1 = \cdots = \tau_{s_0-1} &:= z_0, \\ \tau_{s_0} = \tau_{s_0+1} = \cdots = \tau_{s_0+s_1-1} &:= z_1, \\ &\vdots \\ \tau_{s_0+\cdots+s_{k-1}} = \cdots = \tau_{s_0+\cdots+s_k-1} &:= z_k. \end{aligned} \tag{9}$$

Next, let us consider the Newton basis (6) for the above set of nodes $\tau_0, \ldots, \tau_{s_0+\cdots+s_k-1} = \tau_n$ even though these nodes are not distinct. For simplicity, we will denote the so obtained set of polynomials by $\mathscr{B}_i(x)$ and refer to them as Birkhoff polynomials. They are defined recursively

$$\mathscr{B}_{i+1}(x) = (x - \tau_i)\mathscr{B}_i(x), \qquad i = 0, \ldots, n-1,$$

with $\mathscr{B}_0(x) = 1$. Clearly, $\mathscr{B}_i(x) \in \mathbb{P}_i$.

A direct consequence of (9) and Lemma 1 implies that the set of Birkhoff polynomials $\{\mathscr{B}_i(x)\}_{i=0}^n$ be a set of linearly independent polynomials which may be a basis of $\mathbb{P}_n$.

Lemma 4 *The explicit formulation for $\mathscr{B}_i(x)$ in terms of the interpolation nodes $z_i, i = 0, \ldots, k$ is given by*

$$\mathscr{B}_0(x) = 1, \quad \mathscr{B}_\ell(x) = (x - z_0)^\ell, \qquad \ell = 1, \ldots, s_0,$$

and for $j = 0, \ldots, k-2$ and $\ell = p + \sum_{q=0}^{j} s_q$ with $1 \le p \le s_{j+1}$

$$\mathscr{B}_\ell(x) = \prod_{q=0}^{j} (x - z_q)^{s_q} \cdot (x - z_{j+1})^p,$$

and for $\ell = p + \sum_{q=0}^{k-1} s_q$ with $0 < p < s_k$

$$\mathscr{B}_\ell(x) = \prod_{q=0}^{k-1} (x - z_q)^{s_q} \cdot (x - z_k)^p.$$

These polynomials form a sequence of degree-graded polynomials, in which $\alpha_j = 1$, $\beta_j = \tau_j$, and $\gamma_j = 0$, but some of the β_j's are repeated.

The following formulas clarify the explicit formulation of the Birkhoff basis:

$$\mathscr{B}_0(x) = 1,$$

$$
\begin{aligned}
\mathscr{B}_1(x) &= (x - z_0), \\
\mathscr{B}_2(x) &= (x - z_0)^2, \\
&\vdots \\
\mathscr{B}_{s_0}(x) &= (x - z_0)^{s_0}, \\
\mathscr{B}_{s_0+1}(x) &= (x - z_0)^{s_0}(x - z_1), \\
\mathscr{B}_{s_0+2}(x) &= (x - z_0)^{s_0}(x - z_1)^2, \\
&\vdots \\
\mathscr{B}_{s_0+s_1}(x) &= (x - z_0)^{s_0}(x - z_1)^{s_1}, \\
&\vdots \\
\mathscr{B}_{s_0+s_1+\cdots+s_j}(x) &= (x - z_0)^{s_0}(x - z_1)^{s_1} \cdots (x - z_j)^{s_j}, \\
&\vdots \\
\mathscr{B}_{s_0+s_1+\cdots+s_j+p}(x) &= (x - z_0)^{s_0}(x - z_1)^{s_1} \cdots (x - z_j)^{s_j}(x - z_{j+1})^p.
\end{aligned}
$$

Example 5 Consider the Birkhoff interpolation problem discussed in Example 1. The new set of nodes

$$
\tau_0 = \tau_1 = z_0, \quad \tau_2 = z_1, \quad \tau_3 = z_2, \tag{10}
$$

defines the following basis for $\mathbb{P}_3$:

$$
1, \;\; (x - z_0), \;\; (x - z_0)^2, \;\; (x - z_0)^2(x - z_1).
$$

Lemma 3 also holds for the Birkhoff basis as the qth order derivative of $\prod_{i=0}^{j}(x - \tau_i)$, $j = 0, \ldots, n-1$ does not depend on the specific values of τ_i.

Lemma 5 *The qth order derivative of*

$$
\Pi(x) := \begin{pmatrix} \mathscr{B}_0(x) \\ \mathscr{B}_1(x) \\ \vdots \\ \mathscr{B}_n(x) \end{pmatrix},
$$

is given by

$$
\Pi^{(q)}(x) = \mathbf{D}^q \begin{pmatrix} \mathscr{B}_0(x) \\ \mathscr{B}_1(x) \\ \vdots \\ \mathscr{B}_n(x) \end{pmatrix},
$$

where **D** *and* **Q** *are as in* (5) *and* (7), *respectively.*

Example 6 The differentiation matrix for the Birkhoff interpolation problem considered in Example 1 is given by the matrix (8).

Hence, for the set of nodes (10) it is given by

$$\mathbf{D} = \begin{pmatrix} 0 & 0 & 0 & 0 \\ 1 & 0 & 0 & 0 \\ 0 & 2 & 0 & 0 \\ 0 & 2(z_0 - z_1) & 3 & 0 \end{pmatrix}.$$

As the polynomials $\mathscr{B}_i(x)$, $i = 0, \ldots, n$ are degree-graded, they are linearly independent. Hence they are a basis of $\mathbb{P}_n$ and every polynomial $p(x) \in \mathbb{P}_n$ can be written as a linear combination of this basis.

Now we propose the following approach: Assume that the Birkhoff interpolation problem has a unique solution $P(x)$. Then $P(x)$ can be written as

$$P(x) = \begin{pmatrix} a_0 & a_1 & \cdots & a_n \end{pmatrix} \begin{pmatrix} \mathscr{B}_0(x) \\ \mathscr{B}_1(x) \\ \vdots \\ \mathscr{B}_n(x) \end{pmatrix}, \tag{11}$$

for certain a_i, $i = 0, \ldots, n$. Hence, the qth order derivative of $P(x)$ is given by

$$P^{(q)}(x) = \begin{pmatrix} a_0 & a_1 & \cdots & a_n \end{pmatrix} \mathbf{D}^q \begin{pmatrix} \mathscr{B}_0(x) \\ \mathscr{B}_1(x) \\ \vdots \\ \mathscr{B}_n(x) \end{pmatrix}, \tag{12}$$

where **D** is the differentiation matrix introduced in Lemma 5.

The interpolant (11) has to satisfy the interpolation conditions (3), i.e. for fixed i, $i = 0, \ldots, k$, we have

$$P^{(j)}(z_i) = f_{i,j}, \quad \text{if} \quad J_{i,j} = 1 \quad \text{for} \quad j = 0, \ldots, t.$$

Then, using (12) implies that

$$\begin{pmatrix} a_0 & a_1 & \cdots & a_n \end{pmatrix} \mathbf{D}^j \begin{pmatrix} \mathscr{B}_0(z_i) \\ \mathscr{B}_1(z_i) \\ \vdots \\ \mathscr{B}_n(z_i) \end{pmatrix} = f_{i,j}, \quad \text{if} \quad J_{i,j} = 1 \quad \text{for} \quad i = 0, \ldots, k, \;\; j = 0, \ldots, t.$$

These equations can be summarized using the Birkhoff matrix.

Theorem 2 *The unknowns $a_i, i = 0, \ldots, n$, in the Birkhoff interpolation polynomial* (11) *can be found via the $(n+1) \times (n+1)$ linear system*

$$\mathbf{B}\Phi\Gamma\mathbf{a} = \mathbf{BF}, \tag{13}$$

where **B** *is the $(n+1) \times (t+1)(k+1)$ Birkhoff matrix, and the vector* **F** *is of size $(k+1)(t+1) \times 1$*

$$\mathbf{F} = \left(f_{0,0}\ f_{0,1}\ \cdots\ f_{0,t}\ \cdots\ f_{k,0}\ f_{k,1}\ \cdots\ f_{k,t}\right)^T.$$

The matrix Φ is of size $(t+1)(k+1) \times (t+1)(n+1)$, and is constructed as follows

$$\Phi = \begin{pmatrix} \mathbf{V}_0^T \\ \mathbf{V}_1^T \\ \vdots \\ \mathbf{V}_k^T \end{pmatrix}, \tag{14}$$

where for $i = 0, \ldots, k$, $\mathbf{V}_i^T$ is the following $(t+1) \times (t+1)(n+1)$ matrix

$$\mathbf{V}_i^T = \begin{pmatrix} \mathbf{V}^T(z_i) & & & \\ & \mathbf{V}^T(z_i) & & \\ & & \ddots & \\ & & & \mathbf{V}^T(z_i) \end{pmatrix}, \tag{15}$$

in which the column vector of the Birkhoff polynomial basis is

$$\mathbf{V}(x) = \begin{pmatrix} \mathscr{B}_0(x) \\ \mathscr{B}_1(x) \\ \vdots \\ \mathscr{B}_n(x) \end{pmatrix}.$$

Furthermore, the $(t+1)(n+1) \times (n+1)$ matrix Γ is defined as

$$\Gamma = \begin{pmatrix} \mathbf{I} \\ \mathbf{D}^T \\ (\mathbf{D}^2)^T \\ \vdots \\ (\mathbf{D}^t)^T \end{pmatrix}, \tag{16}$$

where **D** *is the differentiation matrix given by* (5), *and finally the $(n+1) \times 1$ vector $\mathbf{a} = \left(a_0\ a_1\ \cdots\ a_n\right)$ is the vector of unknowns.*

To sum up, the Eq. (13) is the key relation of this contribution, which leads to the desired interpolant. Note that the vector **BF** in the right hand side of the equation contains the available derivative information of the Birkhoff data.

5 Special Cases

In this section, we recover elementary but relevant results for Taylor, Lagrange and Hermite interpolation using our results from the previous section. The existence and uniqueness of the solution to these problems are trivial and well-studied in the literature [7].

5.1 Taylor Interpolation

As already noted, for the Taylor interpolation problem (see Definition 3) we have $k = 0, t = n = s_0 - 1$. It is characterized by an $(n+1) \times (n+1)$ Birkhoff matrix which is identical to the $(n+1) \times (n+1)$ identity matrix. Hence, the system (13) reduces to

$$\Phi \Gamma \mathbf{a} = \mathbf{F}, \tag{17}$$

with the right hand side **F**

$$\mathbf{F} = \begin{pmatrix} f_{0,0} & f_{0,1} & \dots & f_{0,n} \end{pmatrix}^T \in \mathbb{C}^{n+1},$$

and

$$\Phi = \mathbf{V}_0^T = \begin{pmatrix} \mathbf{V}^T(z_0) & & \\ & \ddots & \\ & & \mathbf{V}^T(z_0) \end{pmatrix} \in \mathbb{C}^{(n+1)\times(n+1)^2},$$

where

$$\mathbf{V}^T(z_0) = \begin{pmatrix} \mathscr{B}_0(z_0) & \mathscr{B}_1(z_0) & \dots & \mathscr{B}_n(z_0) \end{pmatrix} = \begin{pmatrix} 1 & 0 & \cdots & 0 \end{pmatrix} \in \mathbb{C}^{1\times(n+1)},$$

as the $\mathscr{B}_j$ are given here by

$$\mathscr{B}_j(x) = (x - z_0)^j, \qquad j = 0, \dots, n.$$

Finally, the $(n+1) \times (n+1)$ differentiation matrix **D** as derived in Lemma 5 is given by

$$\mathbf{D} = \left(\begin{array}{cccccc|c} 0 & 0 & \cdots & \cdots & & 0 & 0 \\ \hline 1 & 0 & \cdots & \cdots & & 0 & 0 \\ 0 & 2 & \ddots & & & \vdots & \vdots \\ \vdots & \ddots & \ddots & \ddots & & \vdots & \vdots \\ \vdots & & & \ddots & n-1 & 0 & 0 \\ 0 & 0 & \cdots & & 0 & n & 0 \end{array}\right),$$

that is, $\mathbf{D}$ has nonzero entries only on its first subdiagonal; $\mathbf{D}_{i+1,i} = i, i = 1, \ldots, n$. All other entries are 0. It is easy to see that $\mathbf{D}^2$ has nonzero entries only on its second subdiagonal, $\mathbf{D}^3$ has nonzero entries only on its third subdiagonal, and so on, until $\mathbf{D}^{n+1} = 0$. Hence,

$$\Phi\Gamma = \begin{pmatrix} \mathbf{V}^T(z_0)\mathbf{I} \\ \mathbf{V}^T(z_0)\mathbf{D}^T \\ \mathbf{V}^T(z_0)(\mathbf{D}^2)^T \\ \vdots \\ \mathbf{V}^T(z_0)(\mathbf{D}^n)^T \end{pmatrix} = \begin{pmatrix} 1 & & & & & \\ & 1 & & & & \\ & & 2! & & & \\ & & & 3! & & \\ & & & & \ddots & \\ & & & & & n! \end{pmatrix}.$$

Therefore, solving (17) gives

$$a_j = \frac{f_{0,j}}{j!}, \qquad j = 0, \ldots, n,$$

which corresponds to (2).

5.2 Lagrange Interpolation

As already noted, for the Lagrange interpolation problem we have $t = 0, n = k$. It is characterized by the Birkhoff incidence matrix $\mathbf{J}$ of size $(k+1) \times 1$ containing only ones, while the corresponding Birkhoff matrix is a $(k+1) \times (k+1)$ identity. Moreover, as $t = 0$, the matrix Γ defined in (16) is the identity matrix of size $k+1$. Hence, the system (13) simplifies to

$$\Phi\mathbf{a} = \mathbf{F}, \tag{18}$$

with the right-hand vector $\mathbf{F}$

$$\mathbf{F} = \left(f_0 \; f_1 \; \ldots \; f_n\right)^T,$$

as in our notation $f_{i,0} = f_i, i = 0, \ldots, n$. The matrices $\mathbf{V}_i$ in (15) are $1 \times (k+1)$ vectors

$$\mathbf{V}_i^T = \begin{pmatrix} \mathscr{B}_0(z_i) & \mathscr{B}_1(z_i) & \cdots & \mathscr{B}_k(z_i) \end{pmatrix}, \qquad i = 0, \ldots, k,$$

or, more precisely, as all nodes are distinct,

$$\mathbf{V}_i^T = \begin{pmatrix} \mathscr{N}_0(z_i) & \mathscr{N}_1(z_i) & \cdots & \mathscr{N}_k(z_i) \end{pmatrix}, \qquad i = 0, \ldots, k.$$

Hence, the elements of the $(k+1) \times (k+1)$ matrix Φ are given by

$$\Phi_{i,j} = \mathscr{N}_{j-1}(z_{i-1}) = \prod_{p=0}^{j-2} (z_{i-1} - z_p), \qquad i, j = 1, \ldots, k+1.$$

Clearly, for $j - 1 > q$, we have $\mathscr{N}_{j-1}(z_q) = 0$. Therefore, Φ is a lower triangular matrix

$$\Phi = \begin{pmatrix} \mathscr{N}_0(z_0) & 0 & 0 & \cdots & 0 & 0 \\ \mathscr{N}_0(z_1) & \mathscr{N}_1(z_1) & 0 & \cdots & 0 & 0 \\ \mathscr{N}_0(z_2) & \mathscr{N}_1(z_2) & \mathscr{N}_2(z_2) & \ddots & & 0 \\ \vdots & \vdots & \vdots & \ddots & \ddots & \vdots \\ \mathscr{N}_0(z_{k-1}) & \mathscr{N}_1(z_{k-1}) & \mathscr{N}_2(z_{k-1}) & \cdots & \mathscr{N}_{k-1}(z_{k-1}) & 0 \\ \mathscr{N}_0(z_k) & \mathscr{N}_1(z_k) & \mathscr{N}_2(z_k) & \cdots & \mathscr{N}_{k-1}(z_k) & \mathscr{N}_k(z_k) \end{pmatrix}$$

$$= \begin{pmatrix} 1 & 0 & 0 & \cdots & 0 & 0 \\ 1 & z_1 - z_0 & 0 & \cdots & 0 & 0 \\ 1 & z_2 - z_0 & (z_2 - z_1)(z_2 - z_0) & \ddots & 0 & 0 \\ \vdots & \vdots & \vdots & & \ddots & \vdots \\ 1 & z_{k-1} - z_0 & (z_{k-1} - z_1)(z_{k-1} - z_0) & \cdots & \prod_{q=0}^{k-2}(z_{k-1} - z_q) & 0 \\ 1 & z_k - z_0 & (z_k - z_1)(z_k - z_0) & \cdots & \prod_{q=0}^{k-2}(z_k - z_q) & \prod_{q=0}^{k-1}(z_k - z_q) \end{pmatrix}.$$

As all nodes z_i are distinct, Φ is nonsingular and (18) has a unique solution; $\det\Phi$ is equal to the determinant of the Vandermonde matrix V with $v_{ij} = z_i^{j-1}$ as can be easily seen from the lower triangular structure of Φ. Note that (18) is just the usual linear system one obtains when solving the Lagrange interpolation problem with respect to the Newton basis. Typically, one does not solve the system (18), but uses divided differences and the Aitken-Neville recursion in order to determine the coefficients a_j, $j = 0, \ldots, k$.

5.3 Hermite Interpolation

As already noted, the Hermite interpolation problem (see Definition 2) is characterized by an $(n+1) \times (t+1)(k+1)$ Birkhoff matrix in which each $s_i \times (t+1)$ diagonal block $\mathbf{B}_i = [\mathbf{I}_{s_i} \quad \mathbf{0}_{s_i \times (t+1-s_i)}]$. The right hand side of (13) is

$$\mathbf{BF} = \left(f_{0,0}\ f_{0,1}\ \cdots\ f_{0,s_0-1}\ \cdots\ f_{k,0}\ f_{k,1}\ \cdots\ f_{k,s_k-1} \right).$$

The standard approach for solving the Hermite interpolation problem makes use of divided differences, hence one does not solve the system $\mathbf{B}\Phi\Gamma\mathbf{a} = \mathbf{BF}$ in order to determine the coefficients a_j, $j = 0, \ldots, k$. But, as the next example shows, the unknowns in the linear system which needs to be solved here, are just the divided differences.

Example 7 Consider the Birkhoff interpolation problem given by:

$$\mathbf{B} = \left(\begin{array}{cc|cc|cc} 1 & 0 & 0 & 0 & 0 & 0 \\ \hline 0 & 0 & 1 & 0 & 0 & 0 \\ 0 & 0 & 0 & 1 & 0 & 0 \\ \hline 0 & 0 & 0 & 0 & 1 & 0 \end{array}\right) \in \mathbb{R}^{4\times 6},$$

which corresponds to the Birkhoff incidence matrix:

$$\mathbf{J} = \begin{pmatrix} 1 & 0 \\ 1 & 1 \\ 1 & 0 \end{pmatrix}.$$

A quick observation shows that this corresponds to a Hermite interpolation problem for $k = 2, t = 1, n = 3$. Following the differentiation matrix in (8) for $\tau_0 = z_0,\ \tau_1 = \tau_2 = z_1,\ \tau_3 = z_2$, we have

$$\mathbf{D} = \begin{pmatrix} 0 & 0 & 0 & 0 \\ 1 & 0 & 0 & 0 \\ z_0 - z_1 & 2 & 0 & 0 \\ (z_0 - z_1)^2 & z_0 - z_1 & 3 & 0 \end{pmatrix}.$$

The vector basis

$$\mathbf{V}(x) = \begin{pmatrix} \mathscr{B}_0(x) \\ \mathscr{B}_1(x) \\ \mathscr{B}_2(x) \\ \mathscr{B}_3(x) \end{pmatrix} = \begin{pmatrix} 1 \\ (x - z_0) \\ (x - z_0)(x - z_1) \\ (x - z_0)(x - z_1)^2 \end{pmatrix},$$

gives V_0^T, V_1^T, V_2^T in (15) which completes the matrix Φ in (14). The system (13) is constructed as follows:

$$\begin{pmatrix} 1 & 0 & 0 & 0 \\ 1 & (z_1 - z_0) & 0 & 0 \\ 0 & 1 & (z_1 - z_0) & 0 \\ 1 & (z_2 - z_0) & (z_2 - z_0)(z_2 - z_1) & (z_2 - z_0)(z_2 - z_1)^2 \end{pmatrix} \begin{pmatrix} a_0 \\ a_1 \\ a_2 \\ a_3 \end{pmatrix} = \begin{pmatrix} f_{0,0} \\ f_{1,0} \\ f_{1,1} \\ f_{2,0} \end{pmatrix},$$

and its solutions correspond to the values obtained from the divided differences.

6 Illustrative Examples

In this section, first we reconsider Example 1. Then, two other examples from the literature are discussed.

Example 8 Suppose that the function $f(x)$ is given by the values $f_{0,0}, f_{0,2}, f_{1,1}, f_{2,0}$ at the distinct nodes z_0, z_1, z_2. Then $k = 2, n = 3$ and $t = 2$. The corresponding Birkhoff matrix $\mathbf{B}$ has been considered in Example 3

$$\mathbf{B} = \left(\begin{array}{ccc|ccc|ccc} 1 & 0 & 0 & 0 & 0 & 0 & 0 & 0 & 0 \\ 0 & 0 & 1 & 0 & 0 & 0 & 0 & 0 & 0 \\ \hline 0 & 0 & 0 & 0 & 1 & 0 & 0 & 0 & 0 \\ \hline 0 & 0 & 0 & 0 & 0 & 0 & 1 & 0 & 0 \end{array}\right) \in \mathbb{R}^{4\times 9}.$$

We first set up the system (13). The Birkhoff matrix $\mathbf{B}$ has already been determined, the right-hand side vector is given by

$$\mathbf{BF} = \begin{pmatrix} f_{0,0} \\ f_{0,2} \\ f_{1,1} \\ f_{2,0} \end{pmatrix},$$

as $\mathbf{F} = \begin{pmatrix} f_{0,0} & f_{0,1} & f_{0,2} & f_{1,0} & f_{1,1} & f_{1,2} & f_{2,0} & f_{2,1} & f_{2,2} \end{pmatrix}^T$. Considering Example 6, the differentiation matrix $\mathbf{D}$ is given by

$$\mathbf{D} = \begin{pmatrix} 0 & 0 & 0 & 0 \\ 1 & 0 & 0 & 0 \\ 0 & 2 & 0 & 0 \\ 0 & 2(z_0 - z_1) & 3 & 0 \end{pmatrix} \in \mathbb{C}^{4\times 4}, \tag{19}$$

hence

$$\Gamma = \begin{pmatrix} \mathbf{I} \\ \mathbf{D}^T \\ (\mathbf{D}^2)^T \end{pmatrix} = \left(\begin{array}{cccc} 1 & 0 & 0 & 0 \\ 0 & 1 & 0 & 0 \\ 0 & 0 & 1 & 0 \\ 0 & 0 & 0 & 1 \\ \hline 0 & 1 & 0 & 0 \\ 0 & 0 & 2 & 2(z_0 - z_1) \\ 0 & 0 & 0 & 3 \\ 0 & 0 & 0 & 0 \\ \hline 0 & 0 & 2 & 2(z_0 - z_1) \\ 0 & 0 & 0 & 6 \\ 0 & 0 & 0 & 0 \\ 0 & 0 & 0 & 0 \end{array}\right) \in \mathbb{C}^{12\times 4}.$$

Finally, the matrix Φ is given by

$$\Phi = \begin{pmatrix} \mathbf{V}_0^T \\ \mathbf{V}_1^T \\ \mathbf{V}_2^T \end{pmatrix} = \left(\begin{array}{ccc} \mathbf{V}^T(z_0) & & \\ & \mathbf{V}^T(z_0) & \\ & & \mathbf{V}^T(z_0) \\ \hline \mathbf{V}^T(z_1) & & \\ & \mathbf{V}^T(z_1) & \\ & & \mathbf{V}^T(z_1) \\ \hline \mathbf{V}^T(z_2) & & \\ & \mathbf{V}^T(z_2) & \\ & & \mathbf{V}^T(z_2) \end{array}\right) \in \mathbb{C}^{9\times 12}.$$

As

$$\mathbf{B}\Phi = \left(\begin{array}{c|c|c} \mathbf{V}^T(z_0) & & \\ & & \mathbf{V}^T(z_0) \\ \hline & \mathbf{V}^T(z_1) & \\ \hline \mathbf{V}^T(z_2) & & \end{array}\right),$$

and

$$\mathbf{V}(x) = \begin{pmatrix} \mathscr{B}_0(x) \\ \mathscr{B}_1(x) \\ \mathscr{B}_2(x) \\ \mathscr{B}_3(x) \end{pmatrix} = \begin{pmatrix} 1 \\ (x - z_0) \\ (x - z_0)^2 \\ (x - z_0)^2(x - z_1) \end{pmatrix}, \tag{20}$$

we obtain

$$\mathbf{B}\Phi = \left(\begin{array}{cccc|cccc|cccc} 1 & 0 & 0 & 0 & 0 & 0 & 0 & 0 & 0 & 0 & 0 & 0 \\ 0 & 0 & 0 & 0 & 0 & 0 & 0 & 0 & 1 & 0 & 0 & 0 \\ \hline 0 & 0 & 0 & 0 & 1 & (z_1 - z_0) & (z_1 - z_0)^2 & 0 & 0 & 0 & 0 & 0 \\ \hline 1 & (z_2 - z_0) & (z_2 - z_0)^2 & (z_2 - z_0)^2(z_2 - z_1) & 0 & 0 & 0 & 0 & 0 & 0 & 0 & 0 \end{array}\right).$$

Summing up, the system (13), $\mathbf{B}\Phi\Gamma\mathbf{a} = \mathbf{B}\mathbf{F}$, reads

$$\underbrace{\begin{pmatrix} 1 & 0 & 0 & 0 \\ 0 & 0 & 2 & 2(z_0-z_1) \\ 0 & 1 & 2(z_1-z_0) & (z_1-z_0)^2 \\ 1 & (z_2-z_0) & (z_2-z_0)^2 & (z_2-z_0)^2(z_2-z_1) \end{pmatrix}}_{\mathbf{C}} \begin{pmatrix} a_0 \\ a_1 \\ a_2 \\ a_3 \end{pmatrix} = \begin{pmatrix} f_{0,0} \\ f_{0,2} \\ f_{1,1} \\ f_{2,0} \end{pmatrix}.$$

Since

$$det(\mathbf{C}) = 2(z_0 - z_2)\Big((z_0 - z_2)^2 - 3(z_0 - z_1)^2\Big),$$

the unknowns $a_i, i = 0, \ldots, 3$ can be obtained uniquely from the above system for the interpolation nodes which do not satisfy $z_2 - z_0 = \pm\sqrt{3}(z_1 - z_0)$. Then the interpolation polynomial is given by

$$P(x) = \begin{pmatrix} a_0 & a_1 & a_2 & a_3 \end{pmatrix} \begin{pmatrix} 1 \\ (x-z_0) \\ (x-z_0)^2 \\ (x-z_0)^2(x-z_1) \end{pmatrix}.$$

The next example has been presented in [13] as a solvable Birkhoff interpolation problem.

Example 9 Let the interpolation nodes z_0, z_1, z_2 and the values of $f_{0,0}, f_{1,1}, f_{2,1}$ be given. Hence, we have $k = 2, n = 2$ and $t = 1$. The Birkhoff matrix is given by

$$\mathbf{B} = \begin{pmatrix} 1 & 0 & 0 & 0 & 0 & 0 \\ 0 & 0 & 0 & 1 & 0 & 0 \\ 0 & 0 & 0 & 0 & 0 & 1 \end{pmatrix},$$

while the differentiation matrix is

$$\mathbf{D} = \begin{pmatrix} 0 & 0 & 0 \\ 1 & 0 & 0 \\ z_0 - z_1 & 2 & 0 \end{pmatrix}. \tag{21}$$

Moreover,

$$\mathbf{B}\Phi = \begin{pmatrix} \mathbf{V}^T(z_0) & 0 \\ 0 & \mathbf{V}^T(z_1) \\ 0 & \mathbf{V}^T(z_2) \end{pmatrix} = \begin{pmatrix} 1 & 0 & 0 & 0 & 0 & 0 \\ 0 & 0 & 0 & 1 & z_1 - z_0 & 0 \\ 0 & 0 & 0 & 1 & z_2 - z_0 & (z_2 - z_0)(z_2 - z_1) \end{pmatrix},$$

as

$$\mathbf{V}(x) = \begin{pmatrix} \mathscr{B}_0(x) \\ \mathscr{B}_1(x) \\ \mathscr{B}_2(x) \end{pmatrix} = \begin{pmatrix} 1 \\ (x - z_0) \\ (x - z_0)(x - z_1) \end{pmatrix}.$$

With

$$\Gamma = \begin{pmatrix} \mathbf{I} \\ \mathbf{D}^T \end{pmatrix} = \begin{pmatrix} 1 & 0 & 0 \\ 0 & 1 & 0 \\ 0 & 0 & 1 \\ 0 & 1 & z_0 - z_1 \\ 0 & 0 & 2 \\ 0 & 0 & 0 \end{pmatrix}, \tag{22}$$

we obtain the following system

$$\underbrace{\begin{pmatrix} 1 & 0 & 0 \\ 0 & 1 & (z_1 - z_0) \\ 0 & 1 & (z_2 - z_0) + (z_2 - z_1) \end{pmatrix}}_{\mathbf{C}} \begin{pmatrix} a_0 \\ a_1 \\ a_2 \end{pmatrix} = \begin{pmatrix} f_{0,0} \\ f_{1,1} \\ f_{2,1} \end{pmatrix}.$$

Since $det(\mathbf{C}) = 2(z_2 - z_1)$, the system has a unique solution as $z_2 \neq z_1$. We obtain

$$\begin{pmatrix} a_0 \\ a_1 \\ a_2 \end{pmatrix} = \begin{pmatrix} f_{0,0} \\ f_{1,1} - \frac{(f_{2,1}-f_{1,1})(z_1-z_0)}{2(z_2-z_1)} \\ \frac{f_{2,1}-f_{1,1}}{2(z_2-z_1)} \end{pmatrix},$$

and the unique interpolation polynomial

$$P(x) = (a_0 \;\; a_1 \;\; a_2)\mathbf{V}(x) = f_{0,0} + \left(f_{1,1} - \frac{(f_{2,1} - f_{1,1})(z_1 - z_0)}{2(z_2 - z_1)} \right)(x - z_0) +$$

$$+ \frac{f_{2,1} - f_{1,1}}{2(z_2 - z_1)}(x - z_0)(x - z_1).$$

Now, we present another example which is conditionally solvable.

Example 10 Consider the distinct interpolation nodes z_0, z_1, z_2 and the given information $f(z_0) = f_{0,0}$, $f'(z_1) = f_{1,1}$, $f(z_2) = f_{2,0}$. According to [23], the interpolation polynomial does not exist, when $z_1 = (z_0 + z_2)/2$, and it uniquely exists for any other choice of z_1. The Birkhoff matrix is given by

$$\mathbf{B} = \begin{pmatrix} 1 & 0 & 0 & 0 & 0 & 0 \\ 0 & 0 & 0 & 1 & 0 & 0 \\ 0 & 0 & 0 & 0 & 1 & 0 \end{pmatrix}.$$

Since the set of the nodes for this problem is the same as Example 9, the differentiation matrix $\mathbf{D}$ is the same as (21) and Γ is as in (22). Hence, we obtain the following system for the unknowns:

$$\underbrace{\begin{pmatrix} 1 & 0 & 0 \\ 0 & 1 & (z_1 - z_0) \\ 1 & (z_2 - z_0) & (z_2 - z_0)(z_2 - z_1) \end{pmatrix}}_{\mathbf{C}} \begin{pmatrix} a_0 \\ a_1 \\ a_2 \end{pmatrix} = \begin{pmatrix} f_{0,0} \\ f_{1,1} \\ f_{2,0} \end{pmatrix}.$$

As

$$det(\mathbf{C}) = (z_2 - z_0)(z_0 - 2z_1 + z_2),$$

the system has a unique solution for all distinct interpolation nodes with the exception of the case $z_1 = (z_0 + z_2)/2$. Thus the problem is conditionally solvable.

To conclude, the following example shows a Birkhoff interpolation problem which is not usually solvable.

Example 11 The values of the $f(z_0)$, $f^{(3)}(z_0)$, $f'(z_1)$, $f^{(3)}(z_1)$ for the distinct interpolation nodes z_0, z_1 are given by $f_{0,0}$, $f_{0,3}$, $f_{1,1}$, $f_{1,3}$ respectively. We want to show that this problem is not solvable for any given interpolation data. In this problem $k = 1, n = 3, t = 3$, and the corresponding Birkhoff matrix is as follows:

$$\mathbf{B} = \begin{pmatrix} 1\,0\,0\,0\,0\,0\,0\,0 \\ 0\,0\,0\,1\,0\,0\,0\,0 \\ 0\,0\,0\,0\,0\,1\,0\,0 \\ 0\,0\,0\,0\,0\,0\,0\,1 \end{pmatrix}.$$

The differentiation matrix remains the same as (19), and the matrix Γ is as follows:

$$\Gamma = \begin{pmatrix} \mathbf{I} \\ \mathbf{D}^T \\ (\mathbf{D}^2)^T \\ (\mathbf{D}^3)^T \end{pmatrix} = \left(\begin{array}{cccc} 1 & 0 & 0 & 0 \\ 0 & 1 & 0 & 0 \\ 0 & 0 & 1 & 0 \\ 0 & 0 & 0 & 1 \\ \hline 0 & 1 & 0 & 0 \\ 0 & 0 & 2 & 2(z_0 - z_1) \\ 0 & 0 & 0 & 3 \\ 0 & 0 & 0 & 0 \\ \hline 0 & 0 & 2 & 2(z_0 - z_1) \\ 0 & 0 & 0 & 6 \\ 0 & 0 & 0 & 0 \\ 0 & 0 & 0 & 0 \\ \hline 0 & 0 & 0 & 6 \\ 0 & 0 & 0 & 0 \\ 0 & 0 & 0 & 0 \\ 0 & 0 & 0 & 0 \end{array}\right) \in \mathbb{C}^{16\times 4}.$$

The basis vector $\mathbf{V}(x)$ equals to the one in (20), and eventually, the system (13) gives the following:

$$\underbrace{\begin{pmatrix} 1 & 0 & 0 & 0 \\ 0 & 0 & 0 & 6 \\ 0 & 1 & 2(z_1 - z_0) & (z_1 - z_0)^2 \\ 0 & 0 & 0 & 6 \end{pmatrix}}_{\mathbf{C}} \begin{pmatrix} a_0 \\ a_1 \\ a_2 \\ a_3 \end{pmatrix} = \begin{pmatrix} f_{0,0} \\ f_{0,3} \\ f_{1,1} \\ f_{1,3} \end{pmatrix}.$$

Since clearly $\mathbf{C}$ is not full-rank, the system is not solvable for any given values of $f_{0,0}, f_{0,3}, f_{1,1}, f_{1,3}$, unless $f_{0,3} = f_{1,3}$ in which case it has infinitely many solutions.

Acknowledgements Alexander von Humboldt Foundation has funded the work of third author.

References

1. Amiraslani, A.: Differentiation matrices in polynomial bases, Math. Sci. **10**(1), 47–53 (2016)
2. Amiraslani, A., Corless, R.M., Lancaster, P.: Linearization of matrix polynomials expressed in polynomial bases. IMA J. Numer. Anal. **29**(1), 141–157 (2009)
3. Atkinson, K., Sharma, A.: A partial characterization of poised Hermite-Birkhoff interpolation problems. SIAM J. Numer. Anal. **6**, 230–235 (1969)
4. Birkhoff, G.D.: General mean value theorems with applications to mechanical differentiation and quadrature. Trans. Am. Math. Soc. **7**(1), 107–136 (1906)
5. Butcher, J.C., Corless, R.M., Gonzalez-Vega, L., Shakoori, A.: Polynomial algebra for Birkhoff interpolants. Numer. Algorithms **56**(3), 319–347 (2011)
6. Chirikalov, V.A.: Computation of the differentiation matrix for the Hermite interpolating polynomials. J. Math. Sci. **68**(6), 766–770 (1994)
7. Davis, P.: Interpolation and Approximation. Dover Publications, New York (1975)
8. de Bruin, M.G., Sharma, A.: Birkhoff interpolation on perturbed roots of unity on the unit circle. J. Nat. Acad. Math. India **11**, 83–97 (1997)
9. de Bruin, M.G., Sharma, A.: Birkhoff interpolation on nonuniformly distributed roots of unity. J. Comput. Appl. Math. **133**(1–2), 295–303 (2001)
10. de Bruin, M.G., Dikshit, H.P.: Birkhoff interpolation on nonuniformly distributed points. J. Indian Math. Soc. (NS) **69**(1–4), 81–101 (2002)
11. de Bruin, M.G., Dikshit, H.P., Sharma, A.: Birkhoff interpolation on unity and on Möbius transform of the roots of unity. Numer. Algorithms **23**(1), 115–125 (2002)
12. Dikshit, H.P.: Birkhoff interpolation on some perturbed roots of unity. Nonlinear Anal. Forum **6**(1), 97–102 (2001)
13. Finden, W.F.: An error term and uniqueness for Hermite-Birkhoff interpolation involving only function values and/or first derivative values. J. Comput. Appl. Math. **212**(1), 1–15 (2008)
14. Ferguson, D.: The question of uniqueness for G. D. Birkhoff interpolation problems. J. Approx. Theory **2**, 1–28 (1969)
15. Gonzalez-Vega, L.: Applying quantifier elimination to the Birkhoff interpolation problem. J. Symb. Comput. **22**(1), 83–103 (1996)
16. Higham, D.J.: Runge-Kutta defect control using Hermite-Birkhoff interpolation. SIAM J. Sci. Comput. **12**, 991–999 (1991)
17. Lorentz, G.G.: Birkhoff interpolation problem. Report CNA-103. The Center of Numerical Analysis, The University of Texas (1975)

18. Lorentz G.G., Jetter, K., Riemenschneider, S.D.: Birkhoff Interpolation. Encyclopedia of Mathematics and its Applications, vol. 19. Cambridge University Press, Cambridge (1984)
19. Mühlbach, G.: An algorithmic approach to Hermite-Birkhoff-interpolation. Numer. Math. **37**(3), 339–347 (1981)
20. Polya, G.: Bemerkungen zur Interpolation und der Naherungstheorie der Balkenbiegung. Z. Angew. Math. Mech. **11**, 445–449 (1931)
21. Pathak, A.K.: A Birkhoff interpolation problem on the unit circle in the complex plane. J. Indian Math. Soc. (N.S.) **73**(3–4), 227–233 (2006)
22. Rababah, A., Al-Refai, M., Al-Jarrah, R.: Computing derivatives of Jacobi polynomials using Bernstein transformation and differentiation matrix. Numer. Funct. Anal. Optim. **29**(5–6), 660–673 (2008)
23. Schoenberg, I.J.: On Hermite-Birkhoff interpolation. J. Math. Anal. Appl. **16**, 538–543 (1966)
24. Shi, Y.G.: Theory of Birkhoff Interpolation. Nova Science, New York (2003)
25. Shen, J., Tang, T., Wang, L.: Spectral methods: algorithms, analysis and applications. SIAM Rev. **55**(2), 405–406 (2013)
26. Turan, P.: Some open problems in approximation theory. Mat. Lapok **25**, 21–75 (1974)
27. Turan, P.: On some open problems of approximation theory. J. Approx. Theory **29**(1), 23–85 (1980)
28. Tadmor, E.: Stability of finite-difference, pseudospectral and Fourier-Galerkin approximations for time-dependent problems. SIAM Rev. **29**(4), 525–555 (1987)
29. Zhou, Y., Martin, C.F.: A regularized solution to the Birkhoff interpolation problem. Commun. Inf. Syst. **4**(1), 89–102 (2004)

On Relation Between P-Matrices and Regularity of Interval Matrices

Milan Hladík

Abstract We explore new results between P-matrix property and regularity of interval matrices. In particular, we show that an interval matrix is regular in and only if some special matrices constructed from its center and radius matrices are P-matrices. We also investigate the converse direction. We reduce the problem of checking P-matrix property to regularity of a special interval matrix. Based on this reduction, novel sufficient condition for a P-matrix property is derived, and its strength is inspected. We also state a new observation to interval P-matrices.

Keywords Interval matrix · P-matrix · Interval analysis · Linear complementarity

1 Introduction

Notation. The kth row of a matrix A is denoted as A_{k*}. The sign of a real r is defined as $\mathrm{sgn}(r) = 1$ if $r \geq 0$ and $\mathrm{sgn}(r) = -1$ otherwise; for vectors the sign is meant entrywise. For a vector y, the diagonal matrix with entries $y_1, \ldots, y_n$ is denoted by D_y. Eventually, $e = (1, \ldots, 1)^T$ stands for a vector of ones and $\rho(A)$ for the spectral radius of a matrix A.

Interval computation. An interval matrix is defined as

$$\mathbf{A} := \{A \in \mathbb{R}^{m\times n};\ \underline{A} \leq A \leq \overline{A}\},$$

where $\underline{A}$ and $\overline{A}$, $\underline{A} \leq \overline{A}$, are given matrices. The midpoint and radius matrices are defined as

$$A_c := \frac{1}{2}(\underline{A} + \overline{A}), \quad A_\Delta := \frac{1}{2}(\overline{A} - \underline{A}).$$

M. Hladík (✉)
Faculty of Mathematics and Physics, Department of Applied Mathematics,
Charles University, Malostranské Nám. 25, 11800 Prague, Czech Republic
e-mail: milan.hladik@matfyz.cz

N. Bebiano (ed.), *Applied and Computational Matrix Analysis*,
Springer Proceedings in Mathematics & Statistics 192,
DOI 10.1007/978-3-319-49984-0_2

The set of interval matrices of size $m \times n$ is denoted by $\mathbb{IR}^{m \times n}$. For definition of interval arithmetic see [8, 10], for instance.

We say that $\mathbf{A}$ is regular if every $A \in \mathbf{A}$ is nonsingular. Regularity of interval matrices is dealt with, e.g., in [5, 15, 16]. In particular, Rohn [16] presents forty equivalent characterizations. NP-hardness of checking regularity was proven by Poljak and Rohn [12, 13]. Sufficient conditions for checking regularity are surveyed in Rex and Rohn [14]. We recall the following one, due to Beeck [1].

Theorem 1 (Beeck [1]) *If* $\rho(|(A_c)^{-1}|A_\Delta) < 1$, *then* $\mathbf{A}$ *is regular.*

P-matrices. A square matrix is a P-matrix if all its principal minors are positive. P-matrices play an important role in linear complementarity problems [9, 22]

$$q + Mx \geq 0, \quad x \geq 0, \quad (q + Mx)^T x = 0.$$

Such a complementarity problem has a unique solution for each q if and only if M is a P-matrix. Since linear complementarity problems appear in so many situations (quadratic programming, bimatrix games, equilibria in specific economies, etc.), P-matrix property is of high importance.

Unfortunately, the problem of checking whether a given matrix is a P-matrix is known to be co-NP-hard [3, 7]. That is why diverse polynomially recognizable subclasses of P-matrices were studied; see [11, 24] and the references therein. Some of them are:

- positive definite matrices;
- M-matrices ($a_{ij} \leq 0\ \forall i, j$ and $A^{-1} \geq 0$);
- B-matrices ($\sum_{k=1}^n a_{ik} > 0$ and $\frac{1}{n}\sum_{k=1}^n a_{ik} > a_{ij}$ for $j \neq i$);
- H-matrices with positive diagonal entries (A is an H-matrix if $\langle A \rangle$ is an M-matrix, where $\langle A \rangle_{ii} = |a_{ii}|$ and $\langle A \rangle_{ij} = -|a_{ij}|, i \neq j$).

The related problem how to generate P-matrices was considered in [18, 24].

The following characterization of P-matrices is due to Fiedler and Pták [4].

Theorem 2 (Fiedler and Pták [4]) *A matrix* $A \in \mathbb{R}^{n \times n}$ *is a P-matrix if and only if for each vector* $x \neq 0$ *there is* i *such that* $x_i(Ax)_i > 0$.

The following relations between regularity of interval matrices and P-matrices are by Rohn [15].

Theorem 3 (Rohn [15]) *An interval matrix* $\mathbf{A} \in \mathbb{IR}^{n \times n}$ *is regular if and only if for each* $y \in \{\pm 1\}^n$ *the matrix* $A_c - D_y A_\Delta$ *is nonsingular and* $(A_c - D_y A_\Delta)^{-1}(A_c + D_y A_\Delta)$ *is a P-matrix.*

Theorem 4 (Rohn [15]) *Let* $\mathbf{A} \in \mathbb{IR}^{n \times n}$ *be regular. Then* $A_1^{-1} A_2$ *is a P-matrix for each* $A_1, A_2 \in \mathbf{A}$.

The following reduction of P-matrix property to interval matrix regularity comes from [19, 21].

Theorem 5 (Rump [21]) *Let $A \in \mathbb{IR}^{n\times n}$ with $A - I$ and $A + I$ nonsingular. Then A is a P-matrix if and only if $[(A-I)^{-1}(A+I) - I,\ (A-I)^{-1}(A+I) + I]$ is regular.*

Similar problem with convex combinations of rows or columns instead of full interval matrices was discussed in [6].

2 Results

Lemma 1 *Let $\mathbf{A} \in \mathbb{IR}^{n\times n}$ with A_c nonsingular. Then $\mathbf{A}$ is regular if and only if $I - A_c^{-1}R$ is a P-matrix for each $R \in [-A_\Delta, A_\Delta]$.*

Proof "Only if." Follows from Theorem 4 by choosing $A_1 := A_c$.

"If." Let $A \in \mathbf{A}$ be singular and denote $R := A_c - A \in [-A_\Delta, A_\Delta]$. Then there is $x \neq 0$ such that $Ax = (A_c - R)x = 0$, from which $(I - A_c^{-1}R)x = 0$. Therefore $I - A_c^{-1}R$ is singular and cannot be a P-matrix. □

Theorem 6 *Let $\mathbf{A} \in \mathbb{IR}^{n\times n}$ with A_c nonsingular. Then $\mathbf{A}$ is regular if and only if $I - A_c^{-1}D_yA_\Delta D_z$ is a P-matrix for each $y, z \in \{\pm 1\}^n$.*

Proof "Only if." Follows from Lemma 1.

"If." Suppose to the contrary that $\mathbf{A}$ is not regular. By Lemma 1, there is $R \in [-A_\Delta, A_\Delta]$ such that $I - A_c^{-1}R$ is not a P-matrix. Hence $I - R^TA_c^{-T}$ is not a P-matrix as well. By Theorem 2, there is $x \neq 0$ such that $x_i((I - R^TA_c^{-T})x)_i \leq 0$ for each i. Equivalently, $x_i^2 \leq x_i(R^TA_c^{-T}x)_i$ for each i. Define $y := \operatorname{sgn}(A_c^{-T}x)$ and $z := \operatorname{sgn}(x)$. Then

$$x_i^2 \leq x_i(R^TA_c^{-T}x)_i \leq x_i(z_i|R|^T|A_c^{-T}x|)_i \leq x_i(D_zA_\Delta^T|A_c^{-T}x|)_i = x_i(D_zA_\Delta^TD_yA_c^{-T}x)_i$$

for each i. Thus, $x_i((I - D_zA_\Delta^TD_yA_c^{-T})x)_i \leq 0$ for each i. This means that $I - D_zA_\Delta^TD_yA_c^{-T}$ is not a P-matrix, and also $I - A_c^{-1}D_yA_\Delta D_z$ is not a P-matrix. A contradiction. □

Remark. Since P-property is not changed by multiplying from the left and from the right by D_z, we can formulate the theorem also as follows: Let $\mathbf{A} \in \mathbb{IR}^{n\times n}$ with A_c nonsingular. Then $\mathbf{A}$ is regular if and only if $I - D_zA_c^{-1}D_yA_\Delta$ is a P-matrix for each $y, z \in \{\pm 1\}^n$.

Contrary to the characterization of regularity in Theorem 3, we have to use both diagonal matrices D_y and D_z. The following example illustrates it. Let

$$\mathbf{A} = \begin{pmatrix} [1,2] & [-1,1] \\ 1 & [1,2] \end{pmatrix}.$$

This interval matrix is not regular since it contains the all-one matrix. On the other hand, all matrices of the form $I - A_c^{-1}D_yA_\Delta$, $y \in \{\pm 1\}^n$, or of the form $I - A_c^{-1}A_\Delta D_z$, $z \in \{\pm 1\}^n$, are P-matrices.

Theorem 7 *Let $A \in \mathbb{R}^{n\times n}$. If $\alpha > 0$ is sufficiently small, then $P := \alpha A$ is a P-matrix if and only if $[(I-P)^{-1} - I,\ (I-P)^{-1} + I]$ is regular.*

Proof “If.” By Theorem 4, regularity of $\mathbf{M} := [(I-P)^{-1} - I,\ (I-P)^{-1} + I]$ implies that $M_c^{-1}\underline{M}$ is a P-matrix. This matrix, however, reads $M_c^{-1}\underline{M} = (I - P)((I-P)^{-1} - I) = I - (I-P) = P$.

“Only if.” By Theorem 6, have to verify that $I - (I-P)D_y I D_z$ is a P-matrix for each $y, z \in \{\pm 1\}^n$. Obviously, is it sufficient to verify matrices $I - (I-P)D_y$, $y \in \{\pm 1\}^n$, only. Without loss of generality suppose that $y = (-e^T, e^T)^T$, where the number of minus ones is k. Then $I - (I-P)D_y = PD_y + (I - D_y)$ has the form of

$$\left(\begin{array}{c|c} - & + \\ \hline - & + \end{array}\right) + \left(\begin{array}{c|c} 2I_k & 0 \\ \hline 0 & 0 \end{array}\right).$$

By the column linearity of determinants (applied on the first k columns), we can express the determinant of this matrix as

$$\sum_{J\subseteq\{1,\ldots,k\}} 2^{|J|}(-1)^{k-|J|}\alpha^{n-|J|}\det(A_J), \tag{1}$$

where A_J denotes the principal submatrix of A obtained by removing the rows and columns indexed by J. So, as $\alpha \to 0$, the dominant term in the summation is that for $J = \{1, \ldots, k\}$ and it draws

$$2^k\alpha^{n-k}\det(A_J).$$

Since A is a P-matrix, this term is positive, as well as the whole summation. Thus, $I - (I-P)D_y$ has the positive determinant. Its principal minors are positive for the same reasons. Therefore, $I - (I-P)D_y$ is a P-matrix. □

Remark 1 (*Estimation of α*) Here we estimate from below the sufficient value of α. This value should be small enough to ensure that (1) is positive, where $k > 0$ (case $k = 0$ holds trivially). That is,

$$\sum_{J\subseteq\{1,\ldots,k\}} 2^{|J|}(-1)^{k-|J|}\alpha^{k-|J|}\det(A_J) > 0.$$

This will be satisfied if

$$2^k\det(A_{\{1,\ldots,k\}}) > \sum_{J\subsetneq\{1,\ldots,k\}} 2^{|J|}\alpha^{k-|J|}\det(A_J).$$

Denote

$$m_1 = \min_{J \subsetneq \{1,\dots,k\}} \det(A_J),$$
$$m_2 = \max_{J \subsetneq \{1,\dots,k\}} \det(A_J).$$

Now, we can write a stronger inequality

$$\begin{aligned} 2^k m_1 &> m_2 \sum_{J \subsetneq \{1,\dots,k\}} 2^{|J|} \alpha^{k-|J|} \\ &= m_2(\alpha+2)^k - m_2 2^k. \end{aligned}$$

From this, we have

$$(\alpha+2)^k < 2^k(1 + m_1/m_2),$$

or,

$$\alpha < -2 + 2\sqrt[k]{1 + m_1/m_2}.$$

Due to overestimations, it suffices to take

$$\alpha := -2 + 2\sqrt[n]{1 + m_1/m_2}.$$

This value can be further simplified. By using concavity of log function and $e^x \geq x + 1$, we have

$$\begin{aligned} -2 + 2\sqrt[n]{1 + m_1/m_2} &= -2 + 2\exp\left(\frac{1}{n}\log(1 + m_1/m_2)\right) \\ &\geq -2 + 2\exp\left(\frac{1}{n}\big((1 - m_1/m_2)\log 1 + (m_1/m_2)\log 2\big)\right) \\ &= -2 + 2\exp\left(\frac{1}{n}(m_1/m_2)\log 2\right) \\ &\geq -2 + 2 + \frac{2}{n}(m_1/m_2)\log 2 = \frac{2}{n}(m_1/m_2)\log 2. \end{aligned}$$

The minimal and maximal determinants m_1 and m_2 can be estimated as follows. By Hadamard's inequality, we have

$$m_2 \leq \prod_{i=1}^{n} \|A_{i*}\|_2.$$

To estimate m_1 is a more involved task. For any nonsingular matrix $M \in \mathbb{R}^{n\times n}$, its determinant (and also sub-determinant) is bounded by

$$\det(M) = \det(M^{-1})^{-1} \geq \rho(M^{-1})^{-n} \geq \sigma_{\max}(M^{-1})^{-n} = \sigma_{\min}(M)^n.$$

This bound, however, can be very conservative. Anyway, we arrive at the possible value of

$$\alpha := \frac{2\log 2}{n} \cdot \frac{\sigma_{\min}(M)^n}{\prod_{i=1}^n \|A_{i*}\|_2}.$$

2.1 *Sufficient Conditions for P-Matrices*

Characterizations of P-matrix property from the previous section enables us to derive new sufficient conditions.

Theorem 8 *The matrix $A \in \mathbb{R}^{n\times n}$ is a P-matrix provided $A - I$ and $A + I$ are nonsingular and*

$$\rho(|(A+I)^{-1}(A-I)|) < 1. \tag{2}$$

Proof Let $A - I$ and $A + I$ be nonsingular. By Theorem 5, A is a P-matrix if and only if $[(A-I)^{-1}(A+I) - I,\ (A-I)^{-1}(A+I) + I]$ is regular. By employing the Beeck sufficient condition for regularity (Theorem 1), we arrive at the final form. □

Obviously, this condition is incomparable with positive definiteness. Moreover, it is also incomparable with M-matrix and H-matrix conditions. For example, the matrix

$$\begin{pmatrix} 46 & -19 \\ -33 & 14 \end{pmatrix}$$

is an M-matrix (and thus also H-matrix), but the condition (2) is not satisfied since the spectral radius is greater than 1.084 (verified by `versoft` [17]). On the other hand, the matrix

$$\begin{pmatrix} 1 & 1 & 1 \\ 1 & 2 & 3 \\ 1 & 4 & 10 \end{pmatrix}$$

is neither an M-matrix nor an H-matrix, but (2) is satisfied with the spectral radius less than 0.955.

Theorem 9 *The matrix $A \in \mathbb{R}^{n\times n}$ is a P-matrix provided for $I - \alpha A$ is nonsingular and $\rho(|I - \alpha A|) < 1$ for some $\alpha > 0$.*

Proof It follows again from the Beeck condition applied to $[(I - \alpha A)^{-1} - I, (I - \alpha A)^{-1} + I]$ and using Theorem 7. □

The latter condition is not new in the essence. If $\rho(|I - \alpha A|) < 1$, then $I - |I - \alpha A|$ is an M-matrix, so also $I - |I - \alpha A| - \operatorname{diag}(I - \alpha A) + \operatorname{diag}(|I - \alpha A|)$ is an M-matrix. The matrix $I - |I - \alpha A| - \operatorname{diag}(I - \alpha A) + \operatorname{diag}(|I - \alpha A|)$ is the comparison matrix of $I - (I - \alpha A) = \alpha A$, so αA is an H-matrix. Moreover, αA has positive diagonal since otherwise if $(\alpha A)_{ii} \leq 0$ for some i, then $|I - \alpha A|_{ii} \geq 1$ and so $\rho(|I - \alpha A|) \geq 1$. Therefore, the sufficient condition is weaker than checking if A is an H-matrix.

2.2 *Interval P-Matrices*

An interval matrix $A \in \mathbb{IR}^{n\times n}$ is called *an interval P-matrix* if each $A \in \mathbf{A}$ is a P-matrix [2, 7, 20]. A more general concept of P-matrix sets was investigated by Song and Gowda [23]. The following characterization of interval P-matrices is due to Białas and Garloff [2], see also [7].

Theorem 10 (Białas and Garloff [2]) $\mathbf{A} \in \mathbb{IR}^{n\times n}$ *is an interval P-matrix if and only if* $A_c - D_z A_\Delta D_z$ *is a P-matrix for each* $z \in \{\pm 1\}^n$.

As a direct consequence we have:

Corollary 1 *Let* $\mathbf{A} \in \mathbb{IR}^{n\times n}$ *such that* $A_c = D$ *is diagonal. Then* $\mathbf{A}$ *is an interval P-matrix if and only if* $\underline{A}$ *is a P-matrix.*

Proof We have that $\mathbf{A} \in \mathbb{IR}^{n\times n}$ is an interval P-matrix if and only if for each $z \in \{\pm 1\}^n$ the matrix $A_c - D_z A_\Delta D_z = D - D_z A_\Delta D_z$ is a P-matrix. This matrix is a P-matrix if and only if $D_z D D_z - A_\Delta = D - A_\Delta = \underline{A}$ is. □

Even though the assumption $A_c = D$ is strong, it might possibly help for checking interval P-matrix property. In a similar way, interval linear equation are often preconditioned such that the midpoint matrix becomes an identity matrix since this case is much easier to solve.

Another special case, reducing the interval P-matrix property to P-property of $\underline{A}$ only, is the following.

Corollary 2 *Let* $\mathbf{A} \in \mathbb{IR}^{n\times n}$ *such that* $A_\Delta = D$ *is diagonal. Then* $\mathbf{A}$ *is an interval P-matrix if and only if* $\underline{A}$ *is a P-matrix.*

Proof We have that $\mathbf{A} \in \mathbb{IR}^{n\times n}$ is an interval P-matrix if and only if for each $z \in \{\pm 1\}^n$ the matrix $A_c - D_z A_\Delta D_z = A_c - D_z D D_z = A_c - D = \underline{A}$ is a P-matrix. □

While Theorem 10 presents a reduction of interval to real P-matrix property, in the theorem below, we show a direct reduction to an elementary formula.

Theorem 11 *$\mathbf{A} \in \mathbb{IR}^{n\times n}$ is an interval P-matrix if and only if*

$$\det(D_{e-|y|} + D_{|y|}A_cD_{|z|} - D_yA_\Delta D_z) > 0 \quad (3)$$

for each $y, z \in \{0, \pm 1\}^n$ such that $|y| = |z|$.

Proof "Only if". This is obvious since $D_{e-|y|} + D_{|y|}A_cD_{|z|} - D_zA_\Delta D_z$ is a block diagonal matrix with entries either ones, or a principal submatrix of some $A \in \mathbf{A}$. Due to P-matrix property, this principal minor is positive.

"If". We use the result from Rohn [16] that an interval matrix $\mathbf{M} \in \mathbb{IR}^{k\times k}$ has all determinants positive, that is, $\det(M) > 0\ \forall M \in \mathbf{M}$, if and only if $\det(M_c - D_yA_\Delta D_z) > 0$ for all $y, z \in \{\pm 1\}^k$. Now, $\mathbf{A}$ is an interval P-matrix if and only if for each $A \in \mathbf{A}$, each minor of A is positive. A minor of A can be expressed as $\det(D_{e-s} + D_sAD_s)$ for some $s \in \{0, 1\}^n$. Thus, we have to show that for each $s \in \{0, 1\}^n$, all determinants of $D_{e-s} + D_s\mathbf{A}D_s$ are positive. By the above reasoning, this is equivalent to $\det(D_{e-s} + D_sA_cD_s - D_yD_sA_\Delta D_sD_z) > 0$ for all $y, z \in \{\pm 1\}^n$. When $s_i = 0$, the values of y_i and z_i play no role, so we can set $s = |y|$ and arrive at the resulting form of (3). □

Theorem 12 *The number of determinants in* (3) *is* 5^n.

Proof By the binomial formula, the number of determinants in (3) is

$$\sum_{k=0}^{n}\binom{n}{k}2^k2^k = \sum_{k=0}^{n}\binom{n}{k}4^k1^{n-k} = (4+1)^n = 5^n,$$

where k denotes the number of nonzero entries of y (or z), $\binom{n}{k}$ gives the number of vectors in $\{0, \pm 1\}^n$ with k nonzero entries, and 2^k counts the number of possibilities for y (and z) when the number of nonzero entries is k. □

3 Conclusion

We reviewed relations between P-matrix property and regularity of interval matrices. We also proposed some new observations and links. In particular, a reduction of interval matrix regularity to P-property and vice versa. As a consequence, new sufficient conditions for P-matrices were stated.

Some new open problems arised as well, e.g., determining a sharper estimation of α from Remark 1. Efficient utilization of Corollary 1 for interval P-matrix property checking is a challenging problem, too.

Acknowledgements The author was supported by the Czech Science Foundation Grant P402/13-10660S.

References

1. Beeck, H.: Zur Problematik der Hüllenbestimmung von Intervallgleichungssystemen. In: Nickel, K. (ed.) Interval Mathemantics: Proceedings of the International Symposium on Interval Mathemantics, LNCS, vol. 29, pp. 150–159. Springer, Berlin (1975)
2. Białas, S., Garloff, J.: Intervals of P-matrices and related matrices. Linear Algebra Appl. **58**, 33–41 (1984)
3. Coxson, G.E.: The P-matrix problem is co-NP-complete. Math. Program **64**, 173–178 (1994)
4. Fiedler, M., Pták, V.: On matrices with non-positive off-diagonal elements and positive principal minors. Czechoslov. Math. J. **12**, 382–400 (1962)
5. Fiedler, M., Nedoma, J., Ramík, J., Rohn, J., Zimmermann, K.: Linear Optimization Problems with Inexact Data. Springer, New York (2006)
6. Johnson, C.R., Tsatsomeros, M.J.: Convex sets of nonsingular and P-matrices. Linear Multilinear Algebra **38**(3), 233–239 (1995)
7. Kreinovich, V., Lakeyev, A., Rohn, J., Kahl, P.: Computational Complexity and Feasibility of Data Processing and Interval Computations. Kluwer, Dordrecht (1998)
8. Moore, R.E., Kearfott, R.B., Cloud, M.J.: Introduction to Interval Analysis. SIAM, Philadelphia (2009)
9. Murty, K.G.: Linear Complementarity, Linear and Nonlinear Programming. Heldermann Verlag, Berlin (1988)
10. Neumaier, A.: Interval Methods for Systems of Equations. Cambridge University Press, Cambridge (1990)
11. Peña, J.M.: A class of P-matrices with applications to the localization of the eigenvalues of a real matrix. SIAM J. Matrix Anal. Appl. **22**(4), 1027–1037 (2001)
12. Poljak, S., Rohn, J.: Radius of nonsingularity. Technical report. KAM Series (88-117), Department of Applied Mathematics, Charles University, Prague (1988)
13. Poljak, S., Rohn, J.: Checking robust nonsingularity is NP-hard. Math. Control Signals Syst. **6**(1), 1–9 (1993)
14. Rex, G., Rohn, J.: Sufficient conditions for regularity and singularity of interval matrices. SIAM J. Matrix Anal. Appl. **20**(2), 437–445 (1998)
15. Rohn, J.: Systems of linear interval equations. Linear Algebra Appl. **126**(C), 39–78 (1989)
16. Rohn, J.: Forty necessary and sufficient conditions for regularity of interval matrices: a survey. Electron. J. Linear Algebra **18**, 500–512 (2009)
17. Rohn, J.: VERSOFT: Verification software in MATLAB / INTLAB, version 10 (2009). URL http://www.nsc.ru/interval/Programing/versoft/
18. Rohn, J.: A note on generating P-matrices. Optim. Lett. **6**(3), 601–603 (2012)
19. Rohn, J.: On Rump's characterization of P-matrices. Optim. Lett. **6**(5), 1017–1020 (2012)
20. Rohn, J., Rex, G.: Interval P-matrices. SIAM J. Matrix Anal. Appl. **17**(4), 1020–1024 (1996)
21. Rump, S.M.: On P-matrices. Linear Algebra Appl. **363**, 237–250 (2003)
22. Schäfer, U.: A linear complementarity problem with a P-matrix. SIAM Rev. **46**(2), 189–201 (2004)
23. Song, Y., Gowda, M.S., Ravindran, G.: On some properties of P-matrix sets. Linear Algebra Appl. **290**(1–3), 237–246 (1999)
24. Tsatsomeros, M.J.: Generating and detecting matrices with positive principal minors. In: Li, L. (ed.) Focus on Computational Neurobiology, pp. 115–132. Nova Science Publishers, Commack (2004)

Interval Linear Algebra and Computational Complexity

Jaroslav Horáček, Milan Hladík and Michal Černý

Abstract This work connects two mathematical fields – computational complexity and interval linear algebra. It introduces the basic topics of interval linear algebra – regularity and singularity, full column rank, solving a linear system, deciding solvability of a linear system, computing inverse matrix, eigenvalues, checking positive (semi)definiteness or stability. We discuss these problems and relations between them from the view of computational complexity. Many problems in interval linear algebra are intractable, hence we emphasize subclasses of these problems that are easily solvable or decidable. The aim of this work is to provide a basic insight into this field and to provide materials for further reading and research.

Keywords Computational complexity · Interval linear algebra · Functional problems · Decision problems · NP-hardness · co-NP-hardness

1 Introduction

The purpose of this work is to emphasize relations between the two mathematical fields - interval linear algebra and computational complexity. This is not a pioneer work. Variety of relations between interval problems and computational complexity is covered by many papers. There are also few monographs that are devoted to this topic [4, 23, 48]. Some questions may arise in mind while reading the previous

J. Horáček (✉) · M. Hladík
Faculty of Mathematics and Physics, Department of Applied Mathematics,
Charles University, Malostranské nám. 25, 118 00 Prague, Czech Republic
e-mail: horacek@kam.mff.cuni.cz

M. Hladík
e-mail: hladik@kam.mff.cuni.cz

M. Černý
Faculty of Computer Science and Statistics, University of Economics,
nám. W. Churchilla 4, 13067 Prague, Czech Republic
e-mail: cernym@vse.cz

N. Bebiano (ed.), *Applied and Computational Matrix Analysis*,
Springer Proceedings in Mathematics & Statistics 192,
DOI 10.1007/978-3-319-49984-0_3

works. Among all, it is the question about the equivalence of the notions NP-hardness and co-NP-hardness. Some authors use these notions as synonyms. Some distinguish between them. Another questions that may arise touches the representation and reducibility of interval problems in a given computational model. We would like to shed more light (not only) on these issues.

Many well-known problems of classical linear algebra become intractable when we introduce intervals into matrices and vectors. However, not everything is lost. There are many interesting sub-classes of problems that behave well. We would like to point out these feasible cases, since they are interesting either from the theoretical or the computational point of view.

Our work does not aspire to substitute the classical monographs or handbooks. It lacks many of their details that are cited in the text. Nevertheless, it collects even some recent results that are missing in the monographs. It also provides links and reductions between the various areas of interval linear algebra. It provides a necessary and compact introduction to computational complexity and interval linear algebra. Then it considers complexity and feasibility of various well-known linear algebraic tasks when considered with interval structures – regularity and singularity, full column rank, solving a linear system, deciding solvability of a linear system, computing inverse matrix, eigenvalues, checking positive (semi)definiteness or stability.

We hope this paper should help newcomers to this area to improve her/his orientation in the field or professionals to provide a signpost to more deeper literature.

2 Interval Linear Algebra – Part I

Interval linear algebra is a mathematical field developed from classical linear algebra. The only difference is, that we do not work with real numbers but with real closed intervals

$$\mathbf{a} = [\underline{a}, \overline{a}],$$

where $\underline{a} \leq \overline{a}$. The set of all closed real intervals is denoted $\mathbb{IR}$ (the set of all closed rational intervals is denoted $\mathbb{IQ}$) We can use intervals for many reasons – in applications we sometimes do not know some parameters precisely, that is why, we rather use intervals of possible values; some real numbers are problematic (e.g., π, $\sqrt{2}, \ldots$) because it is not easy to represent them precisely, that is why, we can represent them with rigorous intervals containing them etc. With intervals we can define arithmetic (there are more possible definitions, we chose one of the most basic ones).

Definition 1 Let us have two intervals $\mathbf{x} = [\underline{x}, \overline{x}]$ a $\mathbf{y} = [\underline{y}, \overline{y}]$. The arithmetical operations $+, *, -, /$ are defined as follows

$$\mathbf{x} + \mathbf{y} = [\underline{x} + \underline{y}, \overline{x} + \overline{y}],$$
$$\mathbf{x} - \mathbf{y} = [\underline{x} - \overline{y}, \overline{x} - \underline{y}],$$

$$\mathbf{x} * \mathbf{y} = [\min(S), \max(S)], \text{ where } S = \{\underline{x}\underline{y}, \underline{x}\overline{y}, \overline{x}\underline{y}, \overline{x}\overline{y}\},$$
$$\mathbf{x} / \mathbf{y} = \mathbf{x} * (1/\mathbf{y}), \quad \text{where } 1/\mathbf{y} = [1/\overline{y}, 1/\underline{y}], \ 0 \notin \mathbf{y}.$$

Hence, we can use intervals instead of real numbers in formulas. However, we have to be careful. If there is a multiple occurrence of the same interval in a formula, the interval arithmetic does see them as two different intervals and we get an overestimation in the resulting interval. For example, let us have $\mathbf{x} = [-2, 1]$ and functions $f_1(x) = x^2$ and $f_2(x) = x * x$. Then we get

$$f_1(\mathbf{x}) = f_1([-2, 1]) = [-2, 1]^2 = [0, 4],$$
$$f_2(\mathbf{x}) = f_2([-2, 1]) = [-2, 1] * [-2, 1] = [-2, 4].$$

In the first case we see the optimal result, in the second case we see an overestimation. That is why, the form of our mathematical expression matters. However, we know the cases when the resulting interval is optimal [30].

Theorem 1 *Applying interval arithmetic on expressions in which all variables occur only once gives the optimal resulting interval.*

Using intervals we can build larger structures. In the interval linear algebra the main notion is an interval matrix. It is defined as follows:

$$\mathbf{A} = \{A \mid \underline{A} \leq A \leq \overline{A}\},$$

where $\underline{A}$, $\overline{A}$ are real $m \times n$ matrices called *lower* and *upper* bound and the relation $\leq$ is always understood componentwise. In another words, it is a matrix with coefficients formed by real closed intervals. In the following text, we will denote every interval structure in boldface. Since an interval vector is a special case of an interval matrix, we define it similarly. We can see that if all intervals in the structures are *degenerate*, i.e., $\underline{A} = \overline{A}$, we get a classical linear algebra. Therefore, interval linear algebra is actually a generalization of the previous one.

Another way to define an interval matrix is using its *midpoint* matrix A_c and its *radius* matrix $\Delta \geq 0$ as

$$\mathbf{A} = [A_c - \Delta, A_c + \Delta].$$

In the following text we automatically suppose that A_c, Δ represent corresponding midpoint and radius matrix of $\mathbf{A}$, and b_c, δ represent corresponding midpoint and radius vector of $\mathbf{b}$. When we talk about a general square matrix we automatically assume that it is of size n.

We mention some special structures that we will use quite often. The identity matrix is denoted I, the matrix containing only ones E and the vector containing only ones e. Another useful matrix is $D_y = \text{diag}(y_1, \ldots, y_n)$ a matrix with the vector y as the main diagonal. We often need to describe some properties of interval structures vectors consisting of only ± 1. We denote the set of all n-dimensional ± 1 vectors as Y_n. A useful concept is a matrix A_{yz} defined as

$$A_{yz} = A_c - D_y \Delta D_z,$$

for some given $y, z \in Y_n$. Each its coefficient on the positon (i, j) is an upper or a lower bound of $\mathbf{A}_{ij}$ depending on the sign of $y_i \cdot z_j$. We will sometimes need to check a spectral radius of a real matrix A, we denote it $\rho(A)$.

Many definitions have an intuitive generalization for interval linear algebra:

An interval matrix $\mathbf{A}$ *has a property* $\mathfrak{P}$ *if every* $A \in \mathbf{A}$ *has the property* $\mathfrak{P}$.

This applies to stability, full column rank, inverse nonnegativity, diagonally dominant matrices, M-matrix and H-matrix property, among others.

Many problems in interval linear algebra are very difficult to be computed exactly (e.g., computing the tightest possible verified interval containing eigenvalues of a general matrix). That is why we inspect the possibility of approximation of these bounds. There are several kinds of errors when we approximate a number a – absolute, relative [6] and inverse relative [22] approximation errors.

Definition 2 An algorithm computes a with *absolute approximation error* ε if it computes a^0 such that $a^0 \in [a - \varepsilon, a + \varepsilon]$.

An algorithm computes a with *relative approximation error* ε if it computes a^0 such that $a^0 \in (1 + [-\varepsilon, \varepsilon])a$.

An algorithm computes a with *inverse relative approximation error* ε if it computes a^0 such that $a \in (1 + [-\varepsilon, \varepsilon])a^0$.

At the end we mention a very useful theorem that we will use very often in this text. It originally comes from the area of numerical mathematics [31].

Theorem 2 (Oettli–Prager) *Let us have an interval matrix and vector* $\mathbf{A}, \mathbf{b}$. *For a real vector* $x \in \mathbb{R}^n$ *it holds* $Ax = b$ *for some* $A \in \mathbf{A}, b \in \mathbf{b}$ *if and only if*

$$|A_c x - b_c| \leq \Delta|x| + \delta.$$

This was just a brief introduction to interval analysis. Interval linear algebra has many important applications – system verification, model checking, handling uncertain data. For a huge variety of applications see, e.g., [17–19]. For more information or applications in nonlinear mathematics see [27].

3 Complexity Theory Background

Now, we take a small break and dig deeper into the area of computational complexity. After that we will return to interval linear algebra and introduce some well-known issues from the viewpoint of computational complexity.

3.1 Binary Encoding and Size of an Instance

For complexity-theoretic classification of interval-theoretic problems, it is a standard to use the Turing computation model. We assume that an instance of a computational problem is formalized as a bit-string, i.e., a finite 0-1 sequence. Thus we cannot work with real-valued instances; instead we usually restrict ourselves to *rational numbers* expressed as fractions $\pm\frac{q}{r}$ with $q, r \in \mathbb{N}$ written down in binary in the coprime form. Then, the *size* of a rational number $\pm\frac{q}{r}$ is understood as the number of bits necessary to write down the sign and both q and r (to be precise, one should also take care of delimiters). If an instance of a problem consists of multiple rational numbers $A = (a_1, \ldots, a_n)$ (e.g., when the input is a vector or a matrix), we define $size(A) = \sum_{i=1}^{n} size(a_i)$.

In interval-theoretic problems, inputs of algorithms are usually interval numbers, vectors or matrices. When we say that an algorithm is to process an $m \times n$ interval matrix $\mathbf{A}$, we understand that the algorithm is given the pair $(\underline{A} \in \mathbb{Q}^{m\times n}, \overline{A} \in \mathbb{Q}^{m\times n})$ and that the size of the input is $L := size(\underline{A}) + size(\overline{A})$. Whenever we speak about *complexity* of such algorithm, we mean a function $\phi(L)$ counting the number of steps of the corresponding Turing machine as a function of the bit-size L of the input $(\underline{A}, \overline{A})$.

Although the literature focuses mainly on the Turing model (and here we also do so), it would be interesting to investigate the behavior of interval-theoretic problems in other computational models, such as the Blum–Shub–Smale (BSS) model for real-valued computing [2] or the quantum model [1].

3.2 Functional Problems and Decision Problems

Formally, a *functional problem* F is a function $\mathsf{F} : \{0, 1\}^* \to \{0, 1\}^*$, where $\{0, 1\}^*$ is the set of all finite bit-strings. A *decision problem* (or *YES/NO problem*) A is a function $\mathsf{A} : \{0, 1\}^* \to \{0, 1\}$.[1]

If there exists a Turing machine computing $\mathsf{A}(x)$ for every $x \in \{0, 1\}^*$, we say that the problem A (either decision or functional) is *computable*.

It is well known that many decision problems in mathematics are uncomputable; e.g., deciding whether a given formula is provable in Zermelo–Fraenkel Set Theory is uncomputable by the famous Gödel Incompleteness Theorem. Fortunately, a majority of decision problems in interval linear algebra are computable. Such problems can usually be written down as arithmetic formulas (i.e., quantified formulas containing natural number constants, arithmetical operations $+$, $\times$, relations $=$, $\leq$ and propositional connectives). Such formulas are decidable (over the reals) by Tarski's Quantifier Elimination Method [33–35].

[1] In computer science it is sometimes emphasized that the functions are defined for each input, or *total* for short. This is to distinguish them from partially defined functions which are also studied in this area, namely within logic and recursion theory.

- *Example A*: *Regularity of an interval matrix*. Each matrix $A \in \mathbf{A}$ is nonsingular iff $(\forall A)[\underline{A} \le A \le \overline{A} \rightarrow \det(A) \ne 0]$. This formula is arithmetical since $\det(\cdot)$ is a polynomial, and thus it is expressible in terms of $+$, $\times$.
- *Example B*: *Is a given $\lambda \in \mathbb{Q}$ the largest eigenvalue of some symmetric $A \in \mathbf{A}$?* This question can be written down as $(\exists A)[A = A^T \;\&\; \underline{A} \le A \le \overline{A} \;\&\; (\exists x \ne 0)[Ax = \lambda x] \;\&\; (\forall \lambda')\{(\exists x' \ne 0)[Ax' = \lambda' x'] \rightarrow \lambda' \le \lambda\}]$.

Although Quantifier Elimination proves computability, it is a highly inefficient method from the practical viewpoint—the computation time can be doubly exponential in general. In spite of this, for many problems, reduction to Quantifier Elimination is the only (and thus "the best") known algorithmic result.

3.3 Weak and Strong Polynomiality

It is a usual convention to say that a problem A is "efficiently" solvable if it is solvable in polynomial time, i.e., in at most $p(L)$ steps of the corresponding Turing machine, where p is a polynomial and L is the size of the input. The class of efficiently solvable decision problems is denoted by P.

Taking a more detailed viewpoint, this is a definition of polynomial-time solvability in the *weak* sense. In our context, we are usually processing a family $a_1, \ldots, a_n$ of rational numbers, where $L = \sum_{i=1}^{n} size(a_i)$, performing arithmetical operations $+, -, \times, \div, \le$ with them. The definition of (weak) polynomiality implies that an algorithm *can perform at most $p_1(L)$ arithmetical operations with numbers of size at most $p_2(L)$ during its computation*, where p_1, p_2 are polynomials.

If a polynomial-time algorithm satisfies the stronger property that it *performs at most $p_1(n)$ arithmetical operations with numbers of size at most $p_2(L)$ during its computation*, we say that it is *strongly polynomial*. The difference is whether we can bound the number of arithmetical operations only by a polynomial in L, or by a polynomial in n.

Example Given a rational A and b, the question $(\exists x)[Ax = b]$ can be decided in strongly polynomial time (although it is nontrivial to implement the Gaussian elimination to yield a strongly polynomial algorithm). On the contrary, the question $(\exists x)[Ax \le b]$ (which is a form of linear programming) is known to be solvable in weakly polynomial time only and it is a major open question whether a strongly polynomial algorithm exists (this is Smales's Ninth Millenium Problem, see [54]).

The main message of the previous example is: whenever an interval-algebraic problem is solvable in polynomial time and requires linear programming (which is a frequent case), it is only a weakly polynomial result. This is why the rare cases, when interval-algebraic problems are solvable in strongly polynomial time, are of special interest.

3.4 NP, coNP

Recall that NP is the class of decision problems A with the following property: there is a polynomial p and a decision problem $\mathsf{B}(x, y)$, solvable in time polynomial in $size(x) + size(y)$, such that, for any instance $x \in \{0, 1\}^*$,

$$\mathsf{A}(x) = 1 \text{ iff } (\exists y \in \{0,1\}^*) \underbrace{size(y) \leq p(size(x))}_{(\star)} \text{ and } \mathsf{B}(x, y) = 1. \quad (1)$$

The string y is called *witness* for the $\exists$-quantifier, or also *witness* of the fact that $\mathsf{A}(x) = 1$. The algorithm for $\mathsf{B}(x, y)$ is called *verifier*. For short, we often write $\mathsf{A}(x) = (\exists^p y)\mathsf{B}(x, y)$, showing that A results from the $\exists$-quantification of the efficiently decidable question B (and the quantifier ranges over strings of polynomially bounded size). Observe that the question $(\exists^p y)\mathsf{B}(x, y)$ need not be decidable in polynomial time (in fact, this is the open problem "P $=^?$ NP"), since the quantification range is exponential in $size(x)$.

A lot of $\exists$-problems from various areas of mathematics are in NP: *"does a given boolean formula x have a satisfying assignment y?"*, *"does a given graph x have 3-coloring y?"*, *"does a given system x = 'Ay ≤ b' have an integral solution y?"*, and many others.

The class coNP is characterized by replacement of the quantifier in (1):

$$\mathsf{A}(x) = 1 \text{ iff } (\forall y \in \{0,1\}^*) \; size(y) \leq p(size(x)) \rightarrow \mathsf{B}(x, y) = 1.$$

It is easily seen that the class coNP is formed of complements of NP-problems, and vice versa. (Recall that a decision problem A is a 0-1 function; its *complement* is defined as $\mathsf{coA} = 1 - \mathsf{A}$.)

The prominent example of a coNP-question is deciding whether a boolean formula is a tautology, or in other words, *"given a boolean formula x, is it true that* ***every*** *assignment y makes it true?"*.

It is easy to see again that deciding a coNP-question can take exponential time since the $\forall$-quantifier ranges over a set exponentially large in $size(x)$.

Example Interval linear algebra is not an exception: a lot of $\exists$-questions belong to NP, but we should be careful a bit. As an example, consider the problem SINGULARITY: given $\mathbf{A} \in \mathbb{IQ}^{n \times n}$, $\exists A \in \mathbf{A}$ which is singular? We could expect that SINGULARITY $\in$ NP since the positive answer can be certified by the $\exists$-witness $A_0 =$ *a particular singular matrix in* $\mathbf{A}$. Indeed, the natural verifier $\mathsf{B}(\mathbf{A}, A_0)$, checking whether $A_0 \in \mathbf{A}$ and A_0 is singular, works in polynomial time. But a problem is hidden in the condition $(\star)$ in (1). To be fully correct, we would have to prove: *there exists a polynomial p such that whenever* $\mathbf{A}$ *contains a singular matrix, then it also contains a rational singular matrix* A_0 *such that* $size(A_0) \leq p(L)$, *where* $L = size(\underline{A}) + size(\overline{A})$. Direct proofs of such properties are "uncomfortable". But we can proceed in a more elegant way, using Theorem 2:

$$\begin{aligned}
&\exists A \in \mathbf{A} \text{ s.t. } A \text{is singular} \\
&\quad \Leftrightarrow \exists A \in \mathbf{A},\ \exists x \neq 0 \text{ s.t. } Ax = 0 \\
&\quad \Leftrightarrow \exists x \neq 0 \text{ s.t. } -\Delta|x| \leq A_c x \leq \Delta|x|, \\
&\quad \Leftrightarrow \exists s \in \{\pm 1\}^n \underbrace{\exists x \text{ s.t. } -\Delta D_s x \leq A_c x \leq \Delta D_s x,\ D_s x \geq 0,\ e^T D_s x \geq 1}_{(\dagger)}. \qquad (2)
\end{aligned}$$

Given $s \in \{\pm 1\}^n$, the relation (†) can be checked in polynomial time by linear programming. Thus, we can define the verifier $\mathsf{B}(\mathbf{A}, s)$ as the algorithm checking the validity of (†). In fact, we have reformulated the ∃-question, *"is there a singular $A \in \mathbf{A}$?"*, *into an equivalent* ∃-question, *"is there a sign vector $s \in \{\pm 1\}^n$ s.t.* (†) *holds true?"*, and now $size(s) \leq L$ is obvious.

The method of (2) is known as *orthant decomposition* since it reduces the problem to inspection of orthants $D_s x \geq 0$, for every $s \in \{\pm 1\}^n$, and the work in each orthant is "easy" (here, the work in an orthant amounts to a single linear program). Many properties with interval data are described by sufficient and necessary conditions that use orthant decomposition.

We can also immediately see that REGULARITY = coSINGULARITY (*"given* **A**, *is **every** $A \in \mathbf{A}$ nonsingular?"*) belongs to coNP.

3.5 Decision Problems: NP-, coNP-Completeness

A decision problem A is *reducible* to a decision problem B (denoted A ≤ B) if there exists a polynomial-time computable function $g : \{0, 1\}^* \to \{0, 1\}^*$, called *reduction*, such that for every $x \in \{0, 1\}^*$ we have

$$\mathsf{A}(x) = \mathsf{B}(g(x)). \qquad (3)$$

Said informally, any algorithm for B can also be used for solving A: given an instance x of A, we can efficiently "translate" it into an instance $g(x)$ of the problem B and run the method deciding $\mathsf{B}(g(x))$, yielding the correct answer to $\mathsf{A}(x)$. Thus, any decision method for B is also a valid method for A, if we admit the polynomial time for computation of the reduction g. In this sense we can say that if A ≤ B, then B "as hard as A, or harder". If both A ≤ B and B ≤ A, then problems A, B are called *polynomially equivalent*.

The relation ≤ induces a partial ordering on classes of polynomially equivalent problems in NP (called *NP-degrees*) and this ordering can be shown to have a maximum element. The problems in the maximum class are called *NP-complete* problems. And similarly, coNP has a class of *coNP-complete* problems. They are complementary: a problem A is NP-complete iff its complement is coNP-complete.

Let $\mathscr{X}$ be one of the classes NP or coNP. If a problem B is $\mathscr{X}$-complete, any method for it can be understood as a universal method for any problem $\mathsf{A} \in \mathscr{X}$, modulo polynomial time needed for computing the reduction. Indeed, since B is the maximum element, we have $\mathsf{A} \leq \mathsf{B}$ for any $\mathsf{A} \in \mathscr{X}$. It is generally believed that $\mathscr{X}$ contains problems which are not efficiently decidable. In NP, boolean satisfiability is a prominent example; in coNP, it is the tautology problem. Then, by $\leq$-maximality, no $\mathscr{X}$-complete problem is efficiently decidable. This shows why a proof of $\mathscr{X}$-completeness of a newly studied problem is often understood as proof of its computational *intractability*.

Remark From a practical perspective, a proof of NP- or coNP-completeness is the same bad news, telling us that "nothing better than superpolynomial-time algorithms can be expected". But formally we must distinguish between NP- and co-NP completeness because it is believed that NP-complete problems are not polynomially equivalent with coNP-complete problems. (This is the "NP $=^?$ coNP" open problem).

NP- and coNP-complete problems in interval analysis. A survey of such problems forms the core of this paper. An important example of an NP-complete problem is SINGULARITY of an interval matrix **A**. Its complement, REGULARITY, is thus coNP-complete.

When we know that B is $\mathscr{X}$-complete and we prove $\mathsf{B} \leq \mathsf{C}$ for a problem $\mathsf{C} \in \mathscr{X}$, then C is also $\mathscr{X}$-complete. This is *the* method behind all $\mathscr{X}$-completeness proofs of this paper. For example, let EIGENVALUE be the problem "given a square interval matrix **A** and a number λ, decide whether λ is an eigenvalue of some $A \in \mathbf{A}$". It is easy to prove SINGULARITY $\leq$ EIGENVALUE; indeed, if we are to decide whether there is a singular matrix $A \in \mathbf{A}$, it suffices to use the reduction $g : \mathbf{A} \mapsto (\mathbf{A}, \lambda = 0)$. The proof of EIGENVALUE $\in$ NP can be derived from the orthant decomposition method; this proves that EIGENVALUE is an NP-complete problem.

3.6 Decision Problems: NP-, coNP-Hardness

We restrict ourselves to NP-hard problems; the reasoning for coNP-hard problems is analogous.

In the previous section we spoke about NP-complete problems as the $\leq$-maximum elements in NP. But our reasoning can be more general. We can work on the entire class of decision problems, including those outside NP. We say that a decision problem H, not necessarily in NP, satisfying $\mathsf{C} \leq \mathsf{H}$ for an NP-complete problem C, is *NP-hard*. Clearly: NP-complete problems are exactly those NP-hard problems which are in NP. But we might encounter a problem H for which we do not have the proof $\mathsf{H} \in$ NP, but still it might be possible to prove $\mathsf{C} \leq \mathsf{H}$. Then the bad news for practice is again the same, that the problem H is computationally intractable. (But we might possibly need even worse computation time than for NP-problems; recall that all problems in NP can be solved in exponential time, not worse.)

To summarize: a proof that a decision problem is NP-hard is a weaker theoretical result than a proof that a decision problem is NP-complete; it leads to an immediate research problem to inspect *why it is difficult to prove the presence in NP*. Usually, the reason is that it is not easy (or impossible at all) to write down the $\exists$-definition; recall the example (2), where the proof of presence in NP required the aid of Theorem 2.

Remark If we are unsuccessful in placing the problem in NP or coNP, being unable to write down the $\exists$- or $\forall$-definition, it might be appropriate to place the problem H into higher levels of the Polynomial Time Hierarchy, or even higher, such as the PSPACE-level; for details see [1], Chap. 5.

3.7 Functional Problems: Efficient Solvability and NP-hardness

Functional problems are problems of computing values of general functions, in contrast to decision problems where we expect only YES/NO answers. We also want to classify functional problems from the complexity-theoretic perspective, whether they are "efficiently solvable", or "intractable", as we did with decision problems. Efficient solvability of a functional problem is again generally understood as polynomial-time computability. To define NP-hardness, we need the following notion of reduction: a decision problem A is *reducible* to a functional problem F, if there exist functions $g : \{0, 1\}^* \to \{0, 1\}^*$ and $h : \{0, 1\}^* \to \{0, 1\}$, both computable in polynomial time, such that

$$\mathsf{A}(x) = h(\mathsf{F}(g(x))) \text{ for every } x \in \{0, 1\}^*. \tag{4}$$

The role of g is analogous to (3): it translates an instance x of A into an instance $g(x)$ of F. What is new here is the function h. Since F is a functional problem, the value $\mathsf{F}(g(x))$ can be an arbitrary bitstring (say, a binary representation of a rational number); then we need another efficiently computable function h translating the value $\mathsf{F}(g(x))$ into a 1-0 value giving the YES/NO answer to $\mathsf{A}(x)$. A trivial example: deciding regularity of a rational matrix (decision problem A) is reducible to the computation of rank (functional problem F). It suffices to define $g(A) = A$ and $h(\zeta) = 1 - \min\{n - \zeta, 1\}$.

Now, a functional problem F is *NP-hard* if there is an NP-hard decision problem reducible to F. For example, the functional problem of counting the number of ones in the truth-table of a given boolean formula is NP-hard since this information allows us to decide whether or not the formula is satisfiable.

Remark (*It is not necessary to distinguish between NP-hardness and coNP-hardness for functional problems*) We could also try to define coNP-hardness of a functional problem G in terms of reducibility of a coNP-hard decision problem C to G via (4). But this is superfluous because here NP-hardness and coNP-hardness coincide. Indeed, if we can reduce a coNP-hard problem C to a functional problem G via (g, h), then we can also reduce the NP-hard problem coC to G via $(g, 1 - h)$. Thus, in case of functional problems, we speak about NP-hardness only.

3.8 More General Reductions: Do We Indeed Have to Distinguish Between NP-hardness and coNP-Hardness of Decision Problems?

In literature, the notions of NP-hardness and coNP-hardness are sometimes used quite freely even for *decision problems*. Sometimes we can read that a decision problem is "NP-hard", even if it would qualify as a coNP-hard problem under our definition based on the reduction (3). This is nothing serious as far as we are aware. It depends how the author understands the notion of a reduction between two decision problems. We have used the *many-one* reduction (3), known also as *Karp* reduction, between two decision problems. This is a standard in complexity-theoretic literature.

However, one could use a more general reduction between two decision problems A, B. For example, taking inspiration from (4), we could define "$\mathsf{A} \leq' \mathsf{B}$ iff $\mathsf{A}(x) = h(\mathsf{B}(g(x)))$ for some polynomial-time computable functions g, h". Then the notions of $\leq'$-NP-hardness and $\leq'$-coNP-hardness coincide and need not be distinguished. (Observe that h must be a function from $\{0, 1\}$ to $\{0, 1\}$ and there are only two such nonconstant functions : $h_1(\xi) = \xi$ and $h_2(\xi) = 1 - \xi$. If we admit only h_1, we get the many-one reduction; if we admit also the negation h_2, we have a generalized reduction under which a problem is NP-hard iff it is coNP-hard. Thus: the notions of NP-hardness and coNP-hardness based on many-one reductions do not coincide just because many-one reductions *do not admit the negation* of the output of $\mathsf{B}(g(x))$.)

To be fully precise, one should always say "a problem A is $\mathscr{X}$-hard w.r.t. a particular reduction $\preceq$". For example, in the previous sections we spoke about $\mathscr{X}$-hard problems for $\mathscr{X} \in \{\text{NP}, \text{coNP}\}$ w.r.t. the many-one reduction (3). If another author uses $\mathscr{X}$-hardness w.r.t. $\leq'$ (e.g., because (s)he considers the ban of negation as too restrictive in her/his context), then (s)he need not distinguish between NP-hardness and coNP-hardness.

For the sake of completeness, we conclude that in literature we can meet the notions of hardness w.r.t. various types of reductions.

Logspace-computable reduction: $\mathsf{A} \leq_{\log} \mathsf{B}$ iff there is a function g computable in memory of size $O(\log \mathit{size}(x))$, such that $\mathsf{A}(x) = \mathsf{B}(g(x))$ for every x. (This reduction in weaker than (3) since every logspace-computable function is also computable in polynomial time.)

Truth-table reduction: $\mathsf{A} \leq_{tt} \mathsf{B}$ iff there is a finite number of polynomial-time computable functions $g_1, \ldots, g_k : \{0, 1\}^* \to \{0, 1\}^*$ and a "truth-table" function $h : \{0, 1\}^k \to \{0, 1\}$ such that $\mathsf{A}(x) = h(\mathsf{B}(g_1(x)), \ldots, \mathsf{B}(g_k(x)))$. This reduction is a generalization of $\leq'$; indeed, $\leq'$ is a restricted truth-table reduction with a two-line truth table. Under $\leq_{tt}$, to decide $\mathsf{A}(x)$ one can compute k instances of B from which the boolean expression h combines the result $\mathsf{A}(x)$.

Turing reduction (or *Cook reduction*): $\mathsf{A} \leq_T \mathsf{B}$ iff there is a polynomial-time algorithm (Turing machine) Q, equipped with a subroutine (an algorithm, *oracle*) computing B, and the entire computation of B is counted as a single step of Q. This is the most general type of reduction: when deciding $\mathsf{A}(x)$, the reduction allows for a polynomial number of computations of $\mathsf{B}(y)$ with $\mathit{size}(y)$ polynomially bounded in *size*(x), and the results can be combined in an arbitrary way; the only limitation is that the overall number of steps is polynomial in *size*(x), assuming that one computation of $\mathsf{B}(y)$ is at the unit cost.

The above mentioned reductions can be ordered in the sequence according to their generality: $\mathsf{A} \leq_{\log} \mathsf{B} \Rightarrow \mathsf{A} \leq \mathsf{B} \Rightarrow \mathsf{A} \leq' \mathsf{B} \Rightarrow \mathsf{A} \leq_{tt} \mathsf{B} \Rightarrow \mathsf{A} \leq_T \mathsf{B}$, where "$\Rightarrow$" means "implies". We know that NP-hardness and coNP-hardness coincide for $\leq'$, and thus also for the generalizations $\leq_{tt}$, $\leq_T$.

3.9 A Reduction-Free Definition of Hardness

For practical purposes, when we do not want to play with properties of particular reductions, we can define the notion of a "hard" problem H (either decision of functional) intuitively as a problem fulfilling this implication: *if* H *is decidable/solvable in polynomial time, then* $\mathrm{P} = \mathrm{NP}$. This is usually satisfactory for the practical understanding of the notion of computational hardness. (Under this definition: if $\mathrm{P} = \mathrm{NP}$, then every decision problem is hard; and if $\mathrm{P} \neq \mathrm{NP}$, then the class of hard decision problems is exactly the class of decision problems not decidable in polynomial time, including all NP-hard and coNP-hard decision problems.)

Even if we accept this definition and do not speak about reductions explicitly, all hardness proofs (at least implicitly) contain some kinds of reductions of previously known hard problems to the newly studied ones.

4 Interval Linear Algebra – Part II

In the following sections we will deal with various problems in interval linear algebra. There are many interesting topics that are unfortunately beyond the scope of this work. We will at least point out some of them in Sect. 4.10. We chose basic topics from introductory courses to linear algebra – regularity and singularity of a matrix, full column rank, solving and solvability of a system of linear equations, matrix inverse, determinant, eigenvalues and eigenvectors, positive (semi)definiteness and stability. The next chapters will offer a great disappointment and also a great challenge, since implanting intervals into a classical linear algebra makes solving most of the problems intractable. That is why, we look for solving relaxed problems, special feasible subclasses of problems or for sufficient conditions checkable in polynomial time. Interval linear algebra still offers many open problems and a lot of place for further research. At the end of each section we present a summary of problems and their complexity. If we only know that a problem is weakly polynomial yet, we just write that it belongs to the class P. When complexity of a problem is not known to our best knowledge (or it is an open problem), we mark it with question mark.

4.1 Regularity and Singularity

Deciding regularity and singularity of an interval matrix is an important task in linear algebra. The definition of interval regularity (and singularity) is intuitive.

Definition 3 A square interval matrix $\mathbf{A}$ is *regular* if every $A \in \mathbf{A}$ is nonsingular. Otherwise, $\mathbf{A}$ is called *singular*.

Considering complexity we can find in the literature the following theorem [42] giving NP-completeness result even for the simple case.

Theorem 3 *Deciding whether an interval matrix* $\mathbf{A} = [A - E, A + E]$ *is singular for some nonnegative symmetric positive definite rational matrix* A *is NP-complete.*

We can prove NP-hardness of this decision problem. Moreover, we get NP-completeness since we know that a singular $\mathbf{A}$ in this form mentioned in the theorem must contain a singular matrix

$$A - \frac{zz^T}{z^T A^{-1} z},$$

for some $z \in \{\pm 1\}^n$ [42] which is a polynomial witness and the above mentioned matrix is checkable in polynomial time (e.g., by Gaussian elimination). This implies that deciding singularity of a general interval matrix is NP-hard. However, in the Sect. 3.4 we saw the construction of a polynomial witness $z \in \{\pm 1\}^n$ certifying that an interval matrix is singular. Hence, we get that checking singularity of a general interval matrix is NP-complete. Clearly, checking regularity as the complement problem to singularity is coNP-complete.

The sufficient and necessary conditions for checking regularity are of exponential nature. In [46] you can see 40 of them. For example, we can use the classical definition of matrix regularity (a matrix A is regular if the system $Ax = 0$ has only trivial solution) and combine it with Oettli–Prager theorem. We get that an interval matrix is regular if and only if the inequality

$$|A_c x| \leq \Delta |x|,$$

has only trivial solution.

Fortunately, there are some sufficient conditions that are computable in polynomial time. It is advantageous to have more conditions, because some of them may suit better to a certain class of matrices or limits of our software tools. Here we present three sufficient conditions for checking regularity and three sufficient conditions for checking singularity.

Theorem 4 (Sufficient conditions for regularity) *An interval matrix* $\mathbf{A} = [A_c - \Delta, A_c + \Delta]$ *is regular if at least one of the following conditions holds*

1. $\rho(|A_c^{-1}|\Delta) < 1$ *[42],*
2. $\sigma_{\max}(\Delta) < \sigma_{\min}(A_c)$ *[50],*

3. $A_c^T A_c - \|\Delta^T \Delta\| I$ *is positive definite for some consistent matrix norm* $\|\cdot\|$ *[36].*

Theorem 5 (Sufficient conditions for singularity) *An interval matrix* $\mathbf{A} = [A_c - \Delta, A_c + \Delta]$ *is singular if at least one of the following conditions holds*

1. $\max_j (|A_c^{-1}|\Delta)_{jj} \geq 1$ *[37],*
2. $(\Delta - |A_c|)^{-1} \geq 0$ *[42],*
3. $\Delta^T \Delta - A_c^T A_c$ *is positive semidefinite [36].*

In the two theorems above, the first condition in the triplet is among the most frequently used sufficient conditions. You can find more sufficient conditions for regularity and singularity in [36].

We can also take a look at the classes of interval matrices that are immediately regular. These are, for example, diagonally dominant matrices [53], M-matrices and H-matrices [30]. These properties are checkable in polynomial time.

Concerning the regularity, in applications we are sometimes interested in *radius of nonsingularity*. This number describes how close is A to a singular matrix – given an $n \times n$ matrix A we are interested in componentwise distance to the nearest singular matrix. This problem is also NP-hard. For more information see e.g., [8, 42].

Summary

Problem	*Complexity*
Is $\mathbf{A}$ regular?	coNP-complete
Is $\mathbf{A}$ singular?	NP-complete
Computing radius of nonsingularity of some A	NP-hard

4.2 Full Column Rank

The definition of the full column rank is natural.

Definition 4 An $m \times n$ interval matrix $\mathbf{A}$ has *full column rank* if every $A \in \mathbf{A}$ has full column rank (i.e., it has rank n).

Deciding whether an interval matrix has full column rank is connected to checking regularity. If an interval matrix $\mathbf{A}$ of size $m \times n$, $m \geq n$, contains a regular submatrix of size n, then obviously $\mathbf{A}$ has a full column rank. What is surprising is that the implication does not hold conversely (in contrast to real matrices). The interval matrix by Irene Sharaya (see [53]) might serve as a counterexample.

$$\mathbf{A} = \begin{pmatrix} 1 & [0, 1] \\ -1 & [0, 1] \\ [-1, 1] & 1 \end{pmatrix}.$$

It has full column rank, but contains no regular submatrix of size 2.

For square matrices, checking regularity can be polynomially reduced to checking full column rank (we just check the matrix $\mathbf{A}$), but the converse is not so easy. Therefore, checking full column rank is coNP-hard. Finding a polynomial certificate for an interval matrix not having full column rank can be done by orthant decomposition similarly as in the case of singularity. That is why, checking full column rank is also coNP-complete.

Again, fortunately, we have some sufficient conditions that are computable in polynomial time.

Theorem 6 *Let $\mathbf{A} = [A_c - \Delta, A_c + \Delta]$ be an $m \times n$ interval matrix. This matrix has full column rank if at least one of the following conditions holds*

1. *A_c has full column rank and $\rho(|A_c^\dagger|\Delta) < 1$, [48],*
2. *$\sigma_{\max}(\Delta) < \sigma_{\min}(A_c)$, [53].*

The symbol † stands for Moore–Penrose inverse (for more details see [26]). The first condition is mentioned implicitly in [48], however the explicit proof can be found in [53]. Notice that the second sufficient condition is the same as the sufficient condition for checking regularity. Many problems can be transformed to checking full column rank – e.g., deciding whether a given interval linear system is solvable, deciding whether a solution set of an interval linear system is bounded.

Summary

Problem	*Complexity*
Does $\mathbf{A}$ have full column rank?	coNP-complete

4.3 Solving a System of Linear Equations

To be brief the title of this section contained the word "solving". Nevertheless, this notion could be a little misguiding. Let us explain what do we mean by solving a system of interval linear equations (or interval linear system for short). The solution set of an interval linear system is defined as follows.

Definition 5 Let $\mathbf{A}x = \mathbf{b}$, where $\mathbf{A}$ is an $m \times n$ interval matrix and $\mathbf{b}$ is an m-dimensional right-hand side vector. Then by a *solution set* Σ we mean

$$\Sigma = \{x \mid Ax = b \text{ for some } A \in \mathbf{A},\ b \in \mathbf{b}\}.$$

We could imagine it as a collection of all solutions of all crisp real systems contained within the bounds of an interval system. Unfortunately, this set is of quite a complex shape. For its description we can use the already mentioned Oettli–Prager Theorem 2. A vector $x \in \mathbb{R}^n$ is a *solution* of $\mathbf{A}x = \mathbf{b}$ (i.e., $x \in \Sigma$) if and only if x satisfies

$$|A_c x - b_c| \le \Delta|x| + \delta.$$

We can see that checking whether a vector y is a solution of $\mathbf{A}x = \mathbf{b}$ is strongly polynomial (we just check the inequality for y).

Oettli–Prager theorem implies that the set Σ is generally non-convex but convex in each orthant (for graphical examples of possible shapes of the solution set see e.g., [14, 27, 29]). To describe this set, we usually enclose it by an n-dimensional box (aligned with axes). Notice that we can view an n-dimensional interval vector as an n-dimensional box aligned with axes.

Definition 6 An n-dimensional interval vector $\mathbf{x}$ is called an interval *enclosure* of Σ if $\Sigma \subseteq \mathbf{x}$. If it is the tightest possible enclosure w.r.t. inclusion (there is no interval box $\mathbf{y}$ such that $\Sigma \subseteq \mathbf{y} \subsetneq \mathbf{x}$), we call $\mathbf{x}$ the interval *hull*.

By *solving* an interval linear system we understand computing any enclosure $\mathbf{x}$ of its solution set Σ. To be brief, we call that $\mathbf{x}$ an enclosure (or the hull) of $\mathbf{A}x = \mathbf{b}$. The notion of enclosure is quite intuitive because we are not always able to compute the interval hull. In [23] we can see that computing the exact hull of $\mathbf{A}x = \mathbf{b}$ is NP-hard.

An interval $\mathbf{a} = [a - \Delta, a + \Delta]$ is absolutely δ-narrow if $\Delta \le \delta$ and relatively δ-narrow if $\Delta \le \delta \cdot |a|$. The problem is still NP-hard even if we limit widths of intervals of a matrix in a system with some $\delta > 0$ [23]. We can summarize it in the following theorem.

Theorem 7 *For every $\delta > 0$, the problem of computing the hull of $\mathbf{A}x = \mathbf{b}$, where $\mathbf{a}_{ij}$, $\mathbf{b}_i$ are both absolutely and relatively δ-narrow is NP-hard.*

Unfortunately, even computing various ε-approximations of the hull components is an NP-hard problem [23].

Theorem 8 *For a given $\varepsilon > 0$ computing the relative and absolute ε-approximation of the hull (its components) of $\mathbf{A}x = \mathbf{b}$ are NP-hard problems.*

That is why, we are usually looking for enclosures, not the hull. Of course, the tighter enclosure the better. For computing enclosures of square systems, there have been various methods developed. Some of them extend the traditional algorithms for the real systems, such as the Gaussian elimination, Jacobi or Gauss–Seidel method [27, 30]. Some of them were designed specifically for interval systems; see for instance [4, 9, 13, 21, 27, 30] among many others.

Overdetermined systems. For an *overdetermined* system (where $\mathbf{A}$ is an $m \times n$ matrix with $m > n$) the situation is slightly more difficult. Many people automatically think of solving overdetermined systems via least squares, i.e.,

Definition 7

$$\Sigma^{lsq} = \{x \mid A^T A x = A^T b \text{ for some } A \in \mathbf{A}, b \in \mathbf{b}\}.$$

Obviously, Σ^{lsq} is not the same set as Σ. Nevertheless, it is not difficult to see that $\Sigma \subseteq \Sigma^{lsq}$. Hence, we can use methods computing least squares for enclosing Σ [29]. The problem of computing the interval hull of Σ^{lsq} is NP-hard, since when $\mathbf{A}$ is square and regular, then $\Sigma^{lsq} = \Sigma$ and computing the exact hull of Σ is NP-hard even for $\mathbf{A}$ regular [4].

If we primarily focus on enclosing just Σ there is a variety of methods – modified Gaussian elimination for overdetermined systems [7], method developed by Rohn [43], Popova [32], or a method using square subsystems [15].

We can try to identify some classes of systems with exact hull computation algorithms that run in polynomial time. If we restrict the right hand side $\mathbf{b}$ to contain only degenerate intervals, we have $\mathbf{A}x = b$. Then, this problem is still NP-hard [23]. If we, however, restricts the matrix to be consisting only of degenerate intervals A and we have a system $Ax = \mathbf{b}$, then, computing exact bounds of the solution set is polynomial, since it can be rewritten as a linear program.

However, even if we allow at most one nondegenerate interval coefficient in each equation, the problem becomes again NP-hard, since an arbitrary interval linear system can be rewritten in this form [23].

Structured systems. We can also explore band and sparse matrices.

Definition 8 A matrix $\mathbf{A}$ is a *w-band* matrix if $\mathbf{a}_{ij} = 0$ for $|i - j| \geq w$.

Band matrices with $d = 1$ are diagonal and computing the hull is clearly strongly polynomial. For $d = 2$ (tridiagonal matrix) it is an open problem. And for $d \geq 3$ it is again NP-hard. We inspected the case of bidiagonal matrices. The result is to our best knowledge new.

Theorem 9 *For a bidiagonal matrix (the matrix with only the main diagonal and an arbitrary neighboring diagonal) computing the exact hull of* $\mathbf{A}x = \mathbf{b}$ *is strongly polynomial.*

Proof Without the loss of generality let us suppose that the matrix $\mathbf{A}$ consists of the main diagonal and the one beyond it. By the forward substitution, we have $\mathbf{x}_1 = \frac{\mathbf{b}_1}{\mathbf{a}_{11}}$ and

$$\mathbf{x}_i = \frac{\mathbf{b}_i - \mathbf{a}_{i,i-1}\mathbf{x}_{i-1}}{\mathbf{a}_{ii}}, \quad i = 2, \ldots, n.$$

By induction, $\mathbf{x}_{i-1}$ is optimally computed with no use of interval coefficients of the ith equations. Since an evaluation in interval arithmetic is optimal in the case there are no multiple occurrences of variables (Theorem 1), $\mathbf{x}_i$ is optimal as well. □

Definition 9 A matrix $\mathbf{A}$ is a *d-sparse* matrix if in each row i at most d elements $\mathbf{a}_{ij} \neq 0$.

For sparse matrices with $d = 1$ computing the hull is clearly strongly polynomial. For $d \geq 2$ it is again NP-hard [23]. Nevertheless, if we combine w-band matrix with system coefficient bounds coming from a given finite set of rational numbers, then we have a polynomial algorithm for computing the hull [23].

If an interval system $\mathbf{A}x = \mathbf{b}$ is in a certain form, the hull can be computed in polynomial time using some already introduced algorithms. If the matrix $\mathbf{A}$ has full column rank and A_c is a diagonal matrix with positive entries, then Hansen-Bliek-Rohn prescription for enclosure gives the exact hull [4]. If $\mathbf{A}$ is an M-matrix, then Gauss-Seidel iteration method converges to the exact hull [30]. And if $\mathbf{A}$ is an M-matrix and $\mathbf{b}$ is nonnegative then the interval version of Gaussian elimination yields the exact hull [30].

In this section we silently supposed that the solution set Σ is bounded. This is not always the case. Many mentioned methods can not deal with an unbounded solution set. That is why we usually need to check for boundedness. However, it is an coNP-complete problem since it is identical with checking the full column rank of the interval matrix $\mathbf{A}$.

Remark A natural generalization of an interval linear system is by incorporating linear dependencies. That is, we have a family of linear systems

$$A(p)x = b(p), \quad p \in \mathbf{p}, \tag{5}$$

where $A(p) = \sum_{k=1}^{K} A^k p_k$ and $b(p) = \sum_{k=1}^{K} b^k p_k$. Here, p is a vector of parameters varying in $\mathbf{p}$. Since this concept generalizes the standard interval systems, many related problems are intractable. We point out one particular efficiently solvable problem. Given $x \in \mathbb{R}^n$, deciding whether it is a solution of a standard interval system $\mathbf{A}x = \mathbf{b}$ is strongly polynomial. For systems with linear dependencies, the problem still stays polynomial, but we can show weak polynomiality only; this is achieved by rewriting (5) as a linear program.

Summary

Problem	*Complexity*
Is x a solution of $\mathbf{A}x = \mathbf{b}$?	strongly P
Computing the hull of $\mathbf{A}x = \mathbf{b}$	NP-hard
Computing the hull of $\mathbf{A}x = b$	NP-hard
Computing the hull of $Ax = \mathbf{b}$	P
Computing the hull of $\mathbf{A}x = \mathbf{b}$, where $\mathbf{A}$ is regular	NP-hard
Computing the hull of $\mathbf{A}x = \mathbf{b}$, where $\mathbf{A}$ is M-matrix	P
Computing the hull of $\mathbf{A}x = \mathbf{b}$, where $\mathbf{A}$ is diagonal	strongly P
Computing the hull of $\mathbf{A}x = \mathbf{b}$, where $\mathbf{A}$ is bidiagonal	strongly P
Computing the hull of $\mathbf{A}x = \mathbf{b}$, where $\mathbf{A}$ is tridiagonal	?
Computing the hull of $\mathbf{A}x = \mathbf{b}$, where $\mathbf{A}$ is 3-band	NP-hard
Computing the hull of $\mathbf{A}x = \mathbf{b}$, where $\mathbf{A}$ is 1-sparse	strongly P
Computing the hull of $\mathbf{A}x = \mathbf{b}$, where $\mathbf{A}$ is 2-sparse	NP-hard
Computing the exact least squares hull of $\mathbf{A}x = \mathbf{b}$	NP-hard
Is Σ bounded?	coNP-complete

4.4 Matrix Inverse

Computation of a matrix inverse is usually avoided in applications. Nonetheless, we chose to mention this topic, since it holds a worthy place in interval linear algebra theory. An interval inverse matrix is defined as follows.

Definition 10 Let us have a square regular interval matrix $\mathbf{A}$. We define its interval inverse matrix as $\mathbf{A}^{-1} = [\underline{B}, \overline{B}]$, where $\underline{B} = \min\{A^{-1}, A \in \mathbf{A}\}$ and $\overline{B} = \max\{A^{-1}, A \in \mathbf{A}\}$, where the min and max are understood componentwise.

As usually, the inverse matrix can be computed using knowledge of inverses of boundary matrices A_{yz} [39].

Theorem 10 *Let* $\mathbf{A}$ *be regular. Then its inverse* $\mathbf{A}^{-1} = [\underline{B}, \overline{B}]$ *is described by*

$$\underline{B} = \min_{y,z \in Y_n} A_{yz}^{-1},$$

$$\overline{B} = \max_{y,z \in Y_n} A_{yz}^{-1},$$

where the min *and* max *is understood componentwise.*

The maximum and minimum bound of each component of the interval inverse is attained at one of the inverse of 2^{2n} boundary matrices. No wonder, it can be proved that generally computing exact inverse matrix is NP-hard [3].

When $A_c = I$, we can compute the exact inverse in polynomial time according to the next theorem [47].

Theorem 11 *Let* $\mathbf{A}$ *be a regular interval matrix with* $A_c = I$. *Let* $M = (I - \Delta)^{-1}$. *Then its inverse* $\mathbf{A}^{-1} = [\underline{B}, \overline{B}]$ *is described by*

$$\underline{B} = -M + D_k,$$
$$\overline{B} = M,$$

where $k_j = \frac{2m_{jj}^2}{2m_{jj}-1}$ *for* $j = 1, \ldots, n$, *with* m_{jj} *being diagonal elements of* M.

There also exists a formula for the exact matrix inverse if all intervals have uniform widths, i.e., $\mathbf{A} = [A_c - \alpha E, A_c + \alpha E]$ [49].

If we wish to only compute an enclosure $\mathbf{B}$ of the matrix inverse we can use any method for computing enclosures of interval linear systems. We get the i-th column of $\mathbf{B}$ by solving the systems $\mathbf{A}x = e_i$, where e_i is i-th column of the identity matrix of order n.

As we mentioned, computing the exact interval inverse is NP-hard. We close this section with a surprising result on inverse nonnegativity ($A^{-1} \geq 0$ for every $A \in \mathbf{A}$). It was first proved in slightly different form in [24]. For this form see [30]. It implies that checking inverse nonnegativity and also computing the exact interval inverse of an inverse nonnegative matrix is strongly polynomial.

Theorem 12 *If* $\underline{A}$, $\overline{A}$ *are regular and* $\underline{A}^{-1}$, $\overline{A}^{-1} \geq 0$ *then* $\mathbf{A}$ *is regular and*

$$\mathbf{A}^{-1} = [\overline{A}^{-1}, \underline{A}^{-1}] \geq 0.$$

Summary

Problem	*Complexity*
Computing the exact inverse of $\mathbf{A}$	NP-hard
Is $\mathbf{A}$ inverse nonnegative?	strongly P
Computing the exact inverse of inverse nonnegative $\mathbf{A}$	strongly P

4.5 Solvability of a Linear System

Of course, before solving a linear system we might want to know, whether it is actually solvable. Considering solvability we should distinguish between two types of solvability.

Definition 11 An interval linear system $\mathbf{A}x = \mathbf{b}$ is *(weakly) solvable* if some system $Ax = b$, where $A \in \mathbf{A}, b \in \mathbf{b}$ is solvable.

In another words, its solution set Σ is not empty. Otherwise, we call the system *unsolvable*.

Definition 12 An interval linear system $\mathbf{A}x = \mathbf{b}$ is *strongly solvable* if every system $Ax = b$, where $A \in \mathbf{A}, b \in \mathbf{b}$ is solvable.

The first definition is interesting for model checking. The second for system verification and automated proofs.

Checking whether an interval systems is solvable is an NP-hard problem [23]. The sign coordinates of the orthant containing the solution can serve as a polynomial witness and existence of a solution can be verified by linear programming, hence this problem is NP-complete and checking unsolvability coNP-complete. The problem of deciding strong solvability is coNP-complete. It can be reformulated as checking unsolvability of a certain linear system using the well known Farkas lemma, e.g., [45].

Sometimes, we look only for nonnegative solutions – *nonnegative solvability*. Checking whether an interval linear system has a nonnegative solution is weakly polynomial. We know the orthant in which the solution should lie. Therefore, we can get rid of the absolute values in Oettli–Prager theorem and apply linear programming. However, checking whether a system is nonnengative strongly solvable is still coNP-complete [4]. We summarize the results in the following table.

Theorem 13 *Checking various types of solvability of* $\mathbf{A}\mathbf{x} = \mathbf{b}$ *is of the following complexity.*

	weak	strong
solvability	NP-complete	coNP-complete
nonnegative solvability	P	coNP-complete

It is easy to see that an interval linear system $\mathbf{A}x = \mathbf{b}$ is unsolvable if the matrix $[\mathbf{A}\ \mathbf{b}]$ has full column rank. That is why, we can use sufficient conditions for full column rank to check unsolvability. Moreover, we can also use methods for computing enclosures. If we have some enclosure $\mathbf{x}$, then clearly a system $\mathbf{A}x = \mathbf{b}$ is unsolvable if $\mathbf{A}\mathbf{x} \cap \mathbf{b} = \emptyset$. Many enclosure algorithms enable detection of unsolvability. Generally speaking, they work in iterative stages and when we intersect enclosures of the solution set from the two subsequent stages and get an empty set, we know for sure that the system is unsolvable. These methods are, for example, Gaussian elimination [7], Jacobi method [27], Gauss–Seidel method [27], subsquares method [15].

Linear inequalities. Just for comparison, considering systems of interval linear inequalities, the problems of checking various types of solvability become much easier. The results are resumed in the following table [4].

Theorem 14 *Checking various types of solvability of* $\mathbf{A}x \leq \mathbf{b}$ *is of the following complexity.*

	weak	strong
solvability	NP-complete	P
nonnegative solvability	P	P

We also would like to mention an interesting nontrivial property of strong solvability of systems of interval linear inequalities. When a system $\mathbf{A}x \leq \mathbf{b}$ is strongly solvable (i.e., every $Ax \leq b$ has a solution), then there exists a solution x satisfying $Ax \leq b$ for every $A \in \mathbf{A}$ and $b \in \mathbf{b}$ [4].

∀∃-solutions. Let us come back to interval linear systems. The traditional concept of a solution (Definition 5) employs existential quantifiers: x is a solution if $\exists A \in \mathbf{A}$, $\exists b \in \mathbf{b} : Ax = b$. Nevertheless, in some applications, another quantification makes sense, too. In particular, ∀∃ quantification was deeply studied [52]. For illustration of complexity of such solution, we will focus on two concepts of solutions – tolerance [4] and control solution [4, 51].

Definition 13
A vector x is a *tolerance* solution of $\mathbf{A}x = \mathbf{b}$ if $\forall A \in \mathbf{A}, \exists b \in \mathbf{b} : Ax = b$.

A vector x is a *control* solution of $\mathbf{A}x = \mathbf{b}$ if $\forall b \in \mathbf{b}, \exists A \in \mathbf{A} : Ax = b$,

Notice that a tolerance solution can equivalently be characterized as $\{Ax \mid A \in \mathbf{A}\} \subseteq \mathbf{b}$ and a control solution as $\mathbf{b} \subseteq \{Ax \mid A \in \mathbf{A}\}$.

Both solutions can be described by a slight modification of Oettli–Prager theorem (one sign change in Oettli–Prager formula) [4].

Theorem 15 *Let us have a system* $\mathbf{A}x = \mathbf{b}$*, then x is*

- *a tolerance solution if it satisfies* $|A_c x - b_c| \leq -\Delta|x| + \delta$.
- *a control solution if it satisfies* $|A_c x - b_c| \leq \Delta|x| - \delta$.

In the case of tolerance solution, the sign change has a large impact on complexity. Deciding whether a system has a tolerance solution is weakly polynomial. However, checking whether a system has a control solution remains NP-complete [23].

Summary

Problem	*Complexity*
Is $\mathbf{A}x = \mathbf{b}$ solvable?	NP-complete
Is $\mathbf{A}x = \mathbf{b}$ strongly solvable?	coNP-complete
Is $\mathbf{A}x = \mathbf{b}$ nonnegative solvable?	P
Is $\mathbf{A}x = \mathbf{b}$ nonnegative strongly solvable?	coNP-complete
Is $\mathbf{A}x \leq \mathbf{b}$ solvable?	NP-complete
Is $\mathbf{A}x \leq \mathbf{b}$ strongly solvable?	P
Is $\mathbf{A}x \leq \mathbf{b}$ nonnegative solvable?	P
Is $\mathbf{A}x \leq \mathbf{b}$ nonnegative strongly solvable?	P
Does $\mathbf{A}x = \mathbf{b}$ have a tolerance solution?	P
Does $\mathbf{A}x = \mathbf{b}$ have a control solution?	NP-complete

4.6 Determinant

Determinants of interval matrices are not often studied. However, we included this section for completeness.

Definition 14 A determinant of $\mathbf{A}$ is defined as $\det(\mathbf{A}) = [\underline{d}, \overline{d}]$, where

$$\underline{d} = \min\{\det(A) \mid A \in \mathbf{A}\},$$

$$\overline{d} = \max\{\det(A) \mid A \in \mathbf{A}\}.$$

Its bounds can be computed from 2^{n^2} boundary matrices $A_{ij} \in \{\underline{A}_{ij}, \overline{A}_{ij}\}$ for $i, j = 1, \ldots, n$. We have the following theoretical result [42].

Theorem 16 *Computing interval determinant of* $\mathbf{A} = [A - E, A + E]$*, where A is rational nonnegative is NP-hard.*

It is intractable even in this simplified case. For interesting relations to eigenvalues and singularity see [42].

Summary

Problem	*Complexity*
Computing $\underline{\det}(\mathbf{A})$	NP-hard
Computing $\overline{\det}(\mathbf{A})$	NP-hard

4.7 Eigenvalues

First, we briefly start with general matrices, then we continue with the symmetric case. Checking singularity of $\mathbf{A}$ can be polynomially reduced to checking whether 0 is an eigenvalue of some matrix $A \in \mathbf{A}$. As we saw in Sect. 3.5 checking whether λ is an eigenvalue of some matrix $A \in \mathbf{A}$ is NP-complete problem. Surprisingly, checking for eigenvectors can be done efficiently [38]. It is strongly polynomial.

How is it with Perron-Frobenius theory of nonnegative matrices ([26])? An interval matrix $\mathbf{A} \in \mathbb{IR}^{n \times n}$ is *nonnegative irreducible* if every $A \in \mathbf{A}$ is nonnegative irreducible. For Perron vectors (positive vectors corresponding to the dominant eigenvalues), we have the following result [44].

Theorem 17 *Let* $\mathbf{A}$ *be nonnegative irreducible. Then the problem of deciding whether* x *is a Perron eigenvector of some matrix in* $\mathbf{A}$ *is strongly polynomial.*

For the sake of simplicity we mentioned only some results considering eigenvalues of a general matrix $\mathbf{A}$. We will go into more detail with symmetric matrices, where their eigenvalues are real.

Definition 15 Let $\mathbf{A} \in \mathbb{IR}^{n \times n}$ with Δ, A_c symmetric. Then the corresponding *symmetric interval matrix* is defined as a set of symmetric matrices in $\mathbf{A}$, that is,

$$\mathbf{A}^S := \{A \in \mathbf{A} : A = A^T\}.$$

For a symmetric $A \in \mathbb{R}^{n \times n}$, we use $\lambda_{\min}(A)$ and $\lambda_{\max}(A)$ for its smallest and largest eigenvalue, respectively. For a symmetric interval matrix, we define the smallest and largest eigenvalues respectively as

$$\lambda_{\min}(\mathbf{A}^S) := \min\{\lambda_{\min}(A) : A \in \mathbf{A}^S\},$$
$$\lambda_{\max}(\mathbf{A}^S) := \max\{\lambda_{\max}(A) : A \in \mathbf{A}^S\}.$$

Even if we consider the symmetric case some problems remain intractable [23, 42]. We are yet able to prove the hardness results, since it is usually difficult to find a proper polynomial witness.

Theorem 18 *On a class of problems with* $A_c \in \mathbb{Q}^{n \times n}$ *symmetric positive definite and entrywise nonnegative, and* $\Delta = E$*, the following problems are intractable*

- *checking whether* 0 *is an eigenvalue of some matrix* $A \in \mathbf{A}^S$ *is NP-hard,*
- *checking* $\lambda_{\max}(\mathbf{A}^S) \in (\underline{a}, \overline{a})$ *for a given open interval* $(\underline{a}, \overline{a})$ *is coNP-hard.*

However, there are some known subclasses for which the eigenvalue range or at least one of the extremal eigenvalues can be determined efficiently [11]:

- If A_c is *essentially non-negative*, i.e., $(A_c)_{ij} \geq 0 \, \forall i \neq j$, then $\lambda_{\max}(\mathbf{A}^S) = \lambda_{\max}(\overline{A})$.
- If Δ is *diagonal*, then $\lambda_{\min}(\mathbf{A}^S) = \lambda_{\min}(\underline{A})$ and $\lambda_{\max}(\mathbf{A}^S) = \lambda_{\max}(\overline{A})$.

In contrast to the extremal eigenvalues $\lambda_{\min}(\mathbf{A}^S)$ and $\lambda_{\max}(\mathbf{A}^S)$, the largest of the minimal eigenvalues and the smallest of the largest eigenvalues,

$$\max\{\lambda_{\min}(A) : A \in \mathbf{A}^S\},$$
$$\min\{\lambda_{\max}(A) : A \in \mathbf{A}^S\},$$

can be computed with an arbitrary precision in polynomial time by using semidefinite programming [16]. As in the general case, checking whether a given vector $0 \neq x \in \mathbb{R}^n$ is an eigenvector of some matrix in $\mathbf{A}^S$ is a polynomial time problem. Nevertheless, strong polynomiality has not been proved yet.

We already know that computing exact bounds on many problems with interval data is intractable. Since we can do no better, we can inspect the hardness of various approximations of their solutions. While doing this we use the following assumption: *Throughout this section, we consider a computational model, in which the exact eigenvalues of rational symmetric matrices are polynomially computable.*

The table below from [11] summarizes the main results. We use the symbol ∞ in case there is no finite approximation factor with polynomial complexity.

Theorem 19 *Approximating the extremal eigenvalues of* $\mathbf{A}^S$ *is of the following complexity.*

	abs. error	rel. error	inverse rel. error
NP-hard with error	any	< 1	1
polynomial with error	∞	1	2

The table below gives results for a more specific case of approximating $\lambda_{\max}(\mathbf{A}^S)$ when A_c is positive semi-definite.

Theorem 20 *Approximating the extremal eigenvalues of* $\mathbf{A}^S$ *with* A_c *rational positive semi-definite is of the following complexity.*

	abs. error	rel. error	inverse rel. error
NP-hard with error	any	$1/(32n^4)$	$1/(32n^4)$
polynomial with error	∞	$1/3$	$1/3$

The tables sums up the generalized idea behind several theorems on computing extremal eigenvalues. For more information and formal details see [11].

At the end of this subsection we mention spectral radius.

Definition 16 Let $\mathbf{A} \in \mathbb{IR}^{n \times n}$, we define the range of *spectral radius* naturally as

$$\rho(\mathbf{A}) = \{\rho(A) : A \in \mathbf{A}\}.$$

Notice that $\rho(\mathbf{A})$ is a compact real interval due to continuity of eigenvalues. Similarly we define spectral radius for $\mathbf{A}^S$.

Complexity of computing $\overline{\rho(\mathbf{A})}$ is an open problem (as Schur stability is; see Sect. 4.9), and, to the best of our knowledge, complexity of computing $\underline{\rho(\mathbf{A})}$ has not been investigated yet.

Anyway, the following gives polynomially solvable subclasses:

- If $\underline{A} \geq 0$, then $\rho(\mathbf{A}) = [\rho(\underline{A}), \rho(\overline{A})]$.
- If $\mathbf{A}$ is diagonal, then $\rho(\mathbf{A}) = [\max_i \min_{a \in \mathbf{a}_{ii}} |a|, \max_i\{|\underline{a}_{ii}|, |\overline{a}_{ii}|\}]$.

Summary

Problem	*Complexity*
Is λ eigenvalue of some $A \in \mathbf{A}$?	NP-complete
Is x eigenvector of some $A \in \mathbf{A}$?	strongly P
Is x Perron vector of nonnegative irreducible $\mathbf{A}$?	strongly P
Is 0 eigenvalue of some $A \in \mathbf{A}^S$?	NP-hard
Is x eigenvector of some $A \in \mathbf{A}^S$?	P
Does $\lambda_{\max}(\mathbf{A}^S)$ belong to a given open interval?	coNP-hard
Computing $\overline{\rho(\mathbf{A})}$	?
Computing $\underline{\rho(\mathbf{A})}$	?
Computing exact bounds on $\rho(\mathbf{A})$ with $\mathbf{A}$ nonnegative	strongly P
Computing exact bounds on $\rho(\mathbf{A})$ with $\mathbf{A}$ diagonal	strongly P

4.8 Positive Definiteness and Semidefiniteness

We should not leave out mentioning the positive definiteness and semidefiniteness. Here without the loss of the generality symmetric matrices are of the only interest. We distinguish between weak and strong definiteness.

Definition 17 A symmetric interval matrix $\mathbf{A}^S$ is weakly positive (semi)definite if some $A \in \mathbf{A}^S$ is positive (semi)definite.

Definition 18 A symmetric interval matrix $\mathbf{A}^S$ is strongly positive (semi)definite if every $A \in \mathbf{A}^S$ is positive (semi)definite.

Checking strong positive definiteness [40] and semidefiniteness [28] are both coNP-hard according to the two following theorems.

Theorem 21 *Checking strong positive semidefiniteness of* $\mathbf{A}^S$ *is co-NP-hard on a class of problems with* $A_c \in \mathbb{Q}^{n\times n}$ *symmetric positive definite and entrywise nonnegative, and* $\Delta = E$.

Theorem 22 *Checking strong positive definiteness of* $\mathbf{A}^S$ *is co-NP-hard on a class of problems with* $A_c \in \mathbb{Q}^{n\times n}$ *symmetric positive definite and entrywise nonnegative, and* $\Delta = E$.

Considering positive definiteness, we have some sufficient conditions that can be checked polynomially [41].

Theorem 23 *An interval matrix* $\mathbf{A}^S$ *is strongly positive definite if at least one of the following condition holds*

- $\lambda_n(A_c) > \rho(\Delta)$,
- A_c *is positive definite and* $\rho(|(A_c)^{-1}|\Delta) < 1$.

The second condition can be reformulated as $\mathbf{A}^S$ being regular and A_c positive definite. If the first condition holds with $\geq$ then $\mathbf{A}^S$ is strongly positive semidefinite.

In contrast to checking strong positive definiteness, weak positive definiteness can be checked in polynomial time by using semidefinite programming [16]; this polynomial result holds also for a more general class of symmetric interval matrices with linear dependencies [12]. For positive semidefiniteness it needn't be the case since semidefinite programming methods work only with some given accuracy.

Summary

Problem	*Complexity*
Is $\mathbf{A}^S$ strongly positive definite?	coNP-hard
Is $\mathbf{A}^S$ strongly positive semidefinite?	coNP-hard
Is $\mathbf{A}^S$ weakly positive definite?	P
Is $\mathbf{A}^S$ weakly positive semidefinite?	?

4.9 Stability

The last section is dedicated to an important and more practical problem – deciding a stability of a matrix. There are many types of stabilities. For illustration, we chose two of them – Hurwitz and Schur.

Definition 19 An interval matrix $\mathbf{A}$ is *Hurwitz stable* if every $A \in \mathbf{A}$ is Hurwitz stable (i.e., all eigenvalues have negative real parts).

Similarly, we define Hurwitz stability for symmetric interval matrices. Due to their relation to positive definiteness ($\mathbf{A}^S$ is Hurwitz stable if $-\mathbf{A}^S$ is positive definite) we could presume that the problem is coNP-hard. It is so, even if we limit ourselves to a special case [40].

Theorem 24 *Checking Hurwitz stability of a symmetric interval matrix* $\mathbf{A}^S$ *is coNP-hard on a class of problems with* $A_c \in \mathbb{Q}^{n\times n}$ *symmetric Hurwitz stable and entrywise nonpositive, and* $\Delta = E$.

For general matrices, coNP-hardness holds as well. The problem is still coNP-hard even if we limit the number of interval coefficients in our matrix [28].

Theorem 25 *Checking Hurwitz stability of* $\mathbf{A}$ *is co-NP-hard on a class of interval matrices with intervals in the last row and column only.*

Likewise, as for checking regularity, checking Hurwitz stability of $\mathbf{A}$ can not be done by checking stability of matrices of type A_{yz} (for reductions of other properties see [5]). On the other hand, it can be checked in this way for $\mathbf{A}^S$. For more discussion and historical context see [23] or [48]. As sufficient conditions we can use conditions for positive definiteness applied to $-\mathbf{A}$. For more sufficient conditions see e.g., [25].

Definition 20 An interval matrix $\mathbf{A}$ is *Schur stable* if every $A \in \mathbf{A}$ is Schur stable (i.e., $\rho(A) < 1$).

In a similar way, we define Schur stability for symmetric interval matrices. For general interval matrices, complexity of checking Schur stability is an open problem, however, for the symmetric case the problem is intractable [40].

Theorem 26 *Checking Schur stability of* $\mathbf{A}^S$ *is coNP-hard on a class of problems with* $A_c \in \mathbb{Q}^{n\times n}$ *symmetric Schur stable and offdiagonal entries nonpositive, and* $\Delta = E$.

Summary

Problem	*Complexity*
Is $\mathbf{A}$ Hurwitz stable?	coNP-hard
Is $\mathbf{A}^S$ Hurwitz stable?	coNP-hard
Is $\mathbf{A}$ Schur stable?	?
Is $\mathbf{A}^S$ Schur stable?	coNP-hard

4.10 Further Topics

Due to the limited space, we had to omit many interesting topics. We touched only briefly the complexity issues of interval linear inequalities, but there are more results; see, e.g., [4, 10]. We did not discussed complexity of computing the range of polynomials over intervals [23], too. In short, we mention two particular problems:

- *Matrix power.* Computing the exact bounds on second power of the matrix $\mathbf{A}^2$ is strongly polynomial (just by evaluating by interval arithmetic), but computing the cube $\mathbf{A}^3$ turns out to be NP-hard [20].

- *Matrix norm.* Computing the range of $\|A\|$ when $A \in \mathbf{A}$ is a trivial task for vector ℓ_p-norms applied on matrices (including Frobenius norm or maximum norm) or for induced 1- and ∞-norms. On the other hand, determining the largest value of the spectral norm $\|A\|_2$ (the largest singular value) subject to $A \in \mathbf{A}$ is NP-hard [28].

5 Summary

In this work we explored the fundamental problems of interval linear algebra. Our goal was to:

- provide a basic introduction to interval linear algebra
- answer elementary computational complexity questions linked with interval linear algebra
- discuss the computational complexity of the basic problems
- explain the relations between these problems
- mention relaxations or special classes of these problems that are easily decidable or there exist polynomial algorithms solving them
- provide a basis for further reading and research

At this place we also would like to apologize to those whose results are not mentioned in this work. There are many great achievements, however this work can unfortunately consume only limited amount of space. We provide links to the literature, where you can find much more of them.

Acknowledgements J. Horáček and M. Hladík were supported by GAČR grant P402/13-10660S. M. Černý was supported by the GAČR grant 16-00408S.

References

1. Arora, S., Barak, B.: Computational Complexity: A Modern Approach. Cambridge University Press, Cambridge (2009)
2. Blum, L., Cucker, F., Shub, M., Smale, S.: Complexity and Real Computation. Springer Science & Business Media, New York (2012)
3. Coxson, G.E.: Computing exact bounds on elements of an inverse interval matrix is NP-hard. Reliab. Comput. **5**(2), 137–142 (1999)
4. Fiedler, M., Nedoma, J., Ramik, J., Rohn, J., Zimmermann, K.: Linear Optimization Problems with Inexact Data. Springer, New York (2006)
5. Garloff, J., Adm, M., Titi, J.: A survey of classes of matrices possessing the interval property and related properties. Reliab. Comput. **22**, 1–10 (2016)
6. Golub, G.H., Van Loan, C.F.: Matrix computations, 3rd edn. Johns Hopkins University Press, Baltimore (1996)
7. Hansen, E., Walster, G.: Solving overdetermined systems of interval linear equations. Reliab. Comput. **12**(3), 239–243 (2006)

8. Hartman, D., Hladík, M.: Tight bounds on the radius of nonsingularity. In: Scientific Computing, Computer Arithmetic, and Validated Numerics, pp. 109–115, Springer, Heidelberg (2015)
9. Hladík, M.: New operator and method for solving real preconditioned interval linear equations. SIAM J. Numer. Anal. **52**(1), 194–206 (2014)
10. Hladík, M.: AE solutions and AE solvability to general interval linear systems. Linear Algebra Appl. **465**, 221–238 (2015)
11. Hladík, M.: Complexity issues for the symmetric interval eigenvalue problem. Open Math. **13**(1), 157–164 (2015)
12. Hladík, M.: Positive Semidefiniteness and Positive Definiteness of a Linear Parametric Interval Matrix, to appear in a Springer book series (2016)
13. Hladík, M., Horáček, J.: A shaving method for interval linear systems of equations. In: Wyrzykowski, R. et al. (ed.) Parallel Processing and Applied Mathematics, vol. 8385 of LNCS, pp. 573–581. Springer, Berlin (2014)
14. Horáček, J., Hladík, M.: Computing enclosures of overdetermined interval linear systems. Reliab. Comput. **19**, 143 (2013)
15. Horáček, J., Hladík, M.: Subsquares approach – a simple scheme for solving overdetermined interval linear systems. In Wyrzykowski, R., et al. (ed.) Parallel Processing and Applied Mathematics, vol. 8385 of LNCS, pp. 613–622. Springer, Berlin (2014)
16. Jaulin, L., Henrion, D.: Contracting optimally an interval matrix without loosing any positive semi-definite matrix is a tractable problem. Reliab. Comput. **11**(1), 1–17 (2005)
17. Jaulin, L., Kieffer, M., Didrit, O., Walter, É.: Applied interval analysis. With Examples in Parameter and State Estimation, Robust Control and Robotics. Springer, London (2001)
18. Kearfott, R.: Interval computations: Introduction, uses, and resources. Euromath Bull. **2**(1), 95–112 (1996)
19. Kearfott, R., Kreinovich, V. (eds.): Applications of Interval Computations. Kluwer, Dordrecht (1996)
20. Kosheleva, O., Kreinovich, V., Mayer, G., Nguyen, H.: Computing the cube of an interval matrix is NP-hard. Proc. ACM Symp. Appl. Comput. **2**, 1449–1453 (2005)
21. Krawczyk, R.: Newton-algorithmen zur bestimmung von nullstellen mit fehlerschranken. Computing **4**(3), 187–201 (1969)
22. Kreinovich, V.: How to define relative approximation error of an interval estimate: A proposal. Appl. Math. Sci. **7**(5), 211–216 (2013)
23. Kreinovich, V., Lakeyev, A.V., Rohn, J., Kahl, P.: Computational Complexity and Feasibility of Data Processing and Interval Computations. Kluwer, Dordrecht (1998)
24. Kuttler, J.: A fourth-order finite-difference approximation for the fixed membrane eigenproblem. Math. Comput. **25**(114), 237–256 (1971)
25. Mansour, M.: Robust stability of interval matrices. In: Proceedings of the 28th IEEE Conference on Decision and Control, vol. 1, pp. 46–51, Tampa, Florida (1989)
26. Meyer, C.D.: Matrix Analysis and Applied Linear Algebra 2. SIAM, Philadelphia (2000)
27. Moore, R., Kearfott, R., Cloud, M.: Introduction to interval analysis. Society for Industrial Mathematics, Philadelphia (2009)
28. Nemirovskii, A.: Several NP-hard problems arising in robust stability analysis. Math. Control Signals Syst. **6**(2), 99–105 (1993)
29. Neumaier, A.: Linear interval equations. Interval Mathematics 1985, pp. 109–120. Springer, Heidelberg (1986)
30. Neumaier, A.: Interval Methods for Systems of Equations, vol. 37. Cambridge University Press, Cambridge (1990)
31. Oettli, W., Prager, W.: Compatibility of approximate solution of linear equations with given error bounds for coefficients and right-hand sides. Numer. Math. **6**(1), 405–409 (1964)
32. Popova, E.D.: Improved solution enclosures for over- and underdetermined interval linear systems. In Lirkov, I., et al. (ed.) Large-Scale Scientific Computing, vol. 3743 of LNCS, pp. 305–312 (2006)
33. Renegar, J.: On the computational complexity and geometry of the first-order theory of the reals. Part I: Introduction. Preliminaries. The geometry of semi-algebraic sets. The decision problem for the existential theory of the reals. J. Symb. Comput. **13**(3), 255–299 (1992)

34. Renegar, J.: On the computational complexity and geometry of the first-order theory of the reals. Part II: The general decision problem. Preliminaries for quantifier elimination. J. Symb. Comput. **13**(3), 301–327 (1992)
35. Renegar, J.: On the computational complexity and geometry of the first-order theory of the reals. Part III: Quantifier elimination. J. Symb. Comput. **13**(3), 329–352 (1992)
36. Rex, G., Rohn, J.: Sufficient conditions for regularity and singularity of interval matrices. SIAM J. Matrix Anal. Appl. **20**(2), 437–445 (1998)
37. Rohn, J.: Systems of linear interval equations. Linear Algebra Appl. **126**, 39–78 (1989)
38. Rohn, J.: Interval matrices: singularity and real eigenvalues. SIAM J. Matrix Anal. Appl. **14**(1), 82–91 (1993)
39. Rohn, J.: Inverse interval matrix. SIAM J. Numer. Anal. **30**(3), 864–870 (1993)
40. Rohn, J.: Checking positive definiteness or stability of symmetric interval matrices is NP-hard. Commentat. Math. Univ. Carol. **35**(4), 795–797 (1994)
41. Rohn, J.: Positive definiteness and stability of interval matrices. SIAM J. Matrix Anal. Appl. **15**(1), 175–184 (1994)
42. Rohn, J.: Checking properties of interval matrices. Technical Report 686, Institute of Computer Science, Academy of Sciences of the Czech Republic, Prague (1996)
43. Rohn, J.: Enclosing solutions of overdetermined systems of linear interval equations. Reliable Comput. **2**(2), 167–171 (1996)
44. Rohn, J.: Perron vectors of an irreducible nonnegative interval matrix. Linear Multilinear Algebra **54**(6), 399–404 (2006)
45. Rohn, J.: Solvability of Systems of Interval Linear Equations and Inequalities, pp. 35–77. Springer (2006)
46. Rohn, J.: Forty necessary and sufficient conditions for regularity of interval matrices: a survey. Electron. J Linear Algebra **18**, 500–512 (2009)
47. Rohn, J.: Explicit inverse of an interval matrix with unit midpoint. Electron. J Linear Algebra **22**, 138–150 (2011)
48. Rohn, J.: A handbook of results on interval linear problems. Technical Report 1163, Institute of Computer Science, Academy of Sciences of the Czech Republic, Prague (2012)
49. Rohn, J., Farhadsefat, R.: Inverse interval matrix: a survey. Electron. J Linear Algebra **22**, 704–719 (2011)
50. Rump, S. M.: Verification methods for dense and sparse systems of equations. In Herzberger, J. (ed.) Topics in Validated Computations. Studies in Computational Mathematics, pp. 63–136 (1994)
51. Shary, S.P.: On controlled solution set of interval algebraic systems. Interval Comput. **6**(6) (1992)
52. Shary, S.P.: A new technique in systems analysis under interval uncertainty and ambiguity. Reliab. Comput. **8**(5), 321–418 (2002)
53. Shary, S.P.: On full-rank interval matrices. Numer. Anal. Appl. **7**(3), 241–254 (2014)
54. Smale, S.: Mathematical problems for the next century. Math. Intell. **20**, 7–15 (1998)

On Optimal Extended Row Distance Profile

P. Almeida, D. Napp and R. Pinto

Abstract In this paper, we investigate extended row distances of Unit Memory (UM) convolutional codes. In particular, we derive upper and lower bounds for these distances and moreover present a concrete construction of a UM convolutional code that almost achieves the derived upper bounds. The generator matrix of these codes is built by means of a particular class of matrices, called superregular matrices. We actually conjecture that the construction presented is optimal with respect to the extended row distances as it achieves the maximum extended row distances possible. This in particular implies that the upper bound derived is not completely tight. The results presented in this paper further develop the line of research devoted to the distance properties of convolutional codes which has been mainly focused on the notions of free distance and column distance. Some open problems are left for further research.

Keywords Convolutional codes · Superregular matrices · Unimemory convolutional codes · Maximum Distance Profile (MDP) · Maximum Distance Separable (MDS)

1 Introduction

During the last two decades, renewed efforts were made to investigate the distance properties of convolutional codes, mainly, their free (Hamming) distance and their column distance. In [20] a Singleton bound for convolutional codes was derived (called generalized Singleton bound) and the codes achieving such a bound were called maximum distance separable (MDS). In [23] the first concrete construction of an MDS convolutional code (over the finite field $\mathbb{F}$) of rate $\frac{k}{n}$ and degree δ was presented for every given set of parameters (n, k, δ), (with the characteristic of the

P. Almeida · D. Napp (✉) · R. Pinto
Department of Mathematics, CIDMA - Center for Research and Development in Mathematics and Applications, University of Aveiro, Aveiro, Portugal
e-mail: diego@ua.pt

N. Bebiano (ed.), *Applied and Computational Matrix Analysis*,
Springer Proceedings in Mathematics & Statistics 192,
DOI 10.1007/978-3-319-49984-0_4

finite field $\mathbb{F}$ and the length n of the code being coprime). Bounds and fundamental properties of the column distances of convolutional codes have also been thoroughly investigated, see for instance [7, 8, 11, 18]. Convolutional codes having the largest columns distances for a given rate $\frac{k}{n}$ and degree δ are called maximum distance profile (MDP). Their existence was proven in [8] and concrete constructions were given in [7] when $(n-k)|\delta$ and in [17] for every set of given parameters (n, k, δ).

In contrast to the column distances, the extended row distances grow beyond the free distance and therefore provide additional information about the performance of the code. Hence, the notion of (extended) row distance is often used when more detailed knowledge of the distance structure of a convolutional code is needed [11]. One of the advantages of the row distance is that it is easy to calculate and serves as an excellent rejection rule when encoders are tested in search for convolutional code with large free distance. As opposed to the free distance and column distance the notion of (extended) row distance has not been fully investigated in the literature.

In this paper we shall focus on Unit Memory (UM) convolutional codes [14]. These codes may be an interesting alternative to the usual convolutional codes as their block length can be chosen to coincide with the word length of microprocessors, see [14, 24] for details. Binary (partial) UM convolutional codes were investigated in the literature by Lauer [13] and Justensen [12, 24] who showed that unit memory codes can perform better in some situations than codes having the same rate and degree but with memory larger than 1.

It is the aim of this work to analyze the row distances of Unit Memory (UM) convolutional codes with finite support. In particular we derive *upper* bounds for extended row distances of UM convolutional codes for a given rate $\frac{k}{n}$ and degree δ. Moreover, we show that such a bounds are tight by presenting concrete constructions of convolutional codes achieving this bound. The encoder matrices of these codes are built by means of a very particular type of matrices called superregular matrices.

The paper is organized as follows. In Sect. 2, we introduce the basic material for the development of the paper: it includes the necessary introductory material on UM convolutional codes and on the class of superregular matrices. In Sect. 3, we include the main results of the paper. In particular we establish upper and lower bounds on the extended row distances and moreover show how to construct (n, k, δ) UM convolutional codes that have (nearly) optimal profile of extended row distances. We conclude the paper in Sect. 4 where we resume the results of the paper and point out some aspects of this construction that can be improved in order to make it more attractive for applications. Finally some interesting avenues for research in this direction are indicated.

2 Distances of Convolutional Codes

This section contains the mathematical background needed for the development of our results. First we introduce convolutional codes with finite support and in particular unit memory codes. We conclude this section by recalling the notion superregular matrices [2]. Such matrices have some similarities with the ones introduced in [3, 7]. They have similar entries and, therefore, some properties are the same but the structure of these new matrices may be different.

Let $\mathbb{F}$ be a finite field and $\mathbb{F}[D]$ be the ring of polynomials with coefficients in $\mathbb{F}$.

2.1 Unit Memory Convolutional Codes

A (finite support) *convolutional code* $\mathscr{C}$ of rate k/n is an $\mathbb{F}[D]$-submodule of $\mathbb{F}[D]^n$ of rank k given by a *basic* and *minimal* full-rank polynomial *encoder matrix* $G(D) \in \mathbb{F}[D]^{k\times n}$,

$$\mathscr{C} = \operatorname{Im}_{\mathbb{F}[D]} G(D) = \left\{u(D)G(D) : u(D) \in \mathbb{F}^k[D]\right\},$$

where *basic* means that $G(D)$ has a polynomial right inverse, and *minimal* means that the sum of the row degrees of $G(D)$ attains its minimal possible value δ, called the *degree* of $\mathscr{C}$.[1] The largest row degree of $G(D)$ is called the *memory*. Note that since $G(D)$ is basic the resulting convolutional code if noncatastrophic, and hence we assume that only noncatastrophic codes are of interest [17, 19].

Although this is the general definition of convolutional codes with finite support, in this paper we will focus on a particular subclass of these codes, namely, Unit Memory (UM), i.e., when the encoder matrix $G(D)$ is described by $G(D) = G_0 + G_1 D$, $G_1 \neq 0$ or equivalently when the memory is equal to 1. Following the notation used in [16] a rate k/n UM convolutional code $\mathscr{C}$ of degree δ is called an (n, k, δ)-convolutional code. Note that in this case $1 \leq \delta \leq k$.

If $u(D) \in \mathbb{F}[D]^k$ has degree $j \geq 0$, $u(D) = u_0 + u_1 D + \cdots + u_{j-1} D^{j-1}$, and

$$G(D) = G_0 + G_1 D,$$

the above representation of $u(D)G(D) = v(D)$ can be expanded as

$$\begin{bmatrix} u_0 \; u_1 \; \cdots \; u_{j-1} \end{bmatrix} \underbrace{\begin{bmatrix} G_0 & G_1 & & & \\ & G_0 & G_1 & & \\ & & \ddots & \ddots & \\ & & & G_0 & G_1 \end{bmatrix}}_{=G_j^r} = \begin{bmatrix} v_0 \; v_1 \; \cdots \; v_j \end{bmatrix}, \tag{1}$$

where $G_j^r \in \mathbb{F}^{jk\times(j+1)n}$ is called the *sliding generator matrix*.

[1] Therefore, the *degree* δ of a convolutional code $\mathscr{C}$ is the sum of the row degrees of one, and hence any, minimal basic encoder.

An important distance measure for a convolutional code $\mathscr{C}$ is its *free distance* defined as

$$d_{\text{free}}(\mathscr{C}) = \min\{\text{wt}(v(D)) \mid v(D) \in \mathscr{C} \quad \text{and} \quad v(D) \neq 0\},$$

where $\text{wt}(v(D))$ is the Hamming weight of a polynomial vector

$$v(D) = \sum_{i \in \mathbb{N}} v_i D^i \in \mathbb{F}[D]^n,$$

defined as

$$\text{wt}(v(D)) = \sum_{i \in \mathbb{N}} \text{wt}(v_i),$$

where $\text{wt}(v_i)$ is the number of the nonzero components of v_i.

The *extended row distance* d_j^r is defined [11, 24] to be the minimum Hamming weight of all paths in the minimal code trellis that diverge from the zero state and then return for the first time back in the zero state *only* after j branches. An UM code can be represented by a trellis [4–6] where the state at time t is u_{t-1}. The number of states is $|\mathbb{F}|^k$ and for UM codes the zero state can always be achieved in one step with input $u_t = 0$. Moreover, a path in the trellis is unmerged with the zero path if and only if each information sub-block is nonzero.

For $j \geq 1$, let I_j denote the set of all $u(D)$ such that $u_\lambda \neq 0$ for $\lambda = 0, 1, \ldots, j-1$ and $u_j = 0$. We formally define the extended row distance d_j^r as

$$d_j^r = \min_{u(D) \in I_j} \text{wt}(u(D)G(D))$$

Thus we are considering the minimum weight of subcodewords corresponding to paths in the trellis from the zero state which reach the zero state again for the first time after exactly $j+1$ time instances. Note that $d_{\text{free}} \leq d_{j+1}^r \leq d_j^r$ and moreover for non-catastrophic codes it holds that $d_{\text{free}} = d_\infty^r = \min_{j=0,1,2,\ldots} d_j^r$ and $\alpha = \lim_{j \to +\infty} \frac{d_j^r}{j}$ gives the average linear *slope* of d_j^r.

2.2 Superregular Matrices

Let $A = [\mu_{i\ell}]$ be a square matrix of order m over $\mathbb{F}$ and S_m the symmetric group of order m. The determinant of A is given by

$$|A| = \sum_{\sigma \in S_m} (-1)^{\text{sgn}(\sigma)} \mu_{1\sigma(1)} \cdots \mu_{m\sigma(m)}.$$

A *trivial term* of the determinant is a term $\mu_\sigma = \mu_{1\sigma(1)} \cdots \mu_{m\sigma(m)}$, with at least one component $\mu_{i\sigma(i)}$ equal to zero. If A is a square submatrix of a matrix B with entries in $\mathbb{F}$, and all the terms of the determinant of A are trivial, we say that $|A|$ is a *trivial minor* of B (if $B = A$ we simply say that $|A|$ is a trivial minor). We say that a matrix B is *superregular* if all its nontrivial minors are different from zero.

The next results were derived in [3] and they will be very useful for our purposes in the next section.

Theorem 1 *Let $\mathbb{F}$ be a field and $a, b \in \mathbb{N}$, such that $a \geq b$ and $B \in \mathbb{F}^{a \times b}$. Suppose that $u = [u_i] \in \mathbb{F}^{b \times 1}$ is a row matrix such that $u_i \neq 0$ for all $1 \leq i \leq b$. If B is a superregular matrix and every column of B has at least one nonzero entry then* $\mathrm{wt}(uB) \geq b - a + 1$.

Theorem 2 *Let α be a primitive element of a finite field $\mathbb{F} = \mathbb{F}_{p^N}$ and $B = [\nu_{i\,\ell}]$ be a matrix over $\mathbb{F}$ with the following properties*

1. *if $\nu_{i\,\ell} \neq 0$ then $\nu_{i\,\ell} = \alpha^{\beta_{i\,\ell}}$ for a positive integer $\beta_{i\,\ell}$;*
2. *if $\nu_{i\,\ell} = 0$ then $\nu_{i'\,\ell} = 0$, for any $i' > i$ or $\nu_{i\,\ell'} = 0$, for any $\ell' < \ell$;*
3. *if $\ell < \ell'$, $\nu_{i\ell} \neq 0$ and $\nu_{i\ell'} \neq 0$ then $2\beta_{i\,\ell} \leq \beta_{i\,\ell'}$;*
4. *if $i < i'$, $\nu_{i\,\ell} \neq 0$ and $\nu_{i'\,\ell} \neq 0$ then $2\beta_{i\,\ell} \leq \beta_{i'\,\ell}$.*

Suppose N is greater than any exponent of α appearing as a nontrivial term of any minor of B. Then B is superregular.

We note that there exist several notions of superregular matrices in the literature. The definition given above generalizes all these notions. Frequently, see for instance [22], a superregular matrix is defined to be a matrix for which every square submatrix is nonsingular. Obviously all the entries of these matrices must be nonzero. Also, in [1, 21], several examples of triangular matrices were constructed in such a way that all submatrices inside this triangular configuration were nonsingular. However, all these notions do not apply to our case as they do not consider submatrices that contain zeros. The more recent contributions [7, 9, 10, 25, 26] consider the same notion of superregularity as us, but defined only for lower triangular matrices. Hence, many examples can be found in these references. In the following section we will adapt this general notion of superregularity to the case of interest in this paper, namely, the sliding generator matrices G_j^r.

3 Bounds and Constructions

In this section we present results of upper and lower bounds on extended row distances of UM convolutional codes. Moreover, we show how we can use the notion of superregular matrices to construct codes that achieve these bounds. We also provide a concrete class of superregular matrices that can be used to build UM convolutional codes with good design row extended distance. We point out some of the advantages and disadvantages of this construction in terms of the size of the field $\mathbb{F}$.

Given a generator matrix $G(D) = G_0 + G_1 D$ of $\mathscr{C}$ we shall assume without loss of generality that the zero rows of G_1 are at the top, i.e.,

$$G_0 = \begin{bmatrix} G_0^{(1)} \\ G_0^{(2)} \end{bmatrix} \qquad G_1 = \begin{bmatrix} 0 \\ G_1^{(2)} \end{bmatrix} \tag{2}$$

with $G_i^{(1)} \in \mathbb{F}^{k-\delta \times n}$ and $G_i^{(2)} \in \mathbb{F}^{\delta \times n}$, where δ is the degree of $\mathscr{C}$. We write $u = [u^{(1)} \quad u^{(2)}]$ accordingly. Note that since $G(D)$ is basic and minimal G_0 and $\begin{bmatrix} G_0^{(1)} \\ G_1^{(2)} \end{bmatrix}$ have full row rank.

The following result establishes an upper bound for the extended row distances.

Theorem 3 *Let $\mathscr{C}$ be a UM (n, k, δ)-convolutional code with generator matrix given by $G(D) = G_0 + G_1 D$ as above. Then,*

$$d_j^r \leq (n - k + 1)j + n \tag{3}$$

Proof We want to estimate

$$\min_{u(D) \in I_j} \mathrm{wt}(u(D)G(D)) = \min_{u_i \neq 0} \mathrm{wt}([u_0\, u_1\, \cdots\, u_{j-1}]G_j^r) \tag{4}$$

where G_j^r is the sliding generator matrix defined in (1). Clearly

$$\min_{u_0 \neq 0} \mathrm{wt}(v_0) = \min_{u_0 \neq 0} \mathrm{wt}(u_0 G_0) \leq n - k + 1$$

as $n - k + 1$ is the Singleton bound for (n, k)-block codes.

If $u_0^{(2)} \neq 0$ then $u_0 G_1 \neq 0$ and therefore $\begin{bmatrix} G_1 \\ G_0 \end{bmatrix}$ has at least $k + 1$ rows. Thus, exists u_1 such that

$$\mathrm{wt}(v_1) = \mathrm{wt}\left([u_0\, u_1] \begin{bmatrix} G_1 \\ G_0 \end{bmatrix}\right) \leq n - k. \tag{5}$$

However we may have $u_1 = 0$ which contradicts $u_i \neq 0$, for all i, and $u_0^{(2)} = 0$ which implies $u_0 G_1 = 0$ and therefore

$$\mathrm{wt}(v_1) \leq n - k + 1. \tag{6}$$

Hence, in any case

$$\min_{u_0 \neq 0} \mathrm{wt}(v_1) \leq n - k + 1. \tag{7}$$

Following the same reasoning, for any u_{i-1} there exists u_i such that

$$\min_{u_0\neq 0} \mathrm{wt}(v_i) = \min_{u_0\neq 0} \mathrm{wt}\left([u_{i-1}\ u_i]\begin{bmatrix} G_1 \\ G_0 \end{bmatrix}\right) \leq n-k+1.$$

for $i = 1, \ldots, j-1$, since, if with $u_{i-1}^{(2)} = 0$ then $\mathrm{wt}(v_i) = n-k+1$. Obviously $\mathrm{wt}(v_j) = \mathrm{wt}(u_{j-1}G_1) \leq n$ and hence for $[v_0\ v_1\ \cdots\ v_j] = [u_0\ u_1\ \cdots\ u_{j-1}]G_j^r$ with $u_i \neq 0$, we have that

$$\begin{aligned}\min_{u_i\neq 0} \mathrm{wt}([v_0\ v_1\ \cdots\ v_j]) =& \min_{u_i\neq 0}(\mathrm{wt}(v_0) + \sum_{i=1}^{j-1}\mathrm{wt}(v_i) + \mathrm{wt}(v_j)) \\ \leq& (n-k+1)j + n\end{aligned}$$

□

Remark 1 Taking a closer look at the proof of the previous lemma we see that between the two upper bounds (5) and (6) we had to consider the largest one (6) in order to prove (3). However we believe that (5) will hold for *a* $[u_0\ u_1\ \cdots\ u_{j-1}]$ minimizing (4). Since we failed to come up with a formal proof for this we leave it for future research and conjecture that the actual upper bound in (3) should be slightly smaller, namely,

$$d_j^r \leq (n-k)j + n + 1. \tag{8}$$

In the next section, we will construct a code that achieves the upper bound in (8).

If $\mathscr{C}$ has its extended row distances achieving the bound (8) for every $j \in \mathbb{N}$ we say that $\mathscr{C}$ has an *almost optimal extended row distances profile* (AOEDP). Note that this upper-bound does not depend on the degree δ of $\mathscr{C}$ in contrast to the generalized Singleton bound for the free distance [20]. Also note that the bound given in (8) grows infinitely and in practice one is interested in knowing the values of d_j^r, $1 \leq j \leq J$ for same given integer J.

The assumption that the zero rows of G_1 are at the top implies that the matrix $\begin{bmatrix} G_1 \\ G_0 \end{bmatrix}$ cannot have zero rows between two nonzero rows.

We will construct UM convolutional codes with designed extended row distances and for that we will require the sliding generator matrix G_j^r to be superregular. Next result characterizes and simplifies the conditions such a G_j^r to be superregular.

Lemma 1 *Let G_j^r be a sliding generator matrix as defined above. Then, G_j^r is superregular if and only if every square submatrix of G_j^r that does not contain zeros in the diagonal is invertible.*

Proof The proof amounts to showing that the unique nontrivial minors of G_j^r are exactly the ones that do not contain zeros in their diagonal. Let $A = [a_{ij}] \in \mathbb{F}^{t\times t}$ be a square submatrix of G_j^r. Obviously, if all the elements in the diagonal of A are nonzero then the corresponding minor is nontrivial. Thus, it is left to prove that if contains a zero in the diagonal, say a_{ss}, then the corresponding minor is trivial. In fact only two possibilities can happen due to the particular structure of blocks of zeros of G^r_j. Or there exists a block of zeros in the upper right corner of A, namely, $a_{ij} = 0$ for $0 \leq i \leq s$ and $s \leq j \leq t$ or otherwise there exists a block o zeros in the left bottom corner of A, namely, $a_{ij} = 0$ for $s \leq i \leq t$ and $0 \leq j \leq s$. It is easy to verify that all terms of $|A|$ have components in both blocks which concludes the proof. □

The next result shows how superregular matrices are related to UM convolutional codes that have an AOEDP.

Theorem 4 *Let $\mathscr{C}$ be a UM (n, k, δ)-convolutional code generated by $G(D) = G_0 + G_1 D$. If all the entries of G_0 and $G_1^{(2)}$ are nonzero and the sliding generator matrix G^r_j is superregular then*

$$d^r_j \geq (n-k)j + n + 1,$$

i.e., $\mathscr{C}$ has an AOEDP.

Proof For $j \geq 1$, let $u(D) \in I_j$. Suppose that the weight of $[u_0\, u_1\, \cdots\, u_{j-1}]$ is t and let $\bar{u}$ be the vector formed by the nonzero components of $[u_0\, u_1\, \cdots\, u_{j-1}]$ and B be the matrix formed by the t rows of G^r_j corresponding to $\bar{u}$. Thus B has $(j+1)n$ columns and t rows. Since $u_\lambda \neq 0$ for $\lambda = 0, 1, \ldots, j-1$ then the $(j+1)n$ columns of B are nonzero. The matrix B is superregular as it is assumed that G^r_j is superregular and any submatrix of a superregular matrix is superregular. Then we can apply Theorem 1 to obtain,

$$\mathrm{wt}(\bar{u}B) = \mathrm{wt}(v(D)) \geq (j+1)n - t + 1.$$

Since $t \leq jk$, we have that

$$\mathrm{wt}(v(D)) \geq (j+1)n - jk + 1 = (n-k)j + n + 1.$$

This concludes the proof. □

For a given $J \geq 1$ and a set of parameters (n, k, δ), with $\delta \leq k < n$ we propose a concrete construction of UM convolutional code constructed via the following class of superregular regular matrices.

Let $G(D) = G_0 + G_1 D$, where G_i, with $i = 1, 2$, are described by

$$G_i = [\gamma_{rs}] \text{ for } \gamma_{rs} = \begin{cases} \alpha^{2^{n+r+s-2}} & \text{if } i = 0 \\ \alpha^{2^{r+s-2}} & \text{if } i = 1 \text{ and } r > k - \delta \\ 0 & \text{if } i = 1 \text{ and } r \leq k - \delta \end{cases} \tag{9}$$

where α is a primitive element of the finite field $\mathbb{F} = \mathbb{F}_{p^N}$.

Lemma 2 *Let $G(D)$ be as in (9). Suppose N is greater than any exponent of α appearing as a nontrivial term of any minor of G^r_J. Then assumptions of Theorem 4 hold for $j = 1, \ldots, J$, namely, all the entries of G_0 and $G_1^{(2)}$ are nonzero and the sliding generator matrix G^r_J is superregular.*

Proof The fact that the entries of G_0 and $G_1^{(2)}$ are nonzero is straightforward. To show that the sliding generator matrix G^r_J is superregular permutate the columns of G^r_J to obtain the matrix

$$A = \begin{bmatrix} & & & G_1 \; G_0 \\ & & G_1 \; G_0 & \\ & \iddots \; \iddots & & \\ G_1 \; G_0 & & & \end{bmatrix}. \tag{10}$$

One can check that A satisfies the conditions of Theorem 2 and therefore it is superregular. Since the minors of A are equal (or symmetric) to the minors of G^r_J this implies that G^r_J is also superregular. □

We are now in a position to present a result that readily follows from Theorem 4 and Lemma 2 and states that the construction rendered in (9) gives rise to a UM convolutional code with a designed extended row distance and moreover has a AOEDP.

Corollary 1 *Let $\mathscr{C}$ be a UM (n, k, δ)-convolutional code generated by $G(D) = G_0 + G_1 D \in \mathbb{F}^{k \times n}$, where G_0 and G_1, are described above. Assume that $\mathbb{F} = \mathbb{F}_{p^N}$, for p prime and N sufficiently large, then the sliding generator matrix G^r_J is superregular and*

$$d^r_j = (n-k)j + n + 1,$$

for $j =, 0, 1, \ldots J$, i.e., d^r_j reaches the upper-bound given in (8) for $j =, 0, 1, \ldots J$.

4 Conclusions

A great deal of attention has been devoted in recent years to the study of convolutional codes with good distance properties. In particular, Maximum Distance Profile (MDP) or Maximum Distance Separable (MDS) have been thoroughly investigated. In this paper we have focused our attention to the construction of unit memory convolutional codes with good extended row distance. It turns out that the question of how to construct them can be related to the construction of a class of matrices, called superregular. We have given conditions for the sliding generator matrix of a code to yield UM convolutional codes with nearly optimal extended row distances. A concrete construction have been presented based on a type of superregular matrices that had been recently used for the authors to build MDP [2]. Moreover, it was recently shown [15] that this class of matrices perform very well when considering rank metric instead of the Hamming metric, producing Maximum Sum Rank Distance convolutional codes. It is natural to ask whether also the presented codes have optimal extended row distance with respect to the rank metric (to be formally defined). This opens up a interesting avenue of future research. Finally we remark that one of the disadvantages of the presented constructions is that they require large fields and it would be convenient to come up with new constructions of superregular matrices over smaller fields.

Acknowledgements The authors would like to thank the reviewer for his/her comments that led to improve the quality of the final version. This work was supported by Portuguese funds through the CIDMA - Center for Research and Development in Mathematics and Applications, and the Portuguese Foundation for Science and Technology (FCT-Fundação para a Ciência e a Tecnologia), within project PEst-UID/MAT/04106/2013.

References

1. Aidinyan, A.K.: On matrices with nondegenerate square submatrices. Probl. Peredachi Inf. **22**(4), 106–108 (1986)
2. Almeida, P., Napp, D., Pinto, R.: A new class of superregular matrices and MDP convolutional codes. Linear Algebra Appl. **439**(7), 2145–2157 (2013)
3. Almeida, P., Napp, D., Pinto, R.: Superregular matrices and applications to convolutional codes. Linear Algebra Appl. **499**, 1–25 (2016)
4. Forney Jr., G.D.: Convolutional codes I: algebraic structure. IEEE Trans. Inf. Theory **IT–16**(5), 720–738 (1970)
5. Forney Jr., G.D.: Structural analysis of convolutional codes via dual codes. IEEE Trans. Inf. Theory **IT–19**(5), 512–518 (1973)
6. Forney Jr., G.D.: Convolutional codes II: maximum likelihood decoding. Inf. Control **25**, 222–266 (1974)
7. Gluesing-Luerssen, H., Rosenthal, J., Smarandache, R.: Strongly MDS convolutional codes. IEEE Trans. Inf. Theory **52**(2), 584–598 (2006)
8. Hutchinson, R., Rosenthal, J., Smarandache, R.: Convolutional codes with maximum distance profile. Syst. Control Lett. **54**(1), 53–63 (2005)
9. Hutchinson, R.: The existence of strongly MDS convolutional codes. SIAM J. Control Optim. **47**(6), 2812–2826 (2008)
10. Hutchinson, R., Smarandache, R., Trumpf, J.: On superregular matrices and MDP convolutional codes. Linear Algebra Appl. **428**, 2585–2596 (2008)
11. Johannesson, R., Zigangirov, K.Sh.: Fundamentals of Convolutional Coding. IEEE Press, New York (1999)
12. Justesen, J., Paaske, E., Ballan, M.: Quasi-cyclic unit memory convolutional codes. IEEE Trans. Inf. Theory **IT–36**(3), 540–547 (1990)
13. Lauer, G.S.: Some optimal partial unit-memory codes. IEEE Trans. Inf. Theory **25**, 240–243 (1979)
14. Lee, L.N.: Short unit-memory byte-oriented binary convolutional codes having maximal free distance. IEEE Trans. Inf. Theory **IT–22**, 349–352 (1976)
15. Mahmood, R., Badr, A., Khisti, A.: Convolutional codes in rank metric for network streaming. in IEEE Int. Symp. Inf. Theory (ISIT) (2015)
16. McEliece, R. J.: The algebraic theory of convolutional codes. In: Handbook of Coding Theory, vol. 1, pp. 1065–1138. Elsevier, Amsterdam (1998)
17. Napp, D., Smarandache, R.: Constructing strongly MDS convolutional codes with maximum distance profile. Advances in Mathematics of Communications (to appear)

18. El Oued, M., Sole, P.: MDS convolutional codes over a finite ring. IEEE Trans. Inf. Theory **59**(11), 7305–7313 (2013)
19. Rosenthal, J., Schumacher, J. M., York, E. V.: On behaviors and convolutional codes. IEEE Trans. Autom. Control **42**(6, part 1), 1881–1891 (1996)
20. Rosenthal, J., Smarandache, R.: Maximum distance separable convolutional codes. Appl. Algebra Eng. Commun. Comput. **1**(10), 15–32 (1999)
21. Roth, R.M., Seroussi, G.: On generator matrices of MDS codes. IEEE Trans. Inf. Theory **31**(6), 826–830 (1985)
22. Roth, R.M., Lempel, A.: On MDS codes via Cauchy matrices. IEEE Trans. Inf. Theory **35**(6), 1314–1319 (1989)
23. Smarandache, R., Gluesing-Luerssen, H., Rosenthal, J.: Constructions for MDS-convolutional codes. IEEE Trans. Autom. Control **47**(5), 2045–2049 (2001)
24. Thommesen, C., Justesen, J.: Bounds on distances and error exponents of unit memory codes. IEEE Trans. Inf. Theory **IT–29**(5), 637–649 (1983)
25. Tomás, V.: Complete-MDP convolutional codes over the Erasure Channel. Ph.D. thesis, Departamento de Ciencia de la Computación e Inteligencia Artificial, Universidad de Alicante, España (2010)
26. Tomás, V., Rosenthal, J., Smarandache, R.: Decoding of MDP convolutional codes over the erasure channel. In: Proceedings of the 2009 IEEE International Symposium on Information Theory (ISIT 2009), Seoul, Korea, pp. 556–560 (2009)

The Dual of Convolutional Codes Over $\mathbb{Z}_{p^r}$

Mohammed El Oued, Diego Napp, Raquel Pinto and Marisa Toste

Abstract An important class of codes widely used in applications is the class of convolutional codes. Most of the literature of convolutional codes is devoted to convolutional codes over finite fields. The extension of the concept of convolutional codes from finite fields to finite rings have attracted much attention in recent years due to fact that they are the most appropriate codes for phase modulation. However convolutional codes over finite rings are more involved and not fully understood. Many results and features that are well-known for convolutional codes over finite fields have not been fully investigated in the context of finite rings. In this paper we focus in one of these unexplored areas, namely, we investigate the dual codes of convolutional codes over finite rings. In particular we study the p-dimension of the dual code of a convolutional code over a finite ring. This contribution can be considered a generalization and an extension, to the ring case, of the work done by Forney and McEliece on the dimension of the dual code of a convolutional code over a finite field.

Keywords Convolutional codes over finite rings · Dual code · p-bases

M. El Oued
High Institute of Maths and Computer Science of Monastir, Monastir, Tunisia
e-mail: wadyel@yahoo.fr

D. Napp (✉) · R. Pinto
Department of Mathematics, CIDMA - Center for Research and Development in Mathematics and Applications,
University of Aveiro, Aveiro, Portugal
e-mail: diego@ua.pt

R. Pinto
e-mail: raquel@ua.pt

M. Toste
CIDMA - Center for Research and Development in Mathematics and Applications,
Superior School of Technologies and Management of Oliveira Do Hospital,
Polytechnic Institute of Coimbra, Coimbra, Portugal
e-mail: marisatoste@gmail.com

N. Bebiano (ed.), *Applied and Computational Matrix Analysis*,
Springer Proceedings in Mathematics & Statistics 192,
DOI 10.1007/978-3-319-49984-0_5

1 Introduction

Codes play an important role in our days. They are implemented in most of all communications systems in order to detect and correct errors that can be introduced during the transmission of information. Convolutional codes over finite rings were first introduced by [8] and are becoming more relevant for communication systems that combine coding and modulation.

We will consider convolutional codes constituted by left compact sequences in $\mathbb{Z}_{p^r}$, where p is a prime and r an integer, i.e., the codewords of the code will be of the form

$$\begin{aligned} w : \mathbb{Z} &\rightarrow \mathbb{Z}_{p^r}^n \\ t &\mapsto w_t \end{aligned}$$

where $w_t = 0$ for $t < k$ for some $k \in \mathbb{Z}$. These sequences can be represented by Laurent series, $w(D) = \sum_{t=k}^{\infty} w_t D^t$. Let us denote by $\mathbb{Z}_{p^r}((D))$ the ring of Laurent series over $\mathbb{Z}_{p^r}$. Moreover, we will represent the ring of polynomials over $\mathbb{Z}_{p^r}$ by $\mathbb{Z}_{p^r}[D]$ and the ring of rational matrices over $\mathbb{Z}_{p^r}$ by $\mathbb{Z}_{p^r}(D)$. More precisely, $\mathbb{Z}_{p^r}(D)$ is the set

$$\left\{ \frac{p(D)}{q(D)} : p(D), q(D) \in \mathbb{Z}_{p^r}[D] \text{ and the coefficient of the smallest power of } D \text{ in } q(D) \text{ is a unit} \right\}$$

modulo the equivalence relation

$$\frac{p(D)}{q(D)} \sim \frac{p_1(D)}{q_1(D)} \text{ if and only if } p(D)q_1(D) = p_1(D)q(D).$$

Convolutional codes over finite rings behave very differently from convolutional codes over finite fields due to the existence of zero divisors. One main difference is that a convolutional code over a finite field $\mathbb{F}$ is always a free module over $\mathbb{F}((D))$ which does not happen in the ring case. In order to deal with this problem we will consider a new type of basis, for $\mathbb{Z}_{p^r}[D]$-submodules of $\mathbb{Z}_{p^r}^n[D]$, which will allow us to define a kind of basis for every convolutional code, called p-basis, and a related type of dimension, called p-dimension. This notions have been extensively used in the last decade [6, 7, 10, 13, 14], extending the ideas of p-adic expansion, p-dimension, p-basis, etc., used in the context of $\mathbb{Z}_{p^r}$-submodules of $\mathbb{Z}_{p^r}^n$, [1, 11, 12, 15].

In this paper we will study the dual of a convolutional code over $\mathbb{Z}_{p^r}[D]$. In particular, we will show that the dual of a convolutional code is also a convolutional code and we will relate the p-dimensions of a convolutional code and its dual. In the field case, this result follows immediately from matrix theory and it is mentioned in [3, 9].

2 The Module $\mathbb{Z}_{p^r}^n[D]$

Any element in $\mathbb{Z}_{p^r}^n$ can be written uniquely as a linear combination of $1, p, p^2, \ldots$ $\ldots, p^{r-1}$, with coefficients in $\mathscr{A}_p = \{0, 1, \ldots, p-1\} \subset \mathbb{Z}_{p^r}$ (called the p-adic expansion of the element) [1]. Note that all elements of $\mathscr{A}_p \backslash \{0\}$ are units. This property provides a kind of linear independence on the elements of $\mathscr{A}_p$. In [15], the authors considered this property to define a special type of linear combination of vectors, called *p-linear combination*, which allowed to define the notion of *p-generator sequence*, *p-basis* and *p-dimension* for every submodule of $\mathbb{Z}_{p^r}^n$. These notions were extended for polynomial vectors in [7] and we recall them in this section.

Definition 1 ([7]) Let $v_1(D), \ldots, v_k(D)$ be in $\mathbb{Z}_{p^r}^n[D]$. The vector $\sum_{j=1}^{k} a_j(D)v_j(D)$, with $a_j(D) \in \mathscr{A}_p[D]$, is said to be a ***p*-linear combination** of $v_1(D), \ldots, v_k(D)$ and the set of all p-linear combination of $v_1(D), \ldots, v_k(D)$ is called the ***p*-span** of $\{v_1(D), \ldots, v_k(D)\}$, denoted by p-span $(v_1(D), \ldots, v_k(D))$.

Note that the p-span of a set of vectors is not always a module. We need to introduce an extra condition to be fulfilled by the vectors.

Definition 2 ([7]) An ordered set of vectors $(v_1(D), \ldots, v_k(D))$ in $\mathbb{Z}_{p^r}^n[D]$ is said to be a ***p*-generator sequence** if $p\, v_i(D)$ is a p-linear combination of $v_{i+1}(D), \ldots,$ $v_k(D),\ \ i = 1, \ldots, k-1$, and $p\, v_k(D) = 0$.

Lemma 1 ([7]) *If* $(v_1(D), \ldots, v_k(D))$ *is a p-generator sequence in* $\mathbb{Z}_{p^r}^n[D]$ *then*

$$p\text{-}span(v_1(D), \ldots, v_k(D)) = span(v_1(D), \ldots, v_k(D)).$$

Consequently p-*span*$(v_1(D), \ldots, v_k(D))$ *is a* $\mathbb{Z}_{p^r}$*-submodule of* $\mathbb{Z}_{p^r}^n[D]$.

Note that if $M = span(v_1(D), \ldots, v_k(D))$ is a submodule of $\mathbb{Z}_{p^r}[D]$, then

$$\begin{aligned}(v_1(D), pv_1(D) \ldots, p^{r-1}v_1(D), v_2(D), pv_2(D), \ldots,\\ \ldots, p^{r-1}v_2(D), \ldots, v_l(D), pv_k(D) \ldots, p^{r-1}v_k(D)).\end{aligned} \tag{1}$$

is a p-generator sequence of M.

Definition 3 ([7]) The vectors $v_1(D), \ldots, v_k(D)$ in $\mathbb{Z}_{p^r}^n[D]$ are said to be ***p*-linearly independent** if the only p-linear combination of $v_1(D), \ldots, v_k(D)$ that is equal to 0 is the trivial one.

Definition 4 ([7]) An ordered set of vectors $(v_1(D), \ldots, v_k(D))$ which is a p-linearly independent p-generator sequence of a submodule M of $\mathbb{Z}_{p^r}^n[D]$ is said to be a ***p*-basis** of M.

It is proved in [6] that two p-bases of a $\mathbb{Z}_{p^r}[D]$-submodule M of $\mathbb{Z}^n_{p^r}[D]$ have the same number of elements. This number of elements is called ***p*-dimension** of M and is denoted by $p\text{-dim}(M)$.

We recall that a free module is a module which admits a basis. The cardinality of a basis of a free module M is called the rank of M.

Lemma 2 ([7]) *Let M be a free submodule of $\mathbb{Z}_{p^r}[D]$ of rank m. Then the p-dimension of M is mr. If $(v_1(D), \ldots, v_m(D))$ is basis of M, then, the sequence*

$$(v_1(D), pv_1(D), \ldots, p^{r-1}v_1(D), \ldots, v_m(D), pv_m(D), \ldots, p^{r-1}v_m(D))$$

is a p-basis of M.

The same notions and results are satisfied for the module $\mathbb{Z}^n_{p^r}$ in [15]. In fact, as mentioned before, these notions were first introduced in this paper for such modules and later extended for the module $\mathbb{Z}^n_{p^r}[D]$ in [7].

Finally, we give the following definition which we need in next sections.

Definition 5 ([5]) A module M is said to be semisimple if it is a direct sum of simple modules, where a simple module is a module that has no submodules other than itself and $\{0\}$.

Let M be a semisimple module. Then every submodule of M is a direct summand, i.e., for every submodule N of M, there is a complement P such that $M = N \oplus P$. Moreover, every submodule of M is semisimple.

3 Convolutional Codes Over $\mathbb{Z}_{p^r}$

Definition 6 A **convolutional code** $\mathscr{C}$ of length n over $\mathbb{Z}_{p^r}$ is a $\mathbb{Z}_{p^r}((D))$-submodule of $\mathbb{Z}^n_{p^r}((D))$ for which there exists a polynomial matrix $\widetilde{G}(D) \in \mathbb{Z}^{\tilde{k}\times n}_{p^r}[D]$ such that

$$\begin{aligned}\mathscr{C} &= \mathrm{Im}_{\mathbb{Z}_{p^r}((D))}\widetilde{G}(D)\\ &= \left\{u(D)\widetilde{G}(D) \in \mathbb{Z}^n_{p^r}((D)) : u(D) \in \mathbb{Z}^{\tilde{k}}_p((D))\right\}.\end{aligned}$$

The matrix $\widetilde{G}(D)$ is called a **generator matrix** of $\mathscr{C}$. If $\widetilde{G}(D)$ has full row rank then it is called an **encoder** of $\mathscr{C}$.

The notion of p-basis can be used to define a p-encoder for a convolutional code.

Definition 7 ([6]) Let $\mathscr{C}$ be a convolutional code of length n over $\mathbb{Z}_{p^r}$. Let $G(D)$ in $\mathbb{Z}^{k\times n}_{p^r}[D]$ be a polynomial matrix whose rows are a p-linearly independent p-generator sequence. Then $G(z)$ is a p-encoder of $\mathscr{C}$ if

$$\begin{aligned}\mathscr{C} &= \mathrm{Im}_{\mathscr{A}_p((D))} G(D) \\ &= \left\{ u(D)G(D) \in \mathbb{Z}_{p^r}^n((D)) : u(D) \in \mathscr{A}_p^k((D)) \right\}.\end{aligned}$$

The integer k is called the p-dimension of $\mathscr{C}$. If there exists a constant matrix $\widetilde{G}$ such that

$$\mathscr{C}+ = \left\{ u(D)\widetilde{G} \in \mathbb{Z}_{p^r}^n((D)) : u(D) \in \mathbb{Z}_p^{\tilde{k}}((D)) \right\},$$

then $\mathscr{C}$ is called a **block code**.

Obviously, block codes are a particular case of convolutional codes. Every block code $\mathscr{C}$ admits a generator matrix in standard form [2]

$$\widetilde{G} = \begin{bmatrix} I_{k_0} & A_{1,0}^0 & A_{2,0}^0 & A_{3,0}^0 & \cdots & A_{r-1,0}^0 & A_{r,0}^0 \\ 0 & pI_{k_1} & pA_{2,1}^1 & pA_{3,1}^1 & \cdots & pA_{r-1,1}^1 & pA_{r,1}^1 \\ 0 & 0 & p^2 I_{k_2} & p^2 A_{3,2}^2 & \cdots & p^2 A_{r-1,2}^2 & p^2 A_{r,2}^2 \\ 0 & 0 & 0 & p^3 I_{k_3} & \cdots & 0 & p^3 A_{r,3}^3 \\ \vdots & \vdots & \vdots & \vdots & \ddots & \vdots & \vdots \\ 0 & 0 & 0 & 0 & \cdots & p^{r-1} I_{k_{r-1}} & p^{r-1} A_{r,r-1}^{r-1} \end{bmatrix}. \tag{2}$$

The integers $k_0, k_1, \ldots, k_{r-1}$ are called the **parameters** of $\widetilde{G}$. All encoders of $\mathscr{C}$ in standard form have the same parameters $k_0, k_1, \ldots, k_{r-1}$.

Note that if $G(D)$ is a generator matrix of a convolutional code $\mathscr{C}$ and $X(D)$ is an invertible rational matrix such that $X(D)G(D)$ is polynomial, then $\mathrm{Im}_{\mathbb{Z}_{p^r}((D))} G(D) = \mathrm{Im}_{\mathbb{Z}_{p^r}((D))} X(D)G(D)$, which means that $X(D)G(D)$ is also a generator matrix of $\mathscr{C}$. Thus, the next straightforward result follows. We include its proof for the sake of completeness.

Lemma 3 *Let $\mathscr{C}$ be a submodule of $\mathbb{Z}_{p^r}^n((D))$ given by $\mathscr{C} = Im_{\mathbb{Z}_{p^r}((D))} N(D)$, where $N(D) \in \mathbb{Z}_{p^r}^{\tilde{k}\times n}(D)$. Then $\mathscr{C}$ is a convolutional code, and if $N(D)$ has full row rank, $\mathscr{C}$ is a free code of rank k.*

Proof Write $N(D) = \left[\frac{p_{ij}(D)}{q_{ij}(D)}\right]$, where $p_{ij}(D), q_{ij}(D) \in \mathbb{Z}_{p^r}[D]$, and the coefficient of the smallest power of D in $q_{ij}(D)$ is a unit. Consider the diagonal matrix $Y(D) \in \mathbb{Z}_{p^r}^{\tilde{k}\times\tilde{k}}[D]$ whose element of the row i is the least common multiple of $q_{i1}(D), q_{i2}(D), \ldots, q_{i\tilde{k}}(D)$. Thus $Y(D)$ is invertible and $N(D) = Y(D)^{-1}X(D)$ for some polynomial matrix $X(D) \in \mathbb{Z}_{p^r}^{\tilde{k}\times n}[D]$. Then $\mathrm{Im}_{\mathbb{Z}_{p^r}((D))} N(D) = \mathrm{Im}_{\mathbb{Z}_{p^r}((D))} X(D)$, which means that $X(D)$ is a generator matrix of $\mathscr{C}$. The last statement of the lemma follows from the fact that $N(D)$ is full row rank if and only if $X(D)$ has full row rank. □

Next we will consider a decomposition of a convolutional code into simpler components. For that we need the following lemma.

Lemma 4 *Let M be a submodule of $\mathbb{Z}_{p^r}^n((D))$. Then, there exists a unique family $M_0, \ldots, M_{r-1}$ of free submodules of $\mathbb{Z}_{p^r}^n((D))$ such that*

$$M = M_0 \oplus pM_1 \oplus \ldots \oplus p^{r-1}M_{r-1}. \tag{3}$$

Proof Let $\overline{M}$ be the projection of M over $\mathbb{Z}_p((D))$ and denote its dimension by $k_0(M)$. Let M_0 be the free code over $\mathbb{Z}_{p^r}((D))$ of rank k_0 satisfying $\overline{M} = \overline{M_0}$ and $M_0 \subset M$. As $\mathbb{Z}_{p^r}^n((D))$ is a semisimple module, M_0 admits a complement code M_0' in M. Necessarily, there exists a code M_1' such that $M_0' = pM_1'$. We have $M = M_0 \oplus pM_1'$. Then by induction we have the result. □

Note that if $\mathscr{C}$ is a block code, this decomposition is directly derived from a generator matrix in standard form. In fact, if G, of the form (2), is a generator matrix of $\mathscr{C}$ then $p^i\mathscr{C}_i = \mathrm{Im}_{\mathbb{Z}_{p^r}((D))} p^i G_i$, where $G_i = [0 \cdots 0 \; I_{k_i} \; A_{2,i}^i \cdots A_{r,i}^i]$, $i = 0, \ldots, r-1$.

Next we will show that any convolutional code $\mathscr{C}$ can be decomposed as

$$\mathscr{C} = \mathscr{C}_0 \oplus p\mathscr{C}_1 \oplus \cdots \oplus p^{r-1}\mathscr{C}_{r-1}$$

where $\mathscr{C}_0, \mathscr{C}_1, \ldots, \mathscr{C}_{r-1}$ are free convolutional codes.

Let $G(D)$ be a generator matrix of $\mathscr{C}$. If $G(D)$ is full row rank then $\mathscr{C}$ is free and $\mathscr{C} = \mathscr{C}_0$.

Let us assume now that $G(D)$ is not full row rank. Then the projection of $G(D)$ into $\mathbb{Z}_p[D]$, $\overline{G(D)} \in \mathbb{Z}_p^{k\times n}[D]$, is also not full row rank and there exists a nonsingular matrix $F_0(D) \in \mathbb{Z}_p^{k\times k}[D]$ such that $F_0(D)\overline{G(D)} = \begin{bmatrix} \widetilde{G}_0(D) \\ 0 \end{bmatrix}$ modulo p, where $\widetilde{G}_0(D)$ is full row rank with rank k_0. Considering $F_0(D) \in \mathbb{Z}_{p^r}^{k\times k}[D]$, it follows that $F_0(D)G(D) = \begin{bmatrix} G_0(D) \\ p\widehat{G}_1(D) \end{bmatrix}$, where $G_0(D) \in \mathbb{Z}_{p^r}^{k_0\times n}$ is such that $\overline{G_0(D)} = \widetilde{G}_0(D)$. Moreover, since $F_0(D)$ is invertible, $\begin{bmatrix} G_0(D) \\ p\widehat{G}_1(D) \end{bmatrix}$ is also a generator matrix of $\mathscr{C}$.

Let us now consider $F_1(D) \in \mathbb{Z}_p^{(k-k_0)\times(k-k_0)}[D]$ such that $F_1(D)\overline{\widehat{G}_1(D)} = \begin{bmatrix} \widetilde{G}_1(D) \\ 0 \end{bmatrix}$ modulo p, where $\widetilde{G}_1(D)$ is full row rank with rank k_1. Then, considering $F_1(D) \in \mathbb{Z}_{p^r}^{(k-k_0)\times(k-k_0)}[D]$, it follows that $F_1(D)\widehat{G}_1(D) = \begin{bmatrix} G_1'(D) \\ p\widehat{G}_2(D) \end{bmatrix}$, where $G_1'(D) \in \mathbb{Z}_{p^r}^{\tilde{k}_1\times n}$ is such that $\overline{G_1'(D)} = \widetilde{G}(D)$, and therefore

$$\begin{bmatrix} I_{k_0} & 0 \\ 0 & F_1(D) \end{bmatrix} F_0(D)G(D) = \begin{bmatrix} G_0(D) \\ pG_1'(D) \\ p^2\widehat{G}_2(D) \end{bmatrix}.$$

If $\begin{bmatrix} G_0(D) \\ G_1'(D) \end{bmatrix}$ is not full row rank, then there exists a permutation matrix P and a rational matrix $L(D) \in \mathbb{Z}_{p^r}^{\tilde{k}_1 \times k_0}(D)$ such that

$$P \begin{bmatrix} I_{k_0} & 0 \\ L_1(D) & I_{k_1} \end{bmatrix} \begin{bmatrix} G_0(D) \\ pG_1'(D) \end{bmatrix} = \begin{bmatrix} G_0(D) \\ pG_1''(D) \\ p^2 G_2'(D) \end{bmatrix},$$

where $G_1''(D) \in \mathbb{Z}_{p^r}^{k_1 \times n}(D)$ and $G_2'(D) \in \mathbb{Z}_{p^r}^{(\tilde{k}_1 - k_1) \times n}(D)$ are rational matrices and $\begin{bmatrix} G_0(D) \\ G_1''(D) \end{bmatrix}$ is a full row rank rational matrix. Note that since $P \begin{bmatrix} I_{k_0} & 0 \\ L_1(D) & I_{k_1} \end{bmatrix}$ is nonsingular it follows that

$$\mathrm{Im}_{\mathbb{Z}_{p^r}((D))} \begin{bmatrix} G_0(D) \\ pG_1'(D) \end{bmatrix} = \mathrm{Im}_{\mathbb{Z}_{p^r}((D))} \begin{bmatrix} G_0(D) \\ pG_1''(D) \\ p^2 G_2'(D) \end{bmatrix}.$$

Let $G_1(D) \mathbb{Z}_{p^r}^{k_1 \times n}[D]$ and $G_2''(D) \in \mathbb{Z}_{p^r}^{(\tilde{k}_1 - k_1) \times n}[D]$ be polynomial matrices (see Lemma 3) such that

$$\mathrm{Im}_{\mathbb{Z}_{p^r}((D))} \begin{bmatrix} G_0(D) \\ pG_1''(D) \\ p^2 G_2'(D) \end{bmatrix} = \mathrm{Im}_{\mathbb{Z}_{p^r}((D))} \begin{bmatrix} G_0(D) \\ pG_1(D) \\ p^2 G_2''(D) \end{bmatrix}.$$

Then

$$\begin{bmatrix} G_0(D) \\ pG_1(D) \\ p^2 G_2''(D) \\ p^2 \widehat{G}_2(D) \end{bmatrix}$$

is still a generator matrix of $\mathscr{C}$ such that $\begin{bmatrix} G_0(D) \\ G_1(D) \end{bmatrix}$ is full row rank.

Proceeding in the same way we obtain a generator matrix of $\mathscr{C}$ of the form

$$\begin{bmatrix} G_0(D) \\ pG_1(D) \\ \vdots \\ p^{r-1} G_{r-1}(D) \end{bmatrix},$$

and such that

$$\begin{bmatrix} G_0(D) \\ G_1(D) \\ \vdots \\ G_{r-1}(D) \end{bmatrix}$$

is full row rank. Thus $\mathscr{C}_i := \operatorname{Im} G_i(D)$ is a free convolutional code, $i = 0, 1, \ldots, r-1$, and $\mathscr{C} = \mathscr{C}_0 \oplus p\mathscr{C}_1 \oplus \cdots \oplus p^{r-1}\mathscr{C}_{r-1}$. If we denote by k_i the rank of $\mathscr{C}_i$ then the family $\{k_0, \ldots, k_{r-1}\}$ is a characteristic of the code. Moreover, it's clear that $\mathscr{C}$ is free if and only if $k_i = 0$ for $i = 1 \ldots r-1$.

The following lemmas will be very useful for deriving the results of the remaining sections.

Lemma 5 *Let $\mathscr{C}$ be a free convolutional code of length n over $\mathbb{Z}_{p^r}((D))$ and rank k. Then, p-$\dim(p^i\mathscr{C}) = (r-i)k$.*

Proof Let $G(D) \in \mathbb{Z}_{p^r}^{k\times n}[D]$ be an encoder of $\mathscr{C}$. The result follows from the fact that $\begin{bmatrix} p^i G(D) \\ p^{i+1}G(D) \\ \vdots \\ p^{r-1}G(D) \end{bmatrix}$ is an p-encoder of $\mathscr{C}$, since $G(D)$ is full row rank. □

Lemma 6 *Let $\mathscr{C}_1$ and $\mathscr{C}_2$ be two convolutional codes over $\mathbb{Z}_{p^r}((D))$. Then we have*

$$p\text{-}\dim(\mathscr{C}_1 + \mathscr{C}_2) = p\text{-}\dim \mathscr{C}_1 + p\text{-}\dim \mathscr{C}_2 - p\text{-}\dim(\mathscr{C}_1 \cap \mathscr{C}_2).$$

If the sum is direct, we have

$$p\text{-}\dim(\mathscr{C}_1 \oplus \mathscr{C}_2) = p\text{-}\dim \mathscr{C}_1 + p\text{-}\dim \mathscr{C}_2.$$

Proof Suppose that $\mathscr{C}_1$ and $\mathscr{C}_2$ are in direct sum, i.e., $\mathscr{C}_1 + \mathscr{C}_2 = \mathscr{C}_1 \oplus \mathscr{C}_2$.

If B_1 is a p-basis of $\mathscr{C}_1$ and B_2 is a p-basis of $\mathscr{C}_2$, then (B_1, B_2) is a p-basis of $\mathscr{C}_1 \oplus \mathscr{C}_2$ which gives the result.

For the general case, Let denote by A the complement of $\mathscr{C}_1 \cap \mathscr{C}_2$ in $\mathscr{C}_1$, i.e., $\mathscr{C}_1 = A \oplus \mathscr{C}_1 \cap \mathscr{C}_2$, and let B such that $\mathscr{C}_2 = B \oplus \mathscr{C}_1 \cap \mathscr{C}_2$. Then we have

$$\mathscr{C}_1 + \mathscr{C}_2 = A \oplus \mathscr{C}_1 \cap \mathscr{C}_2 \oplus B$$

and the result is immediate. □

Next corollary follows immediately from Lemmas 5 and 6.

Corollary 1 *Let $\mathscr{C}$ be a convolutional code of length n such that*

$$\mathscr{C} = \mathscr{C}_0 \oplus p\mathscr{C}_1 \oplus \cdots \oplus p^{r-1}\mathscr{C}_{r-1}$$

with $\mathscr{C}_i$ a free convolutional code with rank k_i, $i = 0, 1, \ldots, r-1$. Then

$$p\text{-dim}(\mathscr{C}) = \sum_{i=0}^{r-1}(r-i)k_i.$$

4 Dual Codes

Let $\mathscr{C}$ be a convolutional code of length n over $\mathbb{Z}_{p^r}((D))$. The **orthogonal** of $\mathscr{C}$, denoted by $\mathscr{C}^\perp$, is defined as

$$\mathscr{C}^\perp = \{y \in \mathbb{Z}_{p^r}^n : [y, x] = 0 \text{ for all } x \in \mathscr{C}\},$$

where $[y, x]$ denotes the inner product over $\mathbb{Z}_{p^r}^n$.

In this section we will show that the dual of a convolutional code is still a convolutional code. The next theorem proves this statement for free convolutional codes, and, as field case, the sum of the rank of the code and its dual is n.

Theorem 1 ([4]) *Let $\mathscr{C}$ be a free convolutional code with length n over $\mathbb{Z}_{p^r}((D))$ and rank $\tilde{k}$. Then $\mathscr{C}^\perp$ is also a free convolutional code of length n and rank $n - \tilde{k}$.*

Proof Let $G(D) \in \mathbb{Z}_{p^r}^{\tilde{k}\times n}$ be an encoder of $\mathscr{C}$. Since $G(D)$ is full row rank there exists a polynomial matrix $L(D) \in \mathbb{Z}_{p^r}^{(n-\tilde{k})\times n}[D]$ such that $\begin{bmatrix} G(D) \\ L(D) \end{bmatrix}$ is nonsingular. Let $[X(D)\ Y(D)]$, with $X(D) \in \mathbb{Z}_{p^r}^{n\times\tilde{k}}(D)$ and $Y(D) \in \mathbb{Z}_{p^r}^{n\times(n-\tilde{k})}(D)$, be the inverse of $\begin{bmatrix} G(D) \\ L(D) \end{bmatrix}$. Then $\mathscr{C}^\perp = \text{Im}_{\mathbb{Z}_{p^r}((D))} Y(D)^t$, which means by Lemma 3 that $\mathscr{C}^\perp$ is a convolutional code. Moreover, since $Y(D)$ is full column rank, there exists a full row rank matrix polynomial matrix $G^\perp(D) \in \mathbb{Z}_{p^r}^{(n-\tilde{k})\times n}[D]$ such that $\mathscr{C}^\perp = \text{Im}_{\mathbb{Z}_{p^r}((D))} G^\perp(D)$. Thus $\mathscr{C}^\perp$ is a free convolutional code of rank $n - \tilde{k}$. □

If $\mathscr{C}$ is a free code of rank $\tilde{k}$, then $p\text{-dim}(\mathscr{C}) = \tilde{k}r$. This gives us the next corollary.

Corollary 2 *Let $\mathscr{C}$ be a free convolutional code of length n over $\mathbb{Z}_{p^r}$. Then we have*

$$p\text{-dim}(\mathscr{C}) + p\text{-dim}(\mathscr{C}^\perp) = nr.$$

In the sequel of this work we propose to establish this result for any code over $\mathbb{Z}_{p^r}((D))$.

The following auxiliary lemmas will be fundamental in the proof of next theorem.

Lemma 7 ([13]) *Let $\mathscr{C}$ be a free convolutional code over $\mathbb{Z}_{p^r}((D))$. For any given integer $i \in \{0, \ldots r-1\}$ we have*

$$\mathscr{C} \cap p^i \mathbb{Z}_{p^r}^n((D)) = p^i \mathscr{C}.$$

Proof The inclusion $p^i\mathscr{C} \subset \mathscr{C} \cap p^i\mathbb{Z}_{p^r}^n((D))$ is trivial. For the other direction, let $y \in p^i\mathbb{Z}_{p^r}^n((D)) \cap \mathscr{C}$. Let $\{x_1, \ldots, x_k\}$ be a basis of $\mathscr{C}$ and its projection $\{\overline{x_1}, \ldots, \overline{x_k}\}$ be a basis of $\overline{\mathscr{C}}$. Then, there exists $a_1, \ldots, a_k \in \mathbb{Z}_{p^r}((D))$ such that $y = \sum_{j=1}^{k} a_j x_j$. As $y \in p^i\mathbb{Z}_{p^r}^n((D))$, we have $\overline{y} = \sum_{j=1}^{k} \overline{a_j x_j} = 0$, where $\overline{a_j} = 0,\ \forall j = 1 \ldots k$. Then, for all $j = 1 \ldots k, a_j$ can be written in the form pb_j where $b_j \in \mathbb{Z}_{p^r}((D))$. By repeating the procedure i times, we obtain $a_j = p^i\alpha_j,\ \forall j = 1 \ldots k$, which gives

$$y = p^i \sum_{j=1}^{k} \alpha_j x_j \in p^i\mathscr{C}.$$

□

Lemma 8 ([13]) *Suppose that $\mathscr{C}$ is a free code. Let $y \in \mathbb{Z}_{p^r}((D))^n$ and let i be an integer in $\{0, \ldots, r-1\}$, such that $p^i y \in \mathscr{C}$. Then $y \in \mathscr{C} + p^{r-i}\mathbb{Z}_{p^r}^n((D))$.*

Proof By the preceding lemma, there exists $x \in \mathscr{C}$ such that $p^i y = p^i x$. This implies that $\overline{y} = \overline{x}$. Thus there exists $y_1 \in \mathscr{C}$, $y_2 \in \mathbb{Z}_{p^r}((D))$ satisfying $y = y_1 + py_2$. We have $p^i y = p^i y_1 + p^{i+1} y_2$, then $p^i y - p^i y_1 = p^{i+1} y_2 \in \mathscr{C}$. Then $y_2 = y_3 + py_4$ where $y_3 \in \mathscr{C}$ and $y_4 \in \mathbb{Z}_{p^r}^n((D))$. Then $y = \underbrace{y_1 + py_3}_{\in\mathscr{C}} + p^2 y_4$. By repeating this procedure $r-i$ times, we obtain $y = x_1 + p^{r-i}x_2$ with $x_1 \in \mathscr{C}$. □

Lemma 9 ([13]) *Let $\mathscr{C}$ be a free convolutional code over $\mathbb{Z}_{p^r}((D))$. For all integer $i \in \{0, \ldots r-1\}$ we have*

$$(p^i\mathscr{C})^\perp = \mathscr{C}^\perp + p^{r-i}\mathbb{Z}_{p^r}^n((D)).$$

Proof It's clear that $\mathscr{C}^\perp + p^{r-i}\mathbb{Z}_{p^r}^n((D)) \subset (p^i\mathscr{C})^\perp$. For the other direction, let $y \in (p^i\mathscr{C})^\perp$, then for all $x \in \mathscr{C}$ we have $[y, p^i x] = [p^i y, x] = 0$, thus $p^i y \in \mathscr{C}^\perp$. As $\mathscr{C}^\perp$ is a free code, we conclude by Lemma 8 that $y \in \mathscr{C}^\perp + p^{r-i}\mathbb{Z}_{p^r}^n((D))$. □

Remark 1 The last lemmas are given in [13] for block codes over $\mathbb{Z}_{p^r}$. The proofs here are just adapted to the ring $\mathbb{Z}_{p^r}((D))$.

Theorem 2 *Let $\mathscr{C} = \mathscr{C}_0 \oplus p\mathscr{C}_1 \oplus \cdots \oplus p^{r-1}\mathscr{C}_{r-1}$ be a convolutional code of length n over $\mathbb{Z}_{p^r}((D))$, such that $\mathscr{C}_i$ is free, $i = 0, 1, \ldots, r-1$, with $\mathscr{C}_0 \oplus \mathscr{C}_1 \oplus \cdots \oplus \mathscr{C}_{r-1} = \mathscr{C}_0 + \mathscr{C}_1 + \cdots + \mathscr{C}_{r-1}$ a free convolutional code. Then, there exists a family of free convolutional codes of length n over $\mathbb{Z}_{p^r}((D))$, $B_i, i = 0, \ldots, r-1$, such that $\mathscr{C}^\perp = B_0 \oplus pB_1 \oplus \cdots \oplus p^{r-1}B_{r-1}$, and*

1. *$B_0 = (\mathscr{C}_0 \oplus \cdots \oplus \mathscr{C}_{r-1})^\perp$.*
2. *For $i \in \{1, \ldots, r-1\}$, $\mathrm{rank}(B_i) = \mathrm{rank}(\mathscr{C}_{r-i})$.*

Proof Suppose that $\text{rank}(\mathcal{C}_i) = k_i$ for $i = 0, \ldots, r-1$. We first begin by looking for the dual of $\mathcal{C}_0 \oplus p\mathcal{C}_1$.

$$\begin{aligned}(\mathcal{C}_0 \oplus p\mathcal{C}_1)^{\perp} &= \mathcal{C}_0^{\perp} \cap (p\mathcal{C}_1)^{\perp} = \mathcal{C}_0^{\perp} \cap (\mathcal{C}_1^{\perp} + p^{r-1}\mathbb{Z}_{p^r}^n) \\ &= \mathcal{C}_0^{\perp} \cap \mathcal{C}_1^{\perp} + p^{r-1}\mathcal{C}_0^{\perp} \\ &= (\mathcal{C}_0 \oplus \mathcal{C}_1)^{\perp} + p^{r-1}\mathcal{C}_0^{\perp}.\end{aligned}$$

By Theorem 1, we can conclude that there exists a free code B_{r-1} such that

$$(\mathcal{C}_0 \oplus p\mathcal{C}_1)^{\perp} = (\mathcal{C}_0 \oplus \mathcal{C}_1)^{\perp} \oplus p^{r-1} B_{r-1}.$$

Suppose $\text{rank}(B_{r-1}) = l_{r-1}$, then we have:

$$\begin{aligned}p\text{-dim}[(\mathcal{C}_0 \oplus p\mathcal{C}_1)^{\perp}] &= p\text{-dim}(\mathcal{C}_0 \oplus \mathcal{C}_1)^{\perp} + p\text{-dim}(p^{r-1} B_{r-1}) \\ &= nr - (k_0 + k_1)r + l_{r-1}.\end{aligned}$$

On the other hand, $p\text{-dim}[(\mathcal{C}_0 \oplus p\mathcal{C}_1)^{\perp}] = nr - (k_0 r + (r-1)k_1)$. We conclude that $\text{rank}(B_{r-1}) = k_1$. We repeat the same procedure with $\mathcal{C}_0 \oplus p\mathcal{C}_1 \oplus p^2\mathcal{C}_2$.

$$\begin{aligned}(\mathcal{C}_0 \oplus p\mathcal{C}_1 \oplus p^2\mathcal{C}_2)^{\perp} &= (\mathcal{C}_0 \oplus p\mathcal{C}_1)^{\perp} \cap (p^2\mathcal{C}_2)^{\perp} \\ &= [(\mathcal{C}_0 \oplus \mathcal{C}_1)^{\perp} \oplus p^{r-1} B_{r-1}] \cap (\mathcal{C}_2^{\perp} + p^{r-2}\mathbb{Z}_{p^r}^n) \\ &= (\mathcal{C}_0 \oplus \mathcal{C}_1 \oplus \mathcal{C}_2)^{\perp} \oplus p^{r-1}(B_{r-1} \cap \mathcal{C}_2^{\perp}) + p^{r-2}(\mathcal{C}_0 \oplus \mathcal{C}_1)^{\perp} + p^{r-1} B_{r-1} \\ &= (\mathcal{C}_0 \oplus \mathcal{C}_1 \oplus \mathcal{C}_2)^{\perp} \oplus p^{r-1} B_{r-1} + p^{r-2}(\mathcal{C}_0 \oplus \mathcal{C}_1)^{\perp}.\end{aligned}$$

By Theorem 1, there exists a free convolutional code B_{r-2} such that

$$(\mathcal{C}_0 \oplus p\mathcal{C}_1 \oplus p^2\mathcal{C}_2)^{\perp} = (\mathcal{C}_0 \oplus \mathcal{C}_1 \oplus \mathcal{C}_2)^{\perp} \oplus p^{r-1} B_{r-1} \oplus p^{r-2} B_{r-2}.$$

Suppose that $\text{rank}(B_{r-2}) = l_{r-2}$, then we have

$$\begin{aligned}&p\text{-dim}(\mathcal{C}_0 \oplus p\mathcal{C}_1 \oplus p^2\mathcal{C}_2)^{\perp} = \\ &\quad = p\text{-dim}[(\mathcal{C}_0 \oplus \mathcal{C}_1 \oplus \mathcal{C}_2)^{\perp}] + p\text{-dim}(p^{r-1} B_{r-1}) + p\text{-dim}(p^{r-2} B_{r-2}) \\ &\quad = nr - (k_0 + k_1 + k_2)r + k_1 + 2l_{r-2}\end{aligned}$$

On the other hand

$$\begin{aligned}p\text{-dim}(\mathcal{C}_0 \oplus p\mathcal{C}_1 \oplus p^2\mathcal{C}_2)^{\perp} &= nr - [k_0 r + k_1(r-1) + k_2(r-2)] \\ &= (n - k_0 - k_1)r + k_1 + 2k_2.\end{aligned}$$

We conclude that $\text{rank}(B_{r-2}) = k_2$. We repeat this procedure $r-1$ times, we thus find the desired result. □

The following result is a consequence of this theorem and generalizes the well-known result for the field case: if $\mathscr{C}$ is a convolutional code of length n over $\mathbb{F}((D))$, where $\mathbb{F}$ is a finite field, then $\dim \mathscr{C} + \dim \mathscr{C}^{\perp} = \dim \mathbb{F}((D)) = n$.

Corollary 3 *Let $\mathscr{C}$ be a convolutional code of length n over $\mathbb{Z}_{p^r}^n$. Then*

$$p\text{-dim}(\mathscr{C}) + p\text{-dim}(\mathscr{C}^{\perp}) = p\text{-dim}(\mathbb{Z}_{p^r}^n((D)) = nr.$$

Proof Let $\mathscr{C} = \mathscr{C}_0 \oplus p\mathscr{C}_1 \oplus \cdots \oplus p^{r-1}\mathscr{C}_{r-1}$ where $\mathscr{C}_i$ is free of rank $k_i, i = 0, 1, \ldots, r-1$. Consider also the free convolutional codes of length n over $\mathbb{Z}_{p^r}((D))$, $B_i, i = 0, \ldots, r-1$, such that $\mathscr{C}^{\perp} = B_0 \oplus pB_1 \oplus \ldots \oplus p^{r-1}B_{r-1}$, and

1. $B_0 = (\mathscr{C}_0 \oplus \ldots \oplus \mathscr{C}_{r-1})^{\perp}$.
2. $\text{rank}(B_i) = \text{rank}(\mathscr{C}_{r-i}), i \in \{1, \ldots, r-1\}$.

Note that $p\text{-dim}(\mathscr{C}) = \sum_{i=0}^{r-1}(r-i)k_i$. From 2. and Lemma 5, it follows that $p\text{-dim}(p^i B_i) = (r-i)k_{r-i}$ and from 1. and Corollary 2 it follows that $p\text{-dim}(B_0) = nr - r(k_0 + k_1, \cdots + k_{r-1})$. Thus,

$$\begin{aligned}
p\text{-dim}(\mathscr{C}^{\perp}) &= p\text{-dim}(B_0) + p\text{-dim}(pB_1) + \cdots + p\text{-dim}(p^{r-1}B_{r-1}) \\
&= nr - r(k_0 + k_1 + \cdots + k_{r-1}) + (r-1)k_{r-1} + (r-2)k_{r-2} + \cdots + k_1 \\
&= nr - (k_0 r + k_1(r-1) + \cdots + k_{r-1}) \\
&= nr - p\text{-dim}(\mathscr{C}).
\end{aligned}$$

□

Remark 2 In the case of block code over a finite ring, we can find this result using the theorem of J.Wood in [16]. Indeed, if $\mathscr{C}$ is a block code of length n over $\mathscr{R}$. $\mathscr{R}$ is a Frobenius ring and then we have

$$|\mathscr{C}||\mathscr{C}^{\perp}| = |\mathscr{R}^n|.$$

If $p\text{-dim}(\mathscr{C}) = k$, we have $|\mathscr{C}| = p^k$ and then $|\mathscr{C}^{\perp}| = p^{nr-k}$ which gives

$$p\text{-dim}(\mathscr{C}^{\perp}) = nr - k.$$

Acknowledgements The work of the second, third and fourth authors was supported in part by the Portuguese Foundation for Science and Technology (FCT-Fundação para a Ciência e a Tecnologia), through CIDMA - Center for Research and Development in Mathematics and Applications, within project UID/MAT/04106/2013.

References

1. Calderbank, A.R., Sloane, N.J.A.: Modular and p-adic cyclic codes. Des. Codes Cryptogr. **6**(1), 21–35 (1995)
2. Conway, J.H., Sloane, N.J.A.: Self-dual codes over the integers modulo 4. J. Comb. Theory A **62**, 30–45 (1993)
3. Forney Jr., G.D.: Convolutional codes I: algebraic structure. IEEE Trans. Inf. Theory **IT–16**(5), 720–738 (1970)
4. Forney Jr., G.D.: Structural analysis of convolutional codes via dual codes. Trans. Inf. Theory **IT–19**(5), 512–518 (1973)
5. Jacobson, N.: Basic Algebra II. W. H. Freeman, San Francisco (1989)
6. Kuijper, M., Pinto, R.: On minimality of convolutional ring encoders. IEEE Trans. Autom. Control **55**(11), 4890–4897 (2009)
7. Kuijper, M., Pinto, R., Polderman, J.W.: The predictable degree property and row reducedness for systems over a finite ring. Linear Algebr. Appl. **425**(2–3), 776–796 (2007)
8. Massey, J. L., Mittelholzer, T.: Convolutional codes over rings. In: Proceedings of the 4th Joint Swedish-Soviet International Workshop Information Theory, pp. 14–18 (1989)
9. McEliece, R.J.: The algebraic theory of convolutional codes. Handbook of Coding Theory, vol. 1, pp. 1065–1138. Elsevier Science Publishers, Amsterdam (1998)
10. Napp, D., Pinto, R., Toste, M.: On MDS convolutional codes over $\mathbb{Z}_{p^r}$. http://arxiv.org/abs/1601.04507. (Submitted on 18 Jan 2016)
11. Norton, G., Salagean, A.: On the structure of linear and cyclic codes over a finite chain ring. Appl. Algebr. Eng. Commun. Comput. **10**(6), 489–506 (2000)
12. Norton, G.H., Salagean, A.: On the hamming distance of linear codes over a finite chain ring. IEEE Trans. Inf. Theory **46**(3), 1060–1067 (2001)
13. Oued, M.El: On MDR codes over a finite ring. IJICoT **3**(2), 107–119 (2015)
14. Oued, M.El, Solé, P.: MDS convolutional codes over a finite ring. IEEE Trans. Inf. Theory **59**(11), 7305–7313 (2013)
15. Vazirani, V.V., Saran, H., Rajan, B.S.: An efficient algorithm for constructing minimal trellises for codes over finite abelian groups. IEEE Trans. Inf. Theory **42**, 1839–1854 (1996)
16. Wood, J.: Duality for modules over finite rings and applications to coding theory. Am. J. Math. **121**(3), 555–575 (1999)

On the K-Theory of the Reduced C^*-Algebras of $GL(n, \mathbb{R})$ and $GL(n, \mathbb{C})$

Sérgio Mendes

Abstract Using Harish-Chandra parameter space, an explicit formula for the K-theory of the reduced C^*-algebra of $\mathrm{GL}(n, \mathbb{C})$ is obtained, in analogy with the real case $\mathrm{GL}(n, \mathbb{R})$ [8]. Applying automorphic induction, an instance of Langlands functoriality principle, we then relate the K-theory of $C_r^*\mathrm{GL}(2n, \mathbb{R})$ and $C_r^*\mathrm{GL}(n, \mathbb{C})$.

Keywords K-theory · GL(n) · Functoriality

1 Introduction

The Gelfand–Naimark Theorem is a well known result in functional analysis. It implies that the category of locally compact Hausdorff spaces and continuous proper maps is equivalent to the opposite of the category of commutative C^*-algebras and proper C^*-morphisms. The main idea of noncommutative geometry is to regard noncommutative C^*-algebras as dual of an, otherwise undefined, category of noncommutative spaces, see [4, p. 7]. An important example of the above is group C^*-algebras.

In this note we consider the reduced C^*-algebras of $\mathrm{GL}(n)$ over the archimedean local fields $\mathbb{R}$ and $\mathbb{C}$. We are mainly interested in the K-theory of these noncommutative spaces. Our K-theory computation is based on a suitable parametrization of the tempered dual.

Let $G = G_F = GL(n, F)$ where F is a local field. The unitary dual of G is the set of equivalence classes of irreducible unitary representations of G and is equipped with the Fell topology. It has also a Plancherel measure μ, whose support is called the tempered dual of G and will be denoted by $\mathscr{A}_n^t(F)$.

Let π be a unitary representation of G on a Hilbert space $\mathscr{H}$. Then, π induces a representation of the convolution algebra $L^1(G)$ given by

S. Mendes (✉)
ISCTE - Lisbon University Institute, Av. das Forças Armadas, Lisboa, 1649-026 Lisbon, Portugal
e-mail: sergio.mendes@iscte.pt

N. Bebiano (ed.), *Applied and Computational Matrix Analysis*, Springer Proceedings in Mathematics & Statistics 192,
DOI 10.1007/978-3-319-49984-0_6

$$\pi(f) = \int_G f(g)\pi(g)d\mu(g),$$

with $f \in L^1(G)$. Let λ be the left regular representation of G on the Hilbert space $L^2(G)$

$$\lambda : L^1(G) \longrightarrow \mathscr{B}(L^2(G)),\ f \mapsto (g \mapsto f * g).$$

The reduced group C^*-algebra $C_r^*(G)$ is the completion of $L^1(G)$ in the operator norm of the image of λ

$$C_r^*(G) = \overline{\{\lambda(f) \in \mathscr{B}(L^2(G)) : f \in L^1(G)\}}^{\|\cdot\|_{\mathscr{B}(L^2(G))}}.$$

The C^*-algebra $C_r^*(G)$ is a noncommutative space and is strong Morita equivalent to the commutative C^*-algebra $C_0(\mathscr{A}_n^t(F))$, see [6]. Since K-theory is stable under strong Morita equivalence, we have an isomorphism

$$K_j C_r^*(G) \cong K^j \mathscr{A}_n^t(F)),\ j = 0, 1.$$

The tempered representations of G may be seen as $C_r^*(G)$-modules. Hence, K-theory of group C^*-algebras is an important tool in the classification of the representations of G. The computation of the K-theory of a group is in general a difficult problem. A possible approach is the Baum-Connes correspondence (or Connes-Kasparov correspondence in certain cases), a major achievement of noncommutative geometry. Echteroff and Pfante [1] used precisely the Connes-Kasparov correspondence to compute the K-theory of $C_r^*\mathrm{GL}(n, \mathbb{R})$ via equivariant K-theory.

In [6], Plymen used a parametrization of the tempered dual due to Harish-Chandra [2] to compute the K-theory of $C_r^*\mathrm{GL}(n, F)$ when F is a nonarchimedean local field. The same approach was used by Plymen and the author in [8] to compute the K-theory of $C_r^*\mathrm{GL}(n, \mathbb{R})$. This method is different from [1] since we need to keep track of the Langlands parameters.

The case of complex semisimple Lie groups was handled by Penington and Plymen [9] and includes the K-theory of $C_r^*\mathrm{GL}(n, \mathbb{C})$.

In view of class field theory and local Langlands correspondence, the K-theory groups of $C_r^*\mathrm{GL}(n)$ over $\mathbb{R}$ and $\mathbb{C}$ are ultimately parametrized by characters of the multiplicative group $\mathbb{C}^\times$. However, to fully understand the parametrization, some representation theory is required. Specifically, the parametrization is given by pairs (M, σ), where M is a Levi subgroup of $\mathrm{GL}(n)$ and σ is a discrete series representation of a certain subgroup of M. Such pairs are data from the Langlands classification of the representations of $\mathrm{GL}(n)$.

We now give a brief description of the main results of this note. In Sect. 3.3 we compute the K-theory of $C_r^*\mathrm{GL}(n, \mathbb{C})$, see Theorem 2. Since we are specializing to the group $\mathrm{GL}(n, \mathbb{C})$, the computation is more explicit than the general case in [9] and is analogous to the one obtained for $C_r^*\mathrm{GL}(n, \mathbb{R})$ in [8]. In Table 1 we verify a certain resemblance between the K-theory of $C_r^*\mathrm{GL}(n, \mathbb{C})$ and the K-theory of $C_r^*\mathrm{GL}(2n, \mathbb{R})$

and in Sect. 4 we apply the principle of functoriality in Langlands theory [3, 5, 8] to interpret the above mentioned similarity of the K-groups, see Theorem 3. The above mentioned resemblance in Table 1 would probably remained unnoticed without the explicit formulae for $K_j C_r^* \mathrm{GL}(n, \mathbb{R})$ and $K_j C_r^* \mathrm{GL}(n, \mathbb{R})$.

Although this note is only concerned with archimedean fields it should be mentioned that in [7] Plymen and the author investigated base change at the level of K-theory for $\mathrm{GL}(n)$ over nonarchimedean local fields. The case of base change over $\mathrm{GL}(n, \mathbb{R})$ is studied in [8]. Base change is another example of the Langlands principle of functoriality.

2 The Harish-Chandra Space

Let $G_F = \mathrm{GL}(n, F)$ where F is either $\mathbb{R}$ or $\mathbb{C}$. Let $C_r^*(G_F)$ denote the reduced C^*-algebra of G_F. The noncommutative space $C_r^*(G_F)$ is strongly Morita equivalent to the commutative C^*-algebra $C_0(\mathscr{A}_n^t(F))$, where $\mathscr{A}_n^t(F)$ is the tempered dual of $\mathrm{GL}(n, F)$. The tempered dual has the structure of locally compact, Hausdorff space and is called the Harish-Chandra parameter space. In order to compute the K-theory of $C_r^*(G_F)$ we need to give a precise description of this parameter space.

Let M be a standard Levi subgroup of G_F. Let M^0 be the subgroup of M such that the determinant of each block-diagonal is ± 1. Denote by $X(M) = \widehat{M/M^0}$ the group of *unramified characters* of M, consisting of those characters which are trivial on M^0.

The Weyl group $W(M) = N(M)/M$ of M acts on the discrete series $E_2(M^0)$ of M^0 by permutations. Choose one element $\sigma \in E_2(M^0)$ for each $W(M)$-orbit. The *isotropy subgroup* of $W(M)$ is the stabilizer $W_\sigma(M) = \{\omega \in W(M) : \omega.\sigma = \sigma\}$. Now, form the disjoint union

$$\bigsqcup_{(M,\sigma)} X(M)/W_\sigma(M) = \bigsqcup_{M} \bigsqcup_{\sigma \in E_2(M^0)} X(M)/W_\sigma(M). \tag{1}$$

The characterization of the tempered dual is due to Harish-Chandra, see [2].

Proposition 1 *There exists a bijection*

$$\begin{array}{ccc} \bigsqcup_{(M,\sigma)} X(M)/W_\sigma(M) & \longrightarrow & \mathscr{A}_n^t(\mathbb{R}) \\ \chi^\sigma & \mapsto & i_{GL(n),MN}(\chi^\sigma \otimes 1), \end{array}$$

where $\chi^\sigma(x) := \chi(x)\sigma(x)$ *for all* $x \in M$.

- The case of $\mathrm{GL}(n, \mathbb{R})$.

The discrete series of $\mathrm{GL}(n, \mathbb{R})$ are empty for $n \geq 3$. Therefore, we only need to consider partitions of n into 1's and 2's. We may decompose n as $n = 2q + r$, where

q is the number of 2's and r is the number of 1's in the partition. We associate the Levi subgroup

$$M \cong \mathrm{GL}(2, \mathbb{R})^q \times \mathrm{GL}(1, \mathbb{R})^r$$

and the subgroup

$$M^0 \cong SL^{\pm}(2, \mathbb{R})^q \times SL^{\pm}(1, \mathbb{R})^r,$$

where $SL^{\pm}(m, \mathbb{R}) = \{g \in \mathrm{GL}(m, \mathbb{R}) : |det(g)| = 1\}$ is the *unimodular subgroup* of $\mathrm{GL}(m, \mathbb{R})$. In particular, $SL^{\pm}(1, \mathbb{R}) = \mathbb{Z}/2\mathbb{Z}$ and $\mathrm{GL}(1, \mathbb{R}) = \mathbb{R}^{\times}$.

The representations in the discrete series of $\mathrm{GL}(2, \mathbb{R})$, denoted $\mathscr{D}_\ell$ for $\ell \in \mathbb{N}$ ($\ell \geq 1$), are induced from $SL(2, \mathbb{R})$ [5, p. 399]:

$$\mathscr{D}_\ell = ind_{SL^{\pm}(2,\mathbb{R}), SL(2,\mathbb{R})}(\mathscr{D}_\ell^{\pm}),$$

where $\mathscr{D}_\ell^{\pm}$ acts in the space

$$\left\{ f : \mathscr{H} \rightarrow \mathbb{C} | f \text{ analytic }, \|f\|^2 = \int\int |f(z)|^2 y^{\ell-1}) dxdy < \infty \right\}.$$

Here, $\mathscr{H}$ denotes the Poincaré upper half plane. The action of $g = \begin{pmatrix} a & b \\ c & d \end{pmatrix}$ is given by

$$\mathscr{D}_\ell^{\pm}(g)(f(z)) = (bz + d)^{-(\ell+1)} f\left(\frac{az + c}{bz + d}\right).$$

More generally, an element σ from the discrete series $E_2(M^0)$ is given by

$$\sigma = \mathscr{D}_{\ell_1} \otimes \cdots \otimes \mathscr{D}_{\ell_q} \otimes \tau_1 \otimes \cdots \otimes \tau_r. \tag{2}$$

Here, $\mathscr{D}_{\ell_i}^{\pm}$ ($\ell_i \geq 1$) are the discrete series representations of $SL^{\pm}(2, \mathbb{R})$ and τ_j ($j = 0, 1$) is a representation of $SL^{\pm}(1, \mathbb{R}) = \mathbb{Z}/2\mathbb{Z}$

$$\tau_0 = id = (x \mapsto x) \text{ and } \tau_1 = sgn = (x \mapsto x/|x|).$$

Now, we quote the following result:

Proposition 2 ([8]) *Let M be a Levi subgroup of* $\mathrm{GL}(n, \mathbb{R})$, *associated to the partition $n = 2q + r$. Then,*

$$X(M) \cong \mathbb{R}^{q+r}.$$

- The case of $\mathrm{GL}(n, \mathbb{C})$.

The tempered dual of $\mathrm{GL}(n, \mathbb{C})$ comprises the *unitary principal series* in accordance with Harish-Chandra [2, p. 277]. The corresponding Levi subgroup is a maximal torus $T \cong (\mathbb{C}^{\times})^n$ and T^0 is the compact n-torus $T^0 \cong \mathbb{T}^n$. The principal series representations are given by parabolic induction

$$\pi_{\ell,it} = i_{G,TU}(\sigma \otimes 1), \tag{3}$$

where $\sigma = \sigma_1 \otimes \cdots \otimes \sigma_n$ and $\sigma_j(z) = (\frac{z}{|z|})^{\ell_j} |z|_{\mathbb{C}}^{it_j}$ ($\ell_j \in \mathbb{Z}$, $t_j \in \mathbb{R}$ and $|z|^2 = |z|_{\mathbb{C}}$).
We have the following result:

Proposition 3 ([8]) *Denote by T the standard maximal torus in* $\mathrm{GL}(n, \mathbb{C})$. *Then,*

$$X(T) \cong \mathbb{R}^n.$$

3 K-Theory

The parametrization of $\mathscr{A}_n^t(F)$ obtained in the previous section allows us to compute the K-theory of $C_r^*(G_F)$ for $F = \mathbb{R}$ and $F = \mathbb{C}$. Denote by M_F the Levi subgroup of G_F. In view of the Strong Morita equivalence described in [6, Sect. 1.2] we infer that

$$\begin{aligned} K_j C_r^* \mathrm{GL}(n, F) &= K^j \left(\bigsqcup\nolimits_{(M_F,\sigma)} X(M_F)/W_\sigma(M_F) \right) \\ &= \bigoplus\nolimits_{(M_F,\sigma)} K^j(\mathbb{R}^{n_{M_F}}/W_\sigma(M_F)), \end{aligned} \tag{4}$$

where $n = 2q + r$, $n_{M_{\mathbb{R}}} = q + r$ and $n_{M_{\mathbb{C}}} = n$. Note that $M_{\mathbb{C}} = T_{\mathbb{C}}$ is a maximal torus.

3.1 *Closed Cones*

Let M be a Levi subgroup of $\mathrm{GL}(n)$. The stabilizer $W_\sigma(M)$ is a subgroup of the symmetric group S_n and acts on $\mathbb{R}^{n_M}$ by permutations.

Definition 1 If $W_\sigma(M) \neq \{1\}$, the orbit space $\mathbb{R}^{n_M}/W_\sigma(M)$ is called a closed cone.

The next result shows that K-groups of closed cones both vanish.

Proposition 4 *For $n > 1$, $K^j(\mathbb{R}^{n_M}/W_\sigma(M)) = 0$, $j = 0, 1$.*

Proof We need the following definition. A point $(a_1, \ldots, a_n) \in \mathbb{R}^n$ is called normalized if $a_j \leq a_{j+1}$, for $j = 1, 2, \ldots, n-1$. Therefore, in each orbit there is exactly one normalized point and $\mathbb{R}^n/S_n$ is homeomorphic to the subset of $\mathbb{R}^n$ consisting of all normalized points of $\mathbb{R}^n$. We denote the set of all normalized points of $\mathbb{R}^n$ by $N(\mathbb{R}^n)$.

In the case of $n = 2$, let (a_1, a_2) be a normalized point of $\mathbb{R}^2$. Write

$$X_1 = [0, +\infty[\times [1, +\infty[$$

$$X_2 =]-\infty, 0[\times [0, 1[$$

$$X_3 = [-\infty, 0[\times[-1, 0[$$

and form the disjoint union

$$\mathbb{R} \times [-1, +\infty[= X_1 \sqcup X_2 \sqcup X_3.$$

Clearly, the map $\varphi : \mathbb{R} \times [-1, +\infty[\to N(\mathbb{R}^2)$ defined by

$$\varphi(a, t) = \begin{cases} (a, at) & , (a, t) \in X_1 \sqcup X_2 \\ (a, -at) & , (a, t) \in X_3 \end{cases}$$

is a homeomorphism.

If $n > 2$ then the map

$$N(\mathbb{R}^{n-1}) \times [-1, +\infty[\to N(\mathbb{R}^n), (a_1, \ldots, a_{n-1}, t) \mapsto (a_1, \ldots, a_{n-2}, \varphi(a_{n-1}, t))$$

is a homeomorphism. Since $[-1, +\infty[$ has zero K-theory in all degrees, the result follows by applying Künneth formula. □

3.2 GL(n, $\mathbb{R}$)

The K-theory of $C_r^*\mathrm{GL}(n, \mathbb{R})$ was computed in [8] using the Harish-Chandra parameter space. The following is well known:

$$K^j(\mathbb{R}^n) = \begin{cases} \mathbb{Z} \text{ if } n = j \mod 2 \\ 0 \text{ otherwise} . \end{cases}$$

From the above result and using (4) and Proposition 4, the K-theory of $C_r^*\mathrm{GL}$ $(n, \mathbb{R})$ may be summarized as follows

Theorem 1 ([8]) *Let $C_r^*GL(n, \mathbb{R})$ be the reduced C^*-algebra of* $\mathrm{GL}(n, \mathbb{R})$. *We have:*

(i) Suppose $n = 2q$ is even. Then the K-groups are

$$K_j C_r^* GL(n, \mathbb{R}) \cong \begin{cases} \bigoplus_{\ell_1 > \cdots > \ell_q} \mathbb{Z} & , j \equiv q (mod\, 2) \\ \bigoplus_{\ell_1 > \cdots > \ell_{q-1}} \mathbb{Z} & , otherwise \end{cases}$$

with $\ell_i \in \mathbb{N}$. If $m = 1$ then $K_j C_r^* GL(2, \mathbb{R}) \cong \mathbb{Z}$.

(ii) Suppose $n = 2q + 1$ is odd. Then the K-groups are

$$K_j C_r^* GL(n, \mathbb{R}) \cong \begin{cases} \bigoplus_{\ell_1 > \cdots > \ell_q, \varepsilon} \mathbb{Z} & , j \equiv q + 1 (mod 2) \\ 0 & , otherwise \end{cases}$$

with $\ell_i \in \mathbb{N}$ and $\varepsilon \in \mathbb{Z}/2\mathbb{Z}$. Here, we use the following convention: if $q = 0$ then the direct sum is $\bigoplus_{\mathbb{Z}/2\mathbb{Z}} \mathbb{Z} \cong \mathbb{Z} \oplus \mathbb{Z}$.

Example 1 For $GL(2, \mathbb{R})$ we have two partitions of $n = 2$.

To the partition $2 = 2 + 0$ we associate

$$M = GL(2, \mathbb{R}), \ M^0 = SL^{\pm}(2, \mathbb{R}), \ W(M) = \{1\}, \ X(M) = \mathbb{R}.$$

An element in the discrete series $\sigma \in E_2(M^0)$ is given by

$$\sigma = i_{G,P}(\mathscr{D}_\ell^+), \ \ell \in \mathbb{N}.$$

To the partition $2 = 1 + 1$ associate

$$M = (\mathbb{R}^\times)^2, \ M^0 = (\mathbb{Z}/2\mathbb{Z})^2, \ W(M) = \mathbb{Z}/2\mathbb{Z}, \ X(M) = \mathbb{R}^2.$$

In this case, an element in the discrete series $\sigma \in E_2(M^0)$ is given by

$$\sigma = i_{G,P}(id \otimes sgn).$$

The tempered dual is parameterized as follows

$$\mathscr{A}_2^t(\mathbb{R}) \cong \bigsqcup_{(M,\sigma)} X(M)/W_\sigma(M) = \left(\bigsqcup_{\ell \in \mathbb{N}} \mathbb{R}\right) \sqcup (\mathbb{R}^2/S_2) \sqcup (\mathbb{R}^2/S_2) \sqcup \mathbb{R}^2,$$

and the K-theory groups are given by

$$K_j C_r^* \mathrm{GL}(2, \mathbb{R}) \cong K^j(\mathscr{A}_2^t(\mathbb{R})) \cong (\bigoplus_{\ell \in \mathbb{N}} K^j(\mathbb{R})) \oplus K^j(\mathbb{R}^2) = \begin{cases} \bigoplus_{\ell \in \mathbb{N}} \mathbb{Z} & , j = 1 \\ \mathbb{Z} & , j = 0. \end{cases}$$

3.3 GL(n, $\mathbb{C}$)

The K-theory for complex semisimple Lie groups was computed by Penington and Plymen [9]. When $G = \mathrm{GL}(n, \mathbb{C})$, the computation was recalled in [8, Theorem 3.9]. In analogy with $\mathrm{GL}(n, \mathbb{R})$ in Sect. 3.2, we are looking for a more explicit description of the K-theory groups for $C_r^* \mathrm{GL}(n, \mathbb{C})$.

Theorem 2 *Let $(\ell_1, \ell_2, \cdots, \ell_n) \in \mathbb{Z}^n$. Then,*

$$K_j C_r^* \mathrm{GL}(n, \mathbb{C}) = \begin{cases} \bigoplus_{\ell_1 > \ell_2 > \cdots > \ell_n} \mathbb{Z} , & \text{if } n = j \mod 2 \\ 0 , & \text{otherwise} , \end{cases}$$

Proof Let $\mathscr{A}_n^t(\mathbb{C})$ denote the Harish-Chandra parameter space. We exploit the strong Morita equivalence described in [9, Proposition 4.1]. We have a homeomorphism of locally compact Hausdorff spaces:

$$\mathscr{A}_n^t(\mathbb{C}) = \bigsqcup_{\sigma \in E_2(T^0)} \mathbb{R}^n / W_{\sigma(T)},$$

by the Harish-Chandra Plancherel Theorem for complex reductive groups [2], and the identification of the Fell topology on the left-hand-side with the natural topology on the right-hand-side, as in [9]. Here, T^0 is the maximal compact subgroup of the maximal torus T of $\mathrm{GL}(n, \mathbb{C})$. Hence,

$$T = (\mathbb{C}^\times)^n \text{ and } T^0 = \mathbb{T}^n.$$

In this case, σ is a character of $\mathbb{T}^n$ and is completely determined by an integers $(\ell_1, \ell_2, \ldots, \ell_n) \in \mathbb{Z}^n$.

The Weyl group is the symmetric group S_n and identifies elements

$$(\ell_1, \ell_2, \ldots, \ell_n) \sim (\ell_{\tau(1)}, \ell_{\tau(2)}, \ldots, \ell_{\tau(n)})$$

for every nontrivial $\tau \in S_n$ since they correspond to equivalent representations. Moreover, if $\ell_i = \ell_j$ for $i \neq j$, then $W_\sigma(T) \neq \{1\}$ and $\mathbb{R}^n / W_{\sigma(T)}$ is a closed cone. Therefore, $W_\sigma(T) = \{1\}$ if, and only if, $\ell_1 > \ell_2 > \cdots > \ell_n$ and the result follows. □

Example 2 The tempered dual of $C_r^*\mathrm{GL}(2, \mathbb{C})$, represented as a lattice. Each dot $\bullet$ represents a pair $(\ell_1, \ell_2) \in \mathbb{Z}^2$ with $\ell_1 > \ell_2$ which corresponds to a copy of the plane $\mathbb{R}^2$. The point $(0, 0)$ denotes the origin of the lattice $\mathbb{Z}^2$.

$$\begin{array}{cccccc}
\circ & \circ & \circ & \circ & \circ & \bullet\mathbb{R}^2 \\
\circ & \circ & \circ & \circ & \bullet\mathbb{R}^2 & \bullet\mathbb{R}^2 \\
\circ & \circ & (0,0) & \bullet\mathbb{R}^2 & \bullet\mathbb{R}^2 & \bullet\mathbb{R}^2 \\
\circ & \circ & \bullet\mathbb{R}^2 & \bullet\mathbb{R}^2 & \bullet\mathbb{R}^2 & \bullet\mathbb{R}^2 \\
\circ & \bullet\mathbb{R}^2 & \bullet\mathbb{R}^2 & \bullet\mathbb{R}^2 & \bullet\mathbb{R}^2 & \bullet\mathbb{R}^2 \\
\bullet\mathbb{R}^2 & \bullet\mathbb{R}^2 & \bullet\mathbb{R}^2 & \bullet\mathbb{R}^2 & \bullet\mathbb{R}^2 & \bullet\mathbb{R}^2
\end{array}$$

4 A Functorial Map

The following table contains the K-groups of the reduced C^*-algebras of $\mathrm{GL}(n, \mathbb{C})$ and $\mathrm{GL}(2n, \mathbb{R})$, where used the convention: $\ell'_k \in \mathbb{Z}$ and $\ell_k \in \mathbb{N}$, for $0 \leq k \leq n$.

We conclude that there exists a certain resemblance between the even (respectively, odd) K-groups of $\mathrm{GL}(n, \mathbb{C})$ and $\mathrm{GL}(2n, \mathbb{R})$ when n is even (respectively, odd). In this section we aim to find an interpretation for this result based on representation theory. In order to do that, we need to delve into the local Langlands correspondence for archimedean fields [5] and a particular instance of the *principle of functoriality* known as automorphic induction, see [3]. We give now some background on the local Langlands correspondence for local archimedean fields.

Let F be either $\mathbb{R}$ or $\mathbb{C}$. The Weyl group of F is the group W_F that fits into the following short exact sequence of topological groups

$$1 \longrightarrow F^\times \longrightarrow W_F \longrightarrow Gal(\overline{F}/F) \longrightarrow 1.$$

Specifically,

$$W_{\mathbb{C}} = \mathbb{C}^\times \quad \text{and} \quad W_{\mathbb{R}} = \langle j \rangle \mathbb{C}^\times,$$

where $j^2 = -1 \in \mathbb{C}^\times$ and $jc = \overline{c}j$, for all $c \in \mathbb{C}^\times$. As a disjoint set we have

$$W_{\mathbb{R}} = \mathbb{C}^\times \sqcup j\mathbb{C}^\times.$$

An L-parameter is a continuous homomorphism

$$\phi : W_F \to GL(n, \mathbb{C})$$

such that $\phi(w)$ is semisimple for all $w \in W_F$. L-parameters are also called Langlands parameters. Two L-parameters are equivalent if they are conjugate under $GL(n, \mathbb{C})$. The set of equivalence classes of L-parameters whose image is bounded is denoted by $\mathcal{G}_n^t$. This is the class of L-parameters we are interested in since they parametrize tempered representations. For that reason, they are called tempered L-parameters.

The local Langlands correspondence is a bijection

$$\mathcal{G}_n^t(F) \to \mathcal{A}_n^t(F)$$

which satisfies some identities on L-functions and ε-factors, see [5].

Example 3 Since $W_{\mathbb{C}} = \mathbb{C}^\times$, a 1-dimensional L-parameter of $W_{\mathbb{C}}$ is simply a unitary quasicharacter of $\mathbb{C}^\times$, i.e., a character:

$$\chi(z) = (z/|z|)^\ell \otimes |z|_{\mathbb{C}}^{it}$$

where $|z|^2 = |z|_{\mathbb{C}} = z\overline{z}$, $\ell \in \mathbb{Z}$ and $t \in \mathbb{R}$. To emphasize the dependence on parameters (ℓ, t) we may write $\chi = \chi_{\ell,t}$ or $\chi = \chi_\ell$. An n-dimensional L-parameter can be written as a direct sum of n 1-dimensional characters of $\mathbb{C}^\times$:

$$\phi = \phi_1 \oplus \cdots \oplus \phi_n,$$

with $\phi_k(z) = (z/|z|)^{\ell_k} \otimes |z|_{\mathbb{C}}^{t_k}$, $\ell_k \in \mathbb{Z}$, $t_k \in \mathbb{R}$, $k = 1, \ldots, n$.

For a description of the L-parameters of $W_{\mathbb{R}}$ and more details about the local Langlands correspondence in the archimedean setting see [5].

The group $Gal(\mathbb{C}/\mathbb{R})$ acts on $\mathscr{G}_1^t(\mathbb{C}) = \widehat{\mathbb{C}^\times}$

$$\chi^\tau(z) = \chi(\overline{z}),$$

where τ is generator of $Gal(\mathbb{C}/\mathbb{R})$. It follows that $Gal(\mathbb{C}/\mathbb{R})$ acts on $\mathscr{G}_n^t(\mathbb{C})$ for every n. A simple computation shows that

$$\chi_{\ell,t}^\tau(z) = \chi_{-\ell,t}(z).$$

Therefore,

$$\chi^\tau = \chi \Leftrightarrow \ell = 0.$$

Note that $W_\mathbb{C} \subset W_\mathbb{R}$, with index $[W_\mathbb{R} : W_\mathbb{C}] = 2$. Hence, there is a natural induction map

$$Ind_{\mathbb{C}/\mathbb{R}} : \mathscr{G}_n^t(\mathbb{C}) \to \mathscr{G}_{2n}^t(\mathbb{R}).$$

By the local Langlands correspondence for archimedean fields, there exists an automorphic induction map $\mathscr{A}\mathscr{I}_{\mathbb{C}/\mathbb{R}}$ such that the following diagram commutes

$$\begin{array}{ccc} \mathscr{A}_n^t(\mathbb{C}) & \xrightarrow{\mathscr{A}\mathscr{I}_{\mathbb{C}/\mathbb{R}}} & \mathscr{A}_{2n}^t(\mathbb{R}) \\ {\scriptstyle {}_\mathbb{C}\mathscr{L}_n} \uparrow & & \uparrow {\scriptstyle {}_\mathbb{R}\mathscr{L}_{2n}} \\ \mathscr{G}_n^t(\mathbb{C}) & \xrightarrow[Ind_{\mathbb{C}/\mathbb{R}}]{} & \mathscr{G}_{2n}^t(\mathbb{R}) \end{array}$$

Example 4 (*Automorphic induction for $n = 1$*). Let $\chi = \chi_{\ell,t}$ be an L-parameter of $W_\mathbb{C}$. If $\chi \neq \chi^\tau$ then $\phi_{\ell,t} \simeq \phi_{-\ell,t}$, see [8]. Hence,

$$\mathscr{A}\mathscr{I}_{\mathbb{C}/\mathbb{R}}({}_\mathbb{C}\mathscr{L}_1(\chi_{\ell,t})) = D_{|\ell|} \otimes |det(.)|^{it}.$$

If $\chi = \chi^\tau$ then $\chi = \chi_{0,t}$ and we have

$$\mathscr{A}\mathscr{I}_{\mathbb{C}/\mathbb{R}}({}_\mathbb{C}\mathscr{L}_1(|.|_\mathbb{C}^{it})) = {}_\mathbb{R}\mathscr{L}_2(\rho \oplus sgn.\rho) = \pi(\rho, \rho^{-1}),$$

where $\pi(\rho, \rho^{-1})$ is a reducible principal series and ρ is the character of $\mathbb{R}^\times \simeq W_\mathbb{R}^{ab}$ associated with $\chi_{0,t} = |.|_\mathbb{C}^{it}$ via class field theory, i.e., such that $\rho_{|W_\mathbb{C}} = \chi$.

As a map of topological spaces, automorphic induction for $n = 1$ may be described as follows:

$$(t, \ell) \in \mathbb{R} \times \mathbb{Z} \mapsto (t, |\ell|) \in \mathbb{R} \times \mathbb{N}, \text{ if } \ell \neq 0 \tag{5}$$

$$(t, 0) \in \mathbb{R} \times \mathbb{Z} \mapsto (t, t) \mapsto \mathbb{R}^2, \text{ if } \ell = 0. \tag{6}$$

Table 1 K-theory groups

	$K_0C_r^*\mathrm{GL}(n,\mathbb{C})$	$K_0C_r^*\mathrm{GL}(2n,\mathbb{R})$	$K_1C_r^*\mathrm{GL}(n,\mathbb{C})$	$K_1C_r^*\mathrm{GL}(2n,\mathbb{R})$
n even	$\bigoplus_{\ell_1'>\cdots>\ell_n'}\mathbb{Z}$	$\bigoplus_{\ell_1>\cdots>\ell_n}\mathbb{Z}$	0	$\bigoplus_{\ell_1>\cdots>\ell_{n-1}}\mathbb{Z}$
n odd	0	$\bigoplus_{\ell_1>\cdots>\ell_{n-1}}\mathbb{Z}$	$\bigoplus_{\ell_1'>\cdots>\ell_n'}\mathbb{Z}$	$\bigoplus_{\ell_1>\cdots>\ell_n}\mathbb{Z}$

We may now prove a result which explains the similarity between the K-theory groups of $C_r^*\mathrm{GL}(n,\mathbb{C})$ and $C_r^*\mathrm{GL}(2n,\mathbb{R})$ as shown in Table 1.

Theorem 3 *Let* $\mathscr{A}\mathscr{I}^*: K_jC_r^*\mathrm{GL}(2n,\mathbb{R}) \longrightarrow K_jC_r^*\mathrm{GL}(n,\mathbb{C})$ *denote the functorial map induced by the automorphic induction map* $\mathscr{A}\mathscr{I} = \mathscr{A}\mathscr{I}_{\mathbb{C}/\mathbb{R}}$, *with* $j \equiv n\,(mod\,2)$. *Then,*

$$Im(\mathscr{A}\mathscr{I}^*) \simeq \bigoplus_{|\ell_1|>|\ell_2|>\cdots>|\ell_n|} \mathbb{Z}.$$

Proof By [8, Theorem 6.3], the generator $([D_m^{|\ell_1|}],\ldots,[D_m^{|\ell_n|}])$ of the component $\mathbb{Z}_{(|\ell_1|,\ldots,|\ell_n|)}$ of $K_jC_r^*\mathrm{GL}(2n,\mathbb{R})$ is sent to $(\mathscr{A}\mathscr{I}_1^*([D_m^{|\ell_1|}]),\ldots,\mathscr{A}\mathscr{I}_1^*([D_m^{|\ell_n|}]))$ which lies in $K_jC_r^*\mathrm{GL}(n,\mathbb{C})$ and this class is nontrivial if and only if $\chi_{\ell_k}^{\tau} \neq \chi_{\ell_k}$, for every $1 \le k \le n$. Moreover, by Theorem 2 we may choose a representative such that $|\ell_1| > |\ell_2| > \cdots > |\ell_n|$. This concludes the proof. □

Example 5 The functorial map $\mathscr{A}\mathscr{I}^*$ is not onto. In fact, for $n = 1$ we have

$$\mathscr{A}\mathscr{I}^*: \bigoplus_{\mathbb{N}}\mathbb{Z} \to \bigoplus_{\mathbb{Z}}\mathbb{Z},\ ([D_1],[D_2],\ldots) \mapsto (\ldots,[D_2],[D_1],0,[D_1],[D_2],\ldots)$$

Acknowledgments I would like to thank the organizers of the Mattriad'2015 held in Coimbra, specially Professors Natália Bebiano, Cristina Câmara and Ana Nata. I also thank the anonymous referee whose remarks helped to improve the paper.

References

1. Echterhoff, S., Pfante, O.: Equivariant K-theory of finite-dimensional real vector spaces. Munster J. of Math. **2**, 65–94 (2009)
2. Harish-Chandra: Collected Papers, vol. 4. Springer, Berlin (1984)
3. Henniart, G.: Induction automorphe pour GL$(n,\mathbb{C})$. J. Funct. Anal. **258**(9), 3082–3096 (2010)
4. Khalkhali, M.: Basic Noncommutative Geometry, 2nd edn. EMS, Zürich (2013)
5. Knapp, A.: Local Langlands correspondence: the archimedean case. Proc. Symp. Pure. Math. **55**, 393–410 (1994)
6. Plymen, R.: The reduced C^*-algebra of the p-adic group GL(n). J. Funct. Anal. **72**, 1–12 (1987)
7. Mendes, S., Plymen, R.: Base change and K-theory for GL(n). J. Noncommut. Geom. **1**, 311–331 (2007)
8. Mendes, S., Plymen, R.: Functoriality and K-theory for GL$_n(\mathbb{R})$. arXiv Preprint (2015) (To appear in the Münster Journal of Mathematics)
9. Penington, M., Plymen, R.: The Dirac operator and the principal series for complex semisimple Lie groups. J. Funct. Anal. **53**, 269–286 (1983)

Spectral Bounds for the k-Regular Induced Subgraph Problem

Domingos Moreira Cardoso and Sofia J. Pinheiro

Abstract Many optimization problems on graphs are reduced to the determination of a subset of vertices of maximum cardinality which induces a k-regular subgraph. For example, a maximum independent set, a maximum induced matching and a maximum clique is a maximum cardinality 0-regular, 1-regular and $(\omega(G) - 1)$-regular induced subgraph, respectively, were $\omega(G)$ denotes the clique number of the graph G. The determination of the order of a k-regular induced subgraph of highest order is in general an NP-hard problem. This paper is devoted to the study of spectral upper bounds on the order of these subgraphs which are determined in polynomial time and in many cases are good approximations of the respective optimal solutions. The introduced upper bounds are deduced based on adjacency, Laplacian and signless Laplacian spectra. Some analytical comparisons between them are presented. Finally, all of the studied upper bounds are tested and compared through several computational experiments.

Keywords Spectral graph theory · Maximum k-regular induced subgraphs · Combinatorial optimization

1 Introduction

Throughout the paper, we deal with simple undirected graphs G, with vertex set $V(G) = \{1, \ldots, n\}$ and edge set $E(G) \neq \emptyset$. Since this graph has n vertices, we say that the graph has *order* n. We write $u \sim v$ whenever the vertices u and v are adjacent. The neighborhood of a vertex $i \in V(G)$, that is, the set of vertices adjacent to i, is denoted by $N_G(i)$, the degree of i is $d_G(i) = |N_G(i)|$, $\Delta(G) = \max_{i \in V(G)} d_G(i)$ and

D.M. Cardoso (✉) · S.J. Pinheiro
Center for Research and Development in Mathematics and Applications,
Department of Mathematics, University of Aveiro, 3810-193 Aveiro, Portugal
e-mail: dcardoso@ua.pt

S.J. Pinheiro
e-mail: spinheiro@ua.pt

N. Bebiano (ed.), *Applied and Computational Matrix Analysis*,
Springer Proceedings in Mathematics & Statistics 192,
DOI 10.1007/978-3-319-49984-0_7

$\delta(G) = \min_{i \in V(G)} d_G(i)$. The subgraph of G induced by the vertex subset $S \subset V(G)$ is denoted by $G[S]$. The graph G is p-regular when all vertices have the same degree equal to p. A vertex subset $S \subseteq V(G)$ is (k, τ)-regular if it induces a k-regular subgraph and $\forall v \notin S,\ |N_G(v) \cap S| = \tau$. The adjacency matrix $A_G = \left(a_{i,j}\right)$ is the $n \times n$ matrix defined by

$$a_{i,j} = \begin{cases} 1 \text{ if } i \sim j, \\ 0 \text{ otherwise.} \end{cases}$$

The Laplacian matrix $L_G = \left(l_{i,j}\right)$ and the signless Laplacian matrix $Q_G = \left(q_{i,j}\right)$ of the graph G, are the matrices $L_G = D_G - A_G$ and $Q_G = D_G + A_G$, respectively, where D_G stands for the diagonal matrix of order n with the i-th entry equal to the vertex degree $d_G(i)$. Therefore, A_G, L_G and Q_G are real symmetric matrices and then all their eigenvalues are real. These eigenvalues are herein denoted, in nonincreasing order, respectively by $\lambda_1 \geq \cdots \geq \lambda_n$, $\mu_1 \geq \cdots \geq \mu_n$ and $q_1 \geq \cdots \geq q_n$. If G has at least one edge, then $\lambda_1 > 0 > \lambda_n$. From now on we consider only simple undirected graphs with at least one edge which will be called graphs.

Each adjacency eigenvalue of a graph G is main if the corresponding eigenspace contains an eigenvector which is not orthogonal to the all ones vector, otherwise is non-main. From Geršgorin's theorem, the eigenvalues of L_G and Q_G are nonnegative real numbers and since the entries of each row of L_G sum 0, then the eigenvalue $\mu_n = 0$ is associated to the all ones eigenvector $\hat{e}$. The multiplicity of 0, as an eigenvalue of L_G, is equal to the number of connected components of G. Furthermore, G is bipartite if and only if $q_n = 0$. Further basic details about graph spectra can be found in [6, 8]. A vertex subset inducing a 0-regular subgraph is called an independent (or stable) set. A maximum independent set is an independent set of maximum cardinality and its cardinality is called independence number and it is denoted by $\alpha(G)$.

In [3] it was proved that the problem of finding a maximum cardinality subset of vertices inducing a k-regular subgraph is NP-hard. Throughout this paper, this maximum is denoted by $\alpha_k(G)$. Note that in the particular case of $k = 0$, $\alpha_0(G) = \alpha(G)$.

The study of spectral upper bounds on the order of k-regular induced subgraphs (it should be noted that the independent sets are 0-regular induced subgraphs) appear in [3–5]. In [1] (see also [11]) an upper bound on the order of induced subgraphs with average degree d (based on adjacency eigenvalues) was obtained for regular graphs, extending the ratio bound (7) to the general case of maximum k-regular induced subgraphs (when $k = 0$, this bound coincide with the ratio bound). A similar result was obtained in [3], using convex quadratic programming techniques. In [4, 5] the arbitrary graph case is analyzed and upper bounds on the order of k-regular induced subgraphs are presented. In [4], the upper bounds are obtained using adjacency eigenvalues and eigenvectors, namely the least eigenvalue (whether it is non-main) and the corresponding eigenspace. In [5], the upper bound is obtained using a quadratic programming technique jointly with the main angles (see [8] for details) and the induced subgraph just must have average degree d.

The main goal of this paper is to introduce some new spectral upper bounds on the order of k-regular induced subgraphs, making an analytic comparison between them when possible. These new upper bounds are based on adjacency, Laplacian and signless Laplacian eigenvalues. Finally, a few computational experiments are presented.

2 Concepts and Fundamental Results

In this section, we introduce some definitions and we recall the previously obtained results needed for the deductions in the next section. In particular, we survey results concerning to spectral upper bounds on the order of k-regular induced subgraphs.

For arbitrary graphs, consider a graph G of order n with $V(G) = S \cup S^c$, where $S \subseteq V(G)$ denotes a vertex subset inducing a k-regular subgraph and S^c is its complement. The set of edges with just one end vertex in S, that is, the cut set defined by S is denoted $\partial(S)$. Hence, $|\partial(S)| = |S|(\bar{d}_S - k)$, where $\bar{d}_S = \frac{1}{|S|}\sum_{i \in S} d_G(i)$.

The next result relates the cardinality of the cut set $\partial(S)$ to the largest eigenvalue of the Laplacian matrix of a graph G.

Lemma 1 ([16]) *Let G be a graph of order n and $S \subseteq V(G)$. Then*

$$|\partial(S)| \leq \mu_1 \frac{|S|(n - |S|)}{n}. \tag{1}$$

Another relationship involving the largest Laplacian eigenvalue and the least adjacency eingenvalue of a graph G is (see [8]).

$$\delta(G) - \lambda_n \leq \mu_1 \leq \Delta(G) - \lambda_n. \tag{2}$$

Now we consider some relationships involving signless Laplacian eigenvalues. Assuming that G is a connected graph of order n, according to [7], the least eigenvalue of Q_G is zero if and only if G is bipartite and, in that case, zero is a simple eigenvalue. They also proved that

$$2\delta(G) \leq q_1 \leq 2\Delta(G). \tag{3}$$

Moreover, according to [9],

$$q_n < \delta(G). \tag{4}$$

From Weyl's inequalities we have an improvement of inequalities (3) and we state relationships between signless Laplacian and adjacency eigenvalues.

$$\delta(G) + \lambda_1 \leq q_1 \leq \Delta(G) + \lambda_1 \tag{5}$$

and

$$\delta(G) + \lambda_n \leq q_n \leq \Delta(G) + \lambda_n. \tag{6}$$

We now present some spectral upper bounds on the size of k-regular induced subgraphs starting with the particular case of $k = 0$, for which we consider only the ones most related with this work.

2.1 Bounds on $\alpha(G)$

In the case of regular graphs, the well known ratio bound, obtained by Hoffman (unpublished) and presented by Lovász in [14] can be stated by the following theorem where, for the last statement, the necessary condition was proved in [12] and the sufficient condition was proved in [2].

Theorem 1 ([2, 12, 14]) *If G is a regular graph of order n, then*

$$\alpha(G) \leq n\frac{-\lambda_n}{\lambda_1 - \lambda_n}. \tag{7}$$

Furthermore, the cardinality of an independent set S attains the upper bound if and only if S is $(0, \tau)$-regular, with $\tau = -\lambda_n$.

The ratio bound (7) was extended by Haemers for arbitrary graphs, according to the following theorem.

Theorem 2 ([11]) *If G is a graph of order n, then*

$$\alpha(G) \leq \frac{-n\ \lambda_n\ \lambda_1}{\delta^2(G) - \lambda_n\ \lambda_1}. \tag{8}$$

The next spectral upper bound based on the largest Laplacian eigenvalue was independently deduced in [10, 15].

Theorem 3 ([10, 15]) *If G is a graph of order n, then*

$$\alpha(G) \leq n\frac{\mu_1 - \delta(G)}{\mu_1}. \tag{9}$$

2.2 Bounds on $\alpha_k(G)$

Cardoso, Kamińsky and Lozin in [3] introduced the following family of convex quadratic programming problems:

$$\upsilon_k(G) = \max_{x \geq 0} 2\hat{e}^T x - \frac{\tau}{k+\tau} x^T \left(\frac{A_G}{\tau} + I_n\right) x, \tag{10}$$

where $\hat{e}$ is the all ones vector, I_n the identity matrix of order n, $k \in \mathbb{N} \cup \{0\}$ and $\tau = -\lambda_n$ and they proved that $\alpha_k(G) \leq \upsilon_k(G)$, where $\alpha_k(G)$ is the cardinality of a vertex subset inducing a k-regular subgraph of maximum order. In fact, in [3], the obtained result was stated as follows.

Theorem 4 ([3]) *Let G be a graph and k a non-negative integer. If $S \subseteq V(G)$ induces a subgraph of G with average degree k, then $|S| \leq \upsilon_k(G)$. The equality holds if and only if $\tau + k \leq |N_G(v) \cap S| \;\; \forall v \notin S$.*

Considering the particular case of regular graphs we have the following theorem, where the upper bound was obtained in [11] and the last statement was proved in [3].

Theorem 5 ([3, 11]) *If G is a p-regular graph of order n, then*

$$\alpha_k(G) \leq n\frac{k - \lambda_n}{p - \lambda_n}. \tag{11}$$

Furthermore, the equality holds if and only if there exists $S \subseteq V(G)$ which $(k, k + \tau)$-regular, with $\tau = -\lambda_n$. In this case, $\alpha_k(G) = |S| = n\frac{k-\lambda_n}{p-\lambda_n}$.

In [4], considering the quadratic program not necessary convex (10), with $\tau > 0$, it was proved that

$$\alpha_k(G) \leq \lambda_{max}(A_{G^c}) + k + 1, \tag{12}$$

where G^c denotes the complement of the graph G, that is, the graph such that $V(G^c) = V(G)$ and $E(G^c) = \{ij : ij \notin E(G)\}$. Furthermore, the following upper bound was obtained.

Theorem 6 ([4]) *Consider a graph G such that $\lambda_{min}(A_G) = \lambda_n = \cdots = \lambda_{n-(p-1)}$ is a non-main eigenvalue with multiplicity p. Assuming that the eigenvectors $\hat{u}_1, \ldots, \hat{u}_n$, associated to the eigenvalues $\lambda_1, \ldots, \lambda_n$, respectively, are unitary and pairwise orthogonal, then*

$$\alpha_k(G) \leq \sum_{j=1}^{n-p} \frac{-\lambda_n + k}{-\lambda_n + \lambda_j} (\hat{e}^T \hat{u}_j)^2. \tag{13}$$

Later, in [5], using a quadratic programming technique jointly with the main angles of G, the upper bound (13) was improved as follows.

Theorem 7 ([5]) *Let G be a graph of order n, and let S be a set of vertices which induces a k-regular subgraph of G $(0 \leq k \leq n - 1)$. If $t > -\lambda_n$ then*

$$\alpha_k(G) \leq h_k^G(t), \tag{14}$$

where $h_k^G(t) = (k + t)\left(1 - \frac{P_{G^c}(t-1)}{(-1)^n P_G(-t)}\right)$ and $P_G(x) = \det(xI - A)$.

3 Upper Bounds Based on the Spectrum of A_G, L_G and Q_G

Now it is worth to recall the following theorem obtained by Haemers.

Theorem 8 ([11]) *Let G be a graph on n vertices of average degree d and let the vertex set of G be partitioned into two sets such that G_1 and G_2 are the subgraphs induced by these two sets. For $i = 1, 2$ let n_i be the number of vertices of G_i, d_i be the average of vertex degrees of G_i and let $\bar{d}_i$ be the average of vertex degrees in G over the vertices of G_i. Then*

(i) $\lambda_1\lambda_2 \geq \frac{nd_id-n_i\bar{d}_i^2}{n-n_i} \geq \lambda_1\lambda_n$.
(ii) If the equality holds on one of the sides, then G_1 and G_2 are regular and also the degrees in G are constant over the vertices of G_1 and G_2.

As a consequence of this theorem, we have the following corollary.

Corollary 1 *If G is a graph of order n, then,*

$$\alpha_k(G) \leq \frac{2k|E(G)| - n\lambda_1\lambda_n}{\delta(G)^2 - \lambda_1\lambda_n}. \tag{15}$$

Proof Let us consider the vertex partition $V(G) = S \cup S^c$, where S induces a k regular subgraph of G. Applying Theorem 8-(i), setting $n_1 = |S|$ and $d_1 = k$, we have,

$$\begin{aligned}\frac{nkd - \bar{d_1}^2|S|}{n - |S|} \geq \lambda_1\lambda_n &\Leftrightarrow \lambda_1\lambda_n(n - |S|) \leq nkd - \bar{d_1}^2|S| \\ &\Leftrightarrow |S|(\bar{d_1}^2 - \lambda_1\lambda_n) \leq nkd - n\lambda_1\lambda_n \\ &\Leftrightarrow |S| \leq \frac{nkd - n\lambda_1\lambda_n}{\bar{d_1}^2 - \lambda_1\lambda_n}.\end{aligned}$$

Since $\bar{d_1} \geq \delta$ and $d = \frac{2|E(G)|}{n}$, the inequality (15) follows. □

Notice that, when G is p-regular, $\lambda_1 = \delta(G)$ and $|E(G)| = \frac{np}{2}$ whereby the upper bound (15) is equal to (11).

The next corollary is a consequence of Lemma 1.

Corollary 2 *If G is a graph of order n, then*

$$\alpha_k(G) \leq n\frac{k + \mu_1 - \delta(G)}{\mu_1}. \tag{16}$$

Proof Considering a vertex subset $S \subseteq V(G)$ inducing a k-regular subgraph and taking into account that (as defined before) $\overline{d_S} = \frac{1}{|S|}\sum_{i\in S} d_G(i)$, it follows that $|\partial(S)| = |S|(\bar{d_S} - k)$. Then applying Lemma 1 we have

$$
\begin{aligned}
|S|(\bar{d}_S - k) \le \mu_1 \frac{|S|(n-|S|)}{n} &\Leftrightarrow \frac{n(\bar{d}_S - k)}{n-|S|} \le \mu_1 \\
&\Leftrightarrow \mu_1|S| \le n\mu_1 - n(\bar{d}_S - k) \\
&\Leftrightarrow |S| \le n\,\frac{k+\mu_1-\bar{d}_S}{\mu_1}.
\end{aligned}
$$

Since $\bar{d}_S \ge \delta(G)$, the inequality (16) follows. □

If a graph G is p-regular, from (2) $\mu_1 + \lambda_n = p$ and we may conclude that the upper bound (16) is equal to (11).

Before the introduction of a new upper bound on the order of k-regular induced subgraphs in function of the largest and the least eigenvalues of the signless Laplacian matrix, it is worth to introduce the following lemma.

Lemma 2 *Let G be a graph of order n without isolated vertices. If G is bipartite or $\delta(G) \ge \frac{\Delta(G)}{2}$ or $q_1 < 4\delta(G)$, then $4\delta(G)^2 - q_nq_1 > 0$.*

Proof Let $\delta = \delta(G)$ and $\Delta = \Delta(G)$.

1. If G is bipartite without isolated vertices, then $q_n = 0$, $\delta > 0$ and therefore, $4\delta^2 - q_nq_1 > 0$.
2. If $\delta \ge \frac{\Delta}{2}$, we have $\delta^2 \ge \frac{\delta\Delta}{2} \Leftrightarrow 4\delta^2 \ge 2\delta\Delta$ and, taking into account (3) and (4), since $q_1 \le 2\Delta$ and $\delta > q_n$ it follows $4\delta^2 - q_nq_1 > 0$.
3. Finally, if $q_1 < 4\delta$, then $q_1q_n \le 4\delta q_n < 4\delta^2$, that is, $q_1q_n < 4\delta^2$ and so $4\delta^2 - q_nq_1 > 0$.

□

Notice that there are graphs G, with $\delta = \delta(G)$, such that $4\delta^2 - q_nq_1 \le 0$, as it is the case of the graph depicted in Fig. 1 which has $\delta = 2$, $q_n = 1.4991$ and $q_1 = 10.8517$.

Theorem 9 *Let G be a graph of order n such that $4\delta^2(G) - q_nq_1 > 0$. Then*

$$
\frac{2k|E(G)| - n\lambda_1\lambda_n}{\delta^2(G) - \lambda_1\lambda_n} \le \frac{4|E(G)|(\Delta(G)+k) - nq_nq_1}{4\delta^2(G) - q_nq_1}. \qquad (17)
$$

Proof Considering $\varepsilon = |E(G)|$, $\delta = \delta(G)$, $\Delta = \Delta(G)$ and assuming that the inequality of (17) holds, we have

Fig. 1 Graph G, with $4\delta(G)^2 - q_nq_1 \le 0$

$$\frac{2k\varepsilon - n\lambda_1\lambda_n}{\delta^2 - \lambda_1\lambda_n} - \frac{\varepsilon(\Delta + k) - n\frac{q_1}{4}q_n}{\delta^2 - \frac{q_1}{4}q_n} \leq 0$$

$$\Updownarrow$$

$$2k\varepsilon\delta^2 - \frac{q_1}{2}q_n k\varepsilon - n\delta^2\lambda_1\lambda_n - \delta^2\Delta\varepsilon - \delta^2 k\varepsilon + n\delta^2\frac{q_1}{4}q_n + \lambda_1\lambda_n\varepsilon\Delta + \lambda_1\lambda_n\varepsilon k \leq 0$$

$$\Updownarrow$$

$$k(\delta^2\varepsilon - \frac{q_1}{2}q_n\varepsilon + \lambda_1\lambda_n\varepsilon) - n\delta^2\lambda_1\lambda_n - \delta^2\Delta\varepsilon + n\delta^2\frac{q_1}{4}q_n + \lambda_1\lambda_n\varepsilon\Delta \leq 0$$

Let $f(k) = k(\delta^2\varepsilon - \frac{q_1}{2}q_n\varepsilon + \lambda_1\lambda_n\varepsilon) - n\delta^2\lambda_1\lambda_n - \delta^2\Delta\varepsilon + n\delta^2\frac{q_1}{4}q_n + \lambda_1\lambda_n\varepsilon\Delta$. Then,

$$\begin{aligned} f'(k) &= \delta^2\varepsilon - \frac{q_1}{2}q_n\varepsilon + \lambda_1\lambda_n\varepsilon \\ &= \varepsilon(\delta^2 - \frac{q_1}{2}q_n + \lambda_1\lambda_n). \end{aligned}$$

From (6),

$$\delta + \lambda_n < q_n \Leftrightarrow \delta^2 + \delta\lambda_n < \delta q_n \Leftrightarrow \delta^2 - \delta q_n + \delta\lambda_n < 0.$$

Since, from (3), $\frac{q_1}{2} \geq \delta$ and, as it is well known, $\lambda_1 \geq \delta$, it follows that $\delta^2 - \frac{q_1}{2}q_n + \lambda_1\lambda_n \leq \delta^2 - \delta q_n + \delta\lambda_n < 0$, that is, $f'(k) < 0$. Therefore, $f(k)$ is a decreasing function.

Considering the function $f(k)$ and setting $k = 0$ and $\Delta = \delta + \xi$ with ξ a nonnegative integer we may define the function

$$g(\delta, \xi) = -n\delta^2\lambda_1\lambda_n - \delta^2(\delta + \xi)\varepsilon + n\delta^2\frac{q_1}{4}q_n + \lambda_1\lambda_n\varepsilon(\delta + \xi).$$

Then

$$\begin{aligned} \frac{\partial g(\delta, \xi)}{\partial \xi} &= -\delta^2\varepsilon + \lambda_1\lambda_n\varepsilon \\ &= \varepsilon(-\delta^2 + \lambda_1\lambda_n) \\ &< 0. \end{aligned}$$

Therefore, $g(\delta, \xi)$ is a decreasing function with respect to ξ. Since $g(\delta, 0) = -n\delta^2\lambda_1\lambda_n - \delta^3\varepsilon + n\delta^2\frac{q_1}{4}q_n + \lambda_1\lambda_n\varepsilon\delta$ and $\delta = \Delta$ it follows that $\lambda_1 = \delta$. Furthermore, from (3), $\frac{q_1}{2} = \delta$ and from (6), $q_n = \delta + \lambda_n$. Therefore,

$$\begin{aligned} g(\delta, 0) &= -n\delta^3\lambda_n - \delta^3\varepsilon + n\frac{\delta^3}{2}(\delta + \lambda_n) + \lambda_n\varepsilon\delta^2 \\ &= -n\delta^3\lambda_n - \delta^3\varepsilon + n\frac{\delta^4}{2} + n\frac{\delta^3}{2}\lambda_n + \lambda_n\varepsilon\delta^2. \end{aligned}$$

Finally, since $\varepsilon = \frac{n\delta}{2}$ we obtain $g(\delta, 0) = -n\delta^3\lambda_n - n\frac{\delta^4}{2} + n\frac{\delta^4}{2} + n\frac{\delta^3}{2}\lambda_n + n\frac{\delta^3}{2}\lambda_n = 0$ and thus, for all nonnegative integers δ and ξ, $g(\delta, \xi) \le 0$. Therefore, $f(0) \le 0$ and, since $f(k)$ is a decreasing function, we may conclude that $f(k) \le 0$ for all k. $\square$

As immediate consequence of Corollary 1 and Theorem 9 we have the following corollary.

Corollary 3 *If G is a graph of order n, ε edges, $\Delta = \Delta(G)$ and $\delta = \delta(G)$, such that $4\delta^2 - q_nq_1 > 0$, then*

$$\alpha_k(G) \le \frac{4\varepsilon(\Delta + k) - nq_nq_1}{4\delta^2 - q_nq_1}. \tag{18}$$

According to [7], a graph G with n vertices and ε edges is regular if and only if $4\varepsilon = nq_1$. Furthermore, when G is regular its degree is equal to $\frac{q_1}{2}$. Thus, assuming that G is p-regular, has n vertices and ε edges, by Lemma 2 the hypothesis of Corollary 3 is fulfilled and then we may write

$$\begin{aligned}
\alpha_k(G) &\le \frac{nq_1(p + k - q_n)}{2pq_1 - q_nq_1} \quad \left(\text{since } \Delta(G) = \delta(G) = p = \frac{q_1}{2} \text{ and } 4\varepsilon = nq_1\right) \\
&= \frac{n(p + k - q_n)}{2p - q_n} = n\frac{k - \lambda_n}{p - \lambda_n} \quad (\text{since } q_n - \lambda_n = p).
\end{aligned}$$

Therefore, for regular graphs, all the upper bounds (11) (15), (16) and (18) are equal. Notice that there are graphs for which these upper bounds are tight. For instance, if $G = K_n$ (a complete graph of order n), then $\lambda_1 = n - 1$ and $\lambda_n = -1$. Thus, if $S \subseteq V(K_n)$ induces a k-regular subgraph, then $n\frac{k-\lambda_n}{\lambda_1-\lambda_n} = k + 1 = |S|$. Therefore, when G is a complete graph, for each k, the upper bounds (15), (16) and (18) on the cardinality of vertex subsets inducing k-regular subgraphs are all reached. More generally, according to Theorem 5, if G is a regular graph and $S \subset V(G)$ is a $(k, k + \tau)$-regular set, with $\tau = -\lambda_n$, then all the above referred upper bounds are reached.

Throughout the paper, in all the proofs of the presented results, only the average degree in S is used and then, in all the obtained results we may replace k-regular induced subgraph by induced subgraph with average degree k. Moreover, all the obtained results remain valid when we consider positive weights on the edges, assuming in that case that the degree of a vertex v is then the sum of the weights of the edges incident to v.

4 Computational Experiments and Conclusions

In this section, several computational experiments with the upper bounds (15), (16) and (18) are presented in Table 1. In each row of this table appears the order n, the maximum degree Δ, the minimum degree δ, the degree of a regular induced subgraph

Table 1 Computational experiments with the upper bounds (15), (16) and (18)

Graph	n	$\Delta(G)$	$\delta(G)$	k	(15)	(16)	(18)
c-fat200-1	200	17	14	0	**74.01**	82.31	97.27
				1	**83.87**	90.72	109.28
				2	**93.73**	99.13	121.29
				6	133.18	**132.75**	169.33
				7	143.04	**141.16**	181.34
c-fat200-2	200	34	32	0	**55.72**	57.29	63.19
				1	**60.28**	61.75	67.86
				2	**64.83**	66.21	72.53
				16	128.65	**128.65**	137.88
				17	133.21	**133.10**	142.55
c-fat200-5	200	86	83	0	**45.85**	48.56	50.10
				1	**47.74**	50.39	52.06
				2	**49.64**	52.21	54.01
				39	119.79	**119.72**	126.41
				40	121.69	**121.55**	128.36
MANN-a9	45	41	40	0	**3.76**	4.46	4.23
				1	**4.81**	5.47	5.32
				2	**5.86**	6.48	6.41
				18	**22.69**	22.70	23.84
				19	23.74	**23.72**	24.93
MANN-a27	378	374	364	0	**5.17**	13.43	13.19
				1	**6.22**	14.43	14.27
				2	**7.27**	15.43	15.36
				3	**8.32**	16.44	16.45
				4	**9.37**	17.44	17.53
Keller4	171	124	102	0	**34.76**	45.74	109.56
				1	**36.20**	46.96	110.51
				2	**37.65**	48.19	111.47
				51	108.46	**108.37**	158.11
brock200-1	200	165	130	0	**20.25**	44.83	75.10
				1	**21.82**	46.02	77.09
				2	**23.40**	47.22	79.08
				64	**121.22**	**121.22**	202.28
				65	122.80	**122.41**	204.26
brock200-2	200	114	78	0	**37.48**	69.19	161.29
				1	**40.12**	70.87	165.49
				2	**42.75**	72.54	169.69
				33	124.54	**124.53**	300.04
				34	127.18	**126.21**	304.24

(continued)

Table 1 (continued)

Graph	n	$\Delta(G)$	$\delta(G)$	k	(15)	(16)	(18)
brock200-3	200	134	99	0	**29.41**	57.79	113.35
				1	**31.51**	59.23	116.37
				2	**33.61**	60.66	119.39
				43	119.58	**119.56**	243.18
				44	121.68	**121.00**	246.20
brock200-4	200	147	112	0	**24.94**	51.73	91.73
				1	**26.76**	53.05	94.15
				2	**28.59**	54.38	96.58
				54	123.58	**123.22**	222.61
				55	125.40	**124.54**	225.03

k and the computed upper bounds on the order of this induced subgraphs for some of the graphs of the family considered in the Second DIMACS Implementation Challenge (see [13]).

Notice that for the particular case of regular graphs the upper bounds (15), (16) and (18) are all equal. Moreover since, according to the Theorem 9, the upper bound (15) is less or equal than the upper bound (18), it follows that

$$\frac{4|E(G)|(\Delta(G)+k)-nq_nq_1}{4\delta(G)^2-q_nq_1} \geq \min\left\{\frac{2k|E(G)|-n\lambda_1\lambda_n}{\delta^2-\lambda_1\lambda_n}, n\frac{k+\mu_1-\delta}{\mu_1}\right\}.$$

Concerning the comparison between the upper bounds (15) and (16) and also between (16) and (18), the computational results presented in the Table 1 show that none of them is always better than the others.

In fact, regarding the upper bounds (15) and (16), for $k = 0, 1, 2$, the former is better than the later. However, for much greater values of k, there are several graphs for which the upper bound (16) is better than (15). Finally, it should be noted that for the graphs MANN-a9 and MANN-a27 for $k = 0, 1, 2$ the upper bound (18) is better than the upper bound (16).

Acknowledgements The authors would like to thank Willem Haemers for several insightful comments and suggestions on this work that have helped us to improve the content of this paper. This research was supported by the Portuguese Foundation for Science and Technology ("FCT-Fundação para a Ciência e a Tecnologia"), through the CIDMA - Center for Research and Development in Mathematics and Applications, within project UID/MAT/04106/2013. We are also indebted to the anonymous referee for her/his careful reading and suggestions which have improved the text.

References

1. Bussemaker, F.C., Cvetković, D., Seidel, J.: Graphs related to exceptional root systems, T. H. - Report 76-WSK-05. Technical University Eindhoven (1976)
2. Cardoso, D.M., Cvetković, D.: Graphs with least eigenvalue -2 attaining a convex quadratic upper bound for the stability number. Bull. Acad. Serbe Sci. Arts. CI. Sci. Math. Natur. Sci. Math. **23**, 41–55 (2006)
3. Cardoso, D.M., Kaminski, M., Lozin, V.: Maximum k-regular induced subgraphs. J. Comb. Optim. **14**, 455–463 (2007)
4. Cardoso, D.M., Pinheiro, S.J.: Spectral upper bounds on the size of k-regular induced subgraphs. Electron. Notes Discret. Math. **32**, 3–10 (2009)
5. Cardoso, D.M., Rowlinson, P.: Spectral upper bounds for the order of a k-regular induced subgraph. Linear Algebra Appl. **433**, 1031–1037 (2010)
6. Cvetković, D., Doob, M., Sachs, H.: Spectra of Graphs, Theory and Applications, 3rd edn. Johan Ambrosius Barth Verlag, Heidelberg (1995)
7. Cvetković, D., Rowlinson, P., Simić, S.K.: Signless Laplacians of finite graphs. Linear Algebra Appl. **423**, 155–171 (2007)
8. Cvetković, D., Rowlinson, P., Simić, S.K.: An Introduction to the Theory of Graph Spectra. London Mathematical Society Student Texts, vol. 75. Cambridge University Press, Cambridge (2010)
9. Das, K.C.: On conjectures involving second largest signless Laplacian eigenvalue of graphs. Linear Algebra Appl. **432**, 3018–3029 (2010)
10. Godsil, C.D., Newman, M.W.: Eigenvalue bounds for independent sets. J. Comb. Theory, Ser. B, **98 (4)**, 721–734 (2008)
11. Haemers, W.: Eigenvalue techniques in design and graph theory (thesis Technical University Eindhoven 1979). Math. Centre Tract 121, Mathematical Centre, Amsterdam (1980)
12. Haemers, W.: Interlacing eigenvalues and graphs. Linear Algebra Appl. **226**(228), 593–616 (1995)
13. Johnson, D.S., Trick, M.A.: Cliques, coloring, and satisfiability: second DIMACS challenge. In: DIMACS Series in Discrete Mathematics and Theoretical Computer Science. American Mathematical Society, Providence (1996)
14. Lovász, L.: On the Shannon capacity of a graph. IEEE Trans. Inf. Theory **25**(2), 1–7 (1979)
15. Lu, M., Liu, H., Tian, F.: New Laplacian spectral bounds for clique and independence numbers of graphs. J. Comb. Theory Ser. B **97**, 726–732 (2007)
16. Mohar, B.: Some applications of Laplace eigenvalues of graphs. Notes taken by Martin Juvan. (English). In: Hahn, G., et al.: (eds.) Graph Symmetry: Algebraic Methods and Applications, NATO ASI Ser., Ser. C, Math. Phys. Sci., vol. 497, pp. 225–275. Kluwer Academic Publishers, Dordrecht (1997)

Multiplicities: Adding a Vertex to a Graph

Kenji Toyonaga, Charles R. Johnson and Richard Uhrig

Abstract Given an Hermitian matrix A whose graph G is a simple undirected graph and its eigenvalues, we suppose the status of each vertex in the graph is known for each eigenvalue of A. We investigate the change of the multiplicity of each eigenvalue, when we add a pendent vertex with given value to a particular vertex in the graph via an edge with given weight. It is shown how each multiplicity changes based on this information. The results are applied to show that more than one eigenvalue may increase in multiplicity with the addition of just one vertex. The intended focus is trees, but the analysis is given for general graphs.

Keywords Eigenvalues · Graph · Matrix · Multiplicities · Symmetric

1 Introduction

If G is a simple, undirected graph on n vertices, denote by $\mathscr{H}(G)$ the set of all n-by-n Hermitian matrices, the graph of whose off-diagonal entries is G. There is long-standing interest in the possible lists of multiplicities for the eigenvalues of matrices in $\mathscr{H}(G)$, especially when G is a tree T. There are several papers on the subject, including ones relating the structure of T to eigenvalue multiplicity, Refs. [2, 4, 5, 7–9]. In many papers, the multiplicity of eigenvalues in a tree is considered when a slight change occurs. Here, we deal with a general graph and consider the new,

K. Toyonaga
Department of Integrated Arts and Science, Kitakyushu National College of Technology, Kokuraminami-ku, Kitakyushu 802-0985, Japan
e-mail: toyonaga@kct.ac.jp

C.R. Johnson (✉) · R. Uhrig
Department of Mathematics, College of William and Mary, P.O. Box 8795, Williamsburg, VA 23187-8795, USA
e-mail: crjohnso@math.wm.edu

R. Uhrig
e-mail: rauhrig@email.wm.edu

N. Bebiano (ed.), *Applied and Computational Matrix Analysis*, Springer Proceedings in Mathematics & Statistics 192,
DOI 10.1007/978-3-319-49984-0_8

but natural issue of adding a single vertex. As all necessary information, particularly multiplicities may be updated, the results could be applied sequentially.

If A is Hermitian, denote the multiplicity of an eigenvalue λ of A by $m_A(\lambda)$. When we remove a vertex u from G, the remaining graph is denoted by $G(u)$. Then we denote by $A(u)$ the $(n-1)$-by-$(n-1)$ principal submatrix of $A \in \mathscr{H}(G)$, resulting from deletion of the row and column corresponding to u. $A[S]$ denotes the principal submatrix of A corresponding to the subgraph S of G. For an identified $A \in \mathscr{H}(G)$, we often speak interchangeably about the graph and the matrix, for convenience.

Our interest here is in precisely what happens to the multiplicities when we add a (pendent) vertex v to a tree T at an identified vertex u. Specifically, we show what happens, for each $A \in \mathscr{H}(T)$, to the multiplicities $m_A(\lambda)$, when we pass to the new tree $\widetilde{T}$, for $\widetilde{A} \in \mathscr{H}(\widetilde{T})$ with $\widetilde{A}(v) = A$, eigenvalue by eigenvalue. Since the analysis is only slightly more complicated when G is a general graph, we present our results at that level of generality.

Because of the interlacing inequalities for an Hermitian matrix and a principal submatrix of it [1], a multiplicity may change by at most 1 when we pass from G to $\widetilde{G}$. For trees, the theory of what may happen, when a particular vertex is deleted, was summarized and further developed in [4], but the basic definitions are the same for general graphs G. A vertex u of G is called "*Parter*" (respectively "*neutral*" or "*downer*") for an eigenvalue λ of $A \in \mathscr{H}(G)$ if

$$m_{A(u)}(\lambda) = m_A(\lambda) + 1 \text{ (resp. } m_A(\lambda),\ m_A(\lambda) - 1).$$

The "*status*" of a vertex u is discussed in [4]. It refers to which of these eventualities occurs, and why.

2 Main Results

We denote the characteristic polynomial of a square matrix A by $p_A(x)$. Suppose that G is a graph on n vertices, that $A \in \mathscr{H}(G)$ is given, and that a new vertex v is appended to G at the vertex u of G, resulting in the graph $\widetilde{G}$ with pendent vertex v. If the weight $\alpha \in \mathbb{R}$ is placed on v and the weight $\tilde{a}_{uv} \in \mathbb{C}$ is placed on the new edge, a new matrix $\widetilde{A} \in \mathscr{H}(\widetilde{G})$ results. Of courese $\widetilde{A}(v) = A$, and, we mean that the u, v entry of $\widetilde{A}$ is $\tilde{a}_{uv}$ and $\tilde{a}_{vv} = \alpha$.

The function $f(x) = \frac{p_{A(u)}(x)}{p_A(x)}$ will be important to us. After cancellation of like terms in the numerator and denominator, because of interlacing, it will be a ratio of two products, each of distinct linear terms. In the numerator will be terms of the form $(x - \tau)$ for eigenvalues τ for which u is Parter, along with eigenvalues of $A(u)$ that do not occur in A. In the denominator will be such terms for eigenvalues μ for which u is a downer. The number of μ's is one more than the number of τ's, and the τ's strictly interlace μ's because of the interlacing inequalities. Important for us is that $f(x)$ will be well-defined and nonzero when evaluated at any eigenvalue for which u is neutral.

Lemma 1 *With the conventions mentioned above, we have for any $\lambda \in \mathbb{R}$:*

(a) If u is a Parter vertex for λ in G,

$$m_{\tilde{A}}(\lambda) = \begin{cases} m_A(\lambda) + 1 & \text{if } \alpha = \lambda \\ m_A(\lambda) & \text{if } \alpha \neq \lambda \end{cases};$$

(b) If u is a neutral vertex for λ in G,

$$m_{\tilde{A}}(\lambda) = \begin{cases} m_A(\lambda) + 1 & \text{if } \alpha = \lambda - |\tilde{a}_{uv}|^2 f(\lambda) \\ m_A(\lambda) & \text{otherwise} \end{cases};$$

and

(c) If u is a downer vertex for λ in G,

$$m_{\tilde{A}}(\lambda) = m_A(\lambda) - 1.$$

Proof Given $A \in \mathscr{H}(G)$, let $\sigma(A) = \{\lambda_1, \lambda_2, \ldots, \lambda_l\}$ be the distinct eigenvalues of A, and their multiplicities in A be $\{m_1, m_2, \ldots, m_l\}$. We focus on a specified eigenvalue λ_k, $(1 \leq k \leq l)$, and now we put $\lambda_k = \lambda$ and $m_k = m$. Then, the characteristic polynomial of $\widetilde{A} = (\tilde{a}_{ij})$ can be represented as follows (cf. [8]).

$$p_{\tilde{A}}(x) = (x - \alpha)p_A(x) - |\tilde{a}_{uv}|^2 p_{A(u)}(x), \tag{1}$$

We further let the distinct eigenvalues of $A(u)$ be $\sigma(A(u)) = \{\mu_1, \mu_2, \ldots\}$, and their multiplicities be $\{m'_1, m'_2, \ldots\}$. As we focus upon one eigenvalue $\lambda = \lambda_k$, $p_{\tilde{A}}(x)$ can be written,

$$p_{\tilde{A}}(x) = (x - \alpha)(x - \lambda)^m f_1(x) - |\tilde{a}_{uv}|^2 (x - \lambda)^{m'} f_2(x), \tag{2}$$

in which $f_1(x) = \prod_{i \neq k} (x - \lambda_i)^{m_i}$, $f_2(x) = \prod_{\mu_i \neq \lambda} (x - \mu_i)^{m'_i}$

In (2), if λ is not an eigenvalue of A or $A(u)$, then m or m' is 0.

If u is a Parter vertex for λ in A, then $m' = m + 1$ in (2). Then,

$$p_{\tilde{A}}(x) = (x - \lambda)^m \{(x - \alpha) f_1(x) - |\tilde{a}_{uv}|^2 (x - \lambda) f_2(x)\}.$$

Here we set $g_1(x) = (x - \alpha) f_1(x) - |\tilde{a}_{uv}|^2 (x - \lambda) f_2(x)$. When $\alpha = \lambda$, $g(\lambda) = 0$, thus $m_{\tilde{A}}(\lambda) = m_A(\lambda) + 1$. However when $\alpha \neq \lambda$, $g(\lambda) \neq 0$, then $m_{\tilde{A}}(\lambda) = m_A(\lambda)$.

If u is a neutral vertex for λ in A, then $m' = m$ in (2). Then,

$$p_{\tilde{A}}(x) = (x - \lambda)^m \{(x - \alpha) f_1(x) - |\tilde{a}_{uv}|^2 f_2(x)\}.$$

When we set $g_2(x) = (x - \alpha) f_1(x) - |\tilde{a}_{uv}|^2 f_2(x)$, if α and $\tilde{a}_{uv}$ has the relation such that $\alpha = \lambda - |\tilde{a}_{uv}|^2 \frac{f_2(\lambda)}{f_1(\lambda)} = \lambda^*$, then $g_2(\lambda) = 0$, so $m_{\tilde{A}}(\lambda) = m_A(\lambda) + 1$. Since

$\frac{f_2(\lambda)}{f_1(\lambda)} = [\frac{p_{A(u)}(x)}{p_A(x)}]_\lambda$ holds, if we put $f(x) = \frac{f_2(x)}{f_1(x)}$, then the assertion holds. If $\alpha \neq \lambda^*$, then $g_2(\lambda) \neq 0$, so $m_{\widetilde{A}}(\lambda) = m_A(\lambda)$.

Lastly, If u is a downer vertex for λ in A, then $m' = m - 1$ in (2). Then

$$p_{\widetilde{A}}(x) = (x - \lambda)^{m-1}\{(x - \alpha)(x - \lambda) f_1(x) - |\tilde{a}_{uv}|^2 f_2(x)\}.$$

If we set $g_3(x) = (x - \alpha)(x - \lambda) f_1(x) - |\tilde{a}_{uv}|^2 f_2(x)$, then $g(\lambda) \neq 0$, thus $m_{\widetilde{A}}(\lambda) = m_A(\lambda) - 1$ for any real number α. □

When we focus on an identified real number λ, if a vertex is appended to a Parter vertex for λ, the multiplicity of λ in $\widetilde{A}$ depends only on the value on the pendent vertex. If it is appended to a neutral vertex for λ, the multiplicity of λ depends only on the relation between the value on the pendent vertex and the weight on the new edge. If the relation $\alpha = \lambda - |\tilde{a}_{uv}|^2 f(\lambda)$ holds, then $|\tilde{a}_{uv}|^2 = \frac{\lambda-\alpha}{f(\lambda)}$ must be positive. So, if $f(\lambda) > 0$, then α must be less than λ, and if $f(\lambda) < 0$, then α must be greater than λ.

If a pendent vertex is appended to a downer vertex, the multiplicity of λ decreases whatever the value on the pendent vertex and the weight on the new edge are.

We note that it follows from the lemma that any eigenvalue of multiplicity 1 in A, for which u is a downer, disappears when we pass to $\widetilde{A}$. In particular, any multiplicity 1 eigenvalue, for which every vertex is a downer, disappears. In the case of trees, for every eigenvalue of multiplicity 1 that has no Parter vertex(equivalently, no neutral vertex), every vertex will be a downer [4] and, so, will disappear. Most of these will be replaced by new eigenvalues in $\widetilde{A}$ that also have multiplicity 1 and no Parter vertex. From the above lemma, we can deduce the next theorem.

Theorem 1 *Let G be a general graph, $A \in \mathscr{H}(G)$ and $\lambda \in \mathbb{R}$. Let u be a vertex in G, and $\widetilde{G}$ be a graph obtained by adding a vertex v valued α to the vertex u of G. Let $\widetilde{A} \in \mathscr{H}(\widetilde{G})$, such that $\widetilde{A}(v) = A$. Let m be the multiplicity of λ as an eigenvalue in A, and let n be the multiplicity of λ in $\widetilde{A}$. Then,*

(a) $m - n = -1$ if and only if u is a Parter vertex for λ in A and $\alpha = \lambda$, or u is a neutral vertex in A and $\alpha = \lambda - |\tilde{a}_{uv}|^2 f(\lambda)$.
(b) $m - n = 0$ if and only if u is a Parter vertex for λ in A and $\alpha \neq \lambda$, or u is a neutral vertex for λ in A and $\alpha \neq \lambda - |\tilde{a}_{uv}|^2 f(\lambda)$.
(c) $m - n = 1$ if and only if u is a downer vertex for λ in A.

In Lemma 1, the status of vertex u in A changes to that in $\widetilde{A}$ as follows.

Corollary 1 *Let G be a general graph, $A \in \mathscr{H}(G)$ and $\lambda \in \mathbb{R}$. Let u be a vertex in G and $\widetilde{G}$ be the graph obtained by adding a vertex v valued α to the vertex u in G. Let $\widetilde{A} \in \mathscr{H}(\widetilde{G})$, such that $\widetilde{A}(v) = A$.*

(a) In case u is Parter for λ in A, the status of u for λ in $\widetilde{A}$ is Parter.
(b) In case u is neutral for λ in A, if $\alpha = \lambda - |\tilde{a}_{uv}|^2 f(\lambda)$, then the status of u for λ in $\widetilde{A}$ becomes downer, if $\alpha = \lambda$, then Parter, and, otherwise, neutral.

(c) *In case u is downer for λ in A, if $\alpha = \lambda$, then the status of u for λ in $\widetilde{A}$ becomes Parter, and, otherwise, neutral.*

Proof (a) If $\alpha = \lambda$, then $m_{\widetilde{A}}(\lambda) = m_A(\lambda) + 1$ from Lemma 1. When u is removed from $\widetilde{G}$, $m_{\widetilde{A}(u)}(\lambda) = m_A(\lambda) + 2$, since u is Parter for λ in G and $\alpha = \lambda$, so that u is Parter in $\widetilde{A}$.

If $\alpha \neq \lambda$, then $m_{\widetilde{A}}(\lambda) = m_A(\lambda)$. When u is removed from $\widetilde{G}$,

$$m_{\widetilde{A}(u)}(\lambda) = m_A(\lambda) + 1,$$

so that u is Parter in $\widetilde{A}$.

(b) If $\alpha = \lambda - |\tilde{a}_{uv}|^2 f(\lambda)$, $m_{\widetilde{A}}(\lambda) = m_A(\lambda) + 1$. When u is removed from $\widetilde{G}$, $m_{\widetilde{A}(u)}(\lambda) = m_A(\lambda)$, so that u is downer in $\widetilde{A}$. If $\alpha = \lambda$, then $m_{\widetilde{A}}(\lambda) = m_A(\lambda)$, and $m_{\widetilde{A}(u)}(\lambda) = m_A(\lambda) + 1$, so that u is Parter in $\widetilde{A}$. If otherwise, $m_{\widetilde{A}}(\lambda) = m_A(\lambda)$, and $m_{\widetilde{A}(u)}(\lambda) = m_A(\lambda)$, so that u is neutral in $\widetilde{A}$.

(c) $m_{\widetilde{A}}(\lambda) = m_A(\lambda) - 1$. If $\alpha = \lambda$, then when u is removed from $\widetilde{G}$, $m_{\widetilde{A}(u)}(\lambda) = m_A(\lambda)$, so that $m_{\widetilde{A}(u)}(\lambda) = m_{\widetilde{A}}(\lambda) + 1$ and u is Parter in $\widetilde{A}$. If $\alpha \neq \lambda$, then $m_{\widetilde{A}(u)}(\lambda) = m_A(\lambda) - 1$, so that u is neutral in $\widetilde{A}$. □

Let T_0 be a branch at vertex v in tree T, and let $A_0 \in \mathscr{H}(T_0)$. Let u be the vertex adjacent to v in T_0. If $m_{A_0(u)}(\lambda) = m_{A_0}(\lambda) - 1$, then T_0 is called a downer branch at v for λ in T relative to A. If a downer branch has eigenvalue λ with multiplicity 1, then we call it a *simple downer branch* for λ. Next we consider the change of multiplicity of λ when we add a simple downer branch for λ to a tree T.

Let b be a simple downer branch for λ. Let $\widehat{T}$ be a tree obtained by adding b to the vertex u in T inserting an edge between u and a downer vertex in b. Let $A \in \mathscr{H}(T)$, $\widehat{A} \in \mathscr{H}(\widehat{T})$ in which A is a principal submatrix of $\widehat{A}$ corresponding to T, and $B \in \mathscr{H}(b)$. Since b is a downer branch for λ at u in $\widehat{A}$, and u is a Parter vertex in $\widehat{A}$, if we set $m_{\widehat{A}}(\lambda) = k$, then $m_{\widehat{A}(u)}(\lambda) = k + 1$. Since $m_B(\lambda) = 1$, $m_{A(u)}(\lambda) = k + 1 - 1 = k$. Thus, $m_{\widehat{A}}(\lambda) = m_{A(u)}(\lambda)$. From this argument, the next Corollary follows.

Corollary 2 *Let $\widehat{T}$ be the tree obtained by adding a simple downer branch for λ to the vertex u of a tree T connecting with an edge. Let $A \in \mathscr{H}(T)$, $\widehat{A} \in \mathscr{H}(\widehat{T})$ in which A is a principal submatrix of $\widehat{A}$ corresponding to T, Then if u is a Parter vertex for λ in A, then $m_{\widehat{A}}(\lambda) = m_A(\lambda) + 1$. If u is a neutral vertex for λ in A, then $m_{\widehat{A}}(\lambda) = m_A(\lambda)$. If u is a downer vertex for λ in A, then $m_{\widehat{A}}(\lambda) = m_A(\lambda) - 1$.*

It is well known that when T is a path, either pendent vertex is a downer for every eigenvalue, all of which are multiplicity 1. Thus, when an end vertex is removed, every eigenvalue disappears and all interlacing inequalities are strict. So a path is a simple downer branch for each eigenvalue. Thus, the previous corollary is applicable to the case that a path is appended to G. Furthermore, by Theorem 1, addition of a new vertex at a pendent vertex also makes every original eigenvalue disappear. This is actually a special case of something much more general that also follows from the theorem.

If T is a tree and λ is a multiplicity 1 eigenvalue for which exactly one vertex is Parter, and that vertex is degree 2, then upon appending a new vertex anywhere in T, except at the Parter vertex, will make the multiplicity 1 eigenvalue disappear. Of course any non-upward multiplicity 1 eigenvalue will disappear, as well. For an eigenvalue λ of A, if there is a vertex such that $m_{A(v)}(\lambda) = m_A(\lambda) + 1$, then λ is called an *upward* eigenvalue, and otherwise *non-upward*. Here is the formal statement.

Corollary 3 *Suppose that T is a tree, that $A \in \mathcal{H}(T)$, that $\lambda \in \sigma(A)$ satisfies $m_A(\lambda) = 1$ and that either λ is upward with exactly one Parter vertex that is degree 2, or that λ is non-upward. Then, if $\widetilde{T}$ is the result of appending a new vertex v at any vertex of T (or any vertex other than u in the upward case), then $\lambda \notin \sigma(\widetilde{A})$ for any $\widetilde{A} \in \mathcal{H}(\widetilde{T})$ such that $\widetilde{A}(v) = A$.*

The multiplicity of an eigenvalue λ of A is changeable by adding a pendent vertex to a graph G as Lemma 1 and Theorem 1 show. However, by perturbing some diagonal entries in $\widetilde{A}$, the multiplicity of the eigenvalue can be preserved as it was in A. Before showing that, we need the next lemma from [5, Theorem 5].

The lemma shows how the multiplicity of an eigenvalue λ changes as a result of perturbing the value on a vertex.

Lemma 2 ([5]) *Let G be a graph, and i a vertex in G. For $A \in \mathcal{H}(G)$, let $A' = A + tE_{ii}, t \neq 0$, where E_{ii} denote the same size matrix with A such that(i, i) element is 1 and zeros elsewhere, then*

(a) $m_{A'}(\lambda) = m_A(\lambda)$ if and only if i is Parter in A or i is neutral in A and t is a unique t_0.
(b) $m_{A'}(\lambda) = m_A(\lambda) + 1$ if and only if i is neutral in A, and $t = t_0$.
(c) $m_{A'}(\lambda) = m_A(\lambda) - 1$ if and only if i is downer in A.

From Lemmas 1 and 2, we can observe the next proposition.

Proposition 1 *Let G be a graph. We suppose that $A \in \mathcal{H}(G)$ has an eigenvalue λ with multiplicity m. Let $\widetilde{G}$ be the graph obtained by adding a pendent vertex v valued α to the vertex u of G connecting with an edge weighted $\tilde{a}_{uv}$. Let the matrix $\widetilde{A} \in \mathcal{H}(\widetilde{G})$, such that $\widetilde{A}(v) = A$. Then there is a $\widetilde{B} \in \mathcal{H}(\widetilde{G})$ such that $\widetilde{B}$ has eigenvalue λ with multiplicity m, and it can be obtained by changing the value on v or u in $\widetilde{A}$.*

Proof First, we suppose that a pendent vertex is added to a Parter vertex for λ in A. If $\alpha \neq \lambda$, then the multiplicity of λ stay same, so it does not matter. If $\alpha = \lambda$, then multiplicity of λ is $m + 1$ in $\widetilde{A}$. In $\widetilde{A}$, the status of vertex v is downer for λ. So, if we perturb the value on v slightly and let the matrix B, the multiplicity of λ will go down, then $m_B(\lambda) = m$.

Secondly we suppose that a pendent vertex is added to a neutral vertex for λ in A. If the relation between α and $\tilde{a}_{uv}$ such as $\alpha = \lambda - |\tilde{a}_{uv}|^2 f(\lambda)$ holds, then multiplicity of λ is $m + 1$ in $\widetilde{A}$. Then the status of vertex u is downer in $\widetilde{A}$. So by perturbing the value on u in $\widetilde{A}$ slightly, we get B such that $m_B(\lambda) = m$. If $\alpha \neq \lambda - |\tilde{a}_{uv}|^2 f(\lambda)$, then $m_{\widetilde{A}}(\lambda) = m_A(\lambda)$. So we can set $\widetilde{A} = B$.

Next we suppose that a pendent vertex is added to a downer vertex for λ in A. Then $m_{\widetilde{A}}(\lambda) = m - 1$. If $\alpha \neq \lambda$, then u is neutral in $\widetilde{A}$. So from Lemma 2, by perturbing the value on u, we can get B such that $m_B(\lambda) = m$.

If $\alpha = \lambda$, then u and v are Parter in $\widetilde{A}$. So we can not make the multiplicity of λ increase only by perturbing the value on u. Then we perturb the value on v slightly from λ, then the status of u is neutral. So, similarly by perturbing the value on u, we can get B such that $m_B(\lambda) = m$. □

From the above proposition, we can observe that when we add a pendent vertex v to the vertex u in G, even if the multiplicity of an eigenvalue changes in $\widetilde{A}$, by further perturbing the value on u or v, we can keep the multiplicity of the eigenvalue as it was in A.

Let $m_1, m_2, \ldots, m_k$ be the multiplicities of the distinct eigenvalues of $A \in \mathscr{H}(T)$. Then we order them as $m_1 \geq m_2 \geq \cdots \geq m_k$. This is called the *unordered multiplicity list* for A, because when the eigenvalues corresponding to this multiplicity list are put in order, their multiplicities are not generally in descending order or increasing order. Let $\mathscr{L}(T)$ be the set of unordered multiplicity lists for all $A \in \mathscr{H}(T)$. There are some papers studying $\mathscr{L}(T)$ [3, 6] etc.; however, for trees with many vertices, not all multiplicity lists have yet been determined. Let $M(T)$ be the maximum multiplicity of an eigenvalue of $A \in \mathscr{H}(T)$. $M(T)$ is equal to the path cover number. (cf. [7]).

Theorem 2 *Let T be a tree, and suppose $(m, 1, 1, \ldots, 1) \in L(T)$ for $m \geq 2$. When we add a pendent vertex to a certain vertex in T and construct $\widetilde{T}$, then there is an Hermitian matrix such that $(m + 1, 1, 1, \ldots, 1) \in L(\widetilde{T})$.*

Proof Let A be an Hermitian matrix with unordered multiplicity list $(m, 1, 1, \ldots, 1)$. We suppose $\sigma(A)$ is ordered as

$$\lambda_1 < \lambda_2 < \cdots < \lambda_k < \cdots < \lambda_{n-m+1}$$

Let the multiplicity of λ_i be m_i. We suppose $m_k = m$ for the eigenvalue λ_k, $2 \leq k \leq n - m$. Now we shift A as $A - \lambda_k I = B$. B also has an unordered multiplicity list $(m, 1, 1, \ldots, 1)$ in which m represents the multiplicity of the eigenvalue 0. Here we order $\sigma(B)$ as $\mu_1 < \mu_2 < \cdots < \mu_k = 0 < \cdots < \mu_{n-m+1}$.

Next we add a pendent vertex v with value 0 to a Parter vertex u for 0 in B. Then we assign the weight of edge $\tilde{b}_{uv}$ and $\tilde{b}_{vu}$ to be ε such that $0 < \varepsilon < \min_i\{\dfrac{\mu_{i+1} - \mu_i}{2}\}$. Then we get the tree $\widetilde{T}$ and corresponding matrix $\widetilde{B} \in \mathscr{H}(\widetilde{T})$, in which B is a principal submatrix of $\widetilde{B}$. If the eigenvalues of $\widetilde{B}$ are ordered as $\tilde{\mu}_1 \leq \tilde{\mu}_2 \leq \cdots \leq \tilde{\mu}_k = 0 \leq \cdots \tilde{\mu}_{n-m+1}$, then $\mu_i - \varepsilon \leq \tilde{\mu}_i \leq \mu_i + \varepsilon$, because spectral radius $\rho(\widetilde{B} - B) = \varepsilon$ and $|\tilde{\mu}_i - \mu_i| \leq \varepsilon$. So, $m_{\widetilde{B}}(\tilde{u}_j) = 1$, $j \neq k$, and $m_{\widetilde{B}}(\tilde{u}_k) = m_B(\mu_k) + 1$, because the pendent vertex is added at a Parter vertex in B. From these, the assertion of the theorem holds. □

3 Examples

Example 1 Let A be an Hermitiam matrix as below,

$$A = \begin{bmatrix} 0\,1\,1\,1\,1\,0\,0\,0 \\ 1\,0\,0\,0\,0\,0\,0\,0 \\ 1\,0\,0\,0\,0\,0\,0\,0 \\ 1\,0\,0\,0\,0\,0\,0\,0 \\ 1\,0\,0\,0\,0\,1\,1\,1 \\ 0\,0\,0\,0\,1\,3\,0\,0 \\ 0\,0\,0\,0\,1\,0\,3\,0 \\ 0\,0\,0\,0\,1\,0\,0\,3 \end{bmatrix}$$

The graph of A is represented in Fig. 1. The circled numbers correspond to the index of the vertex. And the numbers outside of circles represent the values assigned on the vertices. The matrix A has eigenvalues 0 and 3 with multiplicity 2 each, among others. When we remove vertex 1 from T, the multiplicities of eigenvalues 0 and 3 become 3 and 2 in $A(1) \in \mathscr{H}(T(1))$, respectively. So vertex 1 is a Parter vertex for 0 and neutral vertex for 3 in A. When we add a pendent vertex at vertex 1, we consider the case in which the multiplicities of the eigenvalues 0 and 3 each go up in the new graph $\widetilde{A}$. To make the multiplicity of 0 go up in $\widetilde{A}$, the value on the added vertex 9 must be 0, because vertex 1 is Parter for 0.

Furthermore, to make the multiplicity of 3 go up, we must set the weight of the edge $\tilde{a}_{19}, \tilde{a}_{91}$ as the next equation dictates by Lemma 1 or Theorem 1.

$$3 - |\tilde{a}_{19}|^2 f(3) = 0, \tag{3}$$

in which $f(3)$ is the value of $f(x) = \frac{p_{A(1)}(x)}{p_A(x)}$ at 3. Since $p_A(x) = x^2(x-3)^2(x^4 - 3x^3 - 7x^2 + 12x + 9)$, and $p_{A(1)}(x) = x^3(x-3)^2(x^2 - 3x - 3)$, the value of $\tilde{a}_{19}$ is $\sqrt{6}e^{i\theta}$. Then, $\widetilde{A}$ is as follows, and $\widetilde{A}$ has eigenvalues 0 and 3 with multiplicity 3 each.

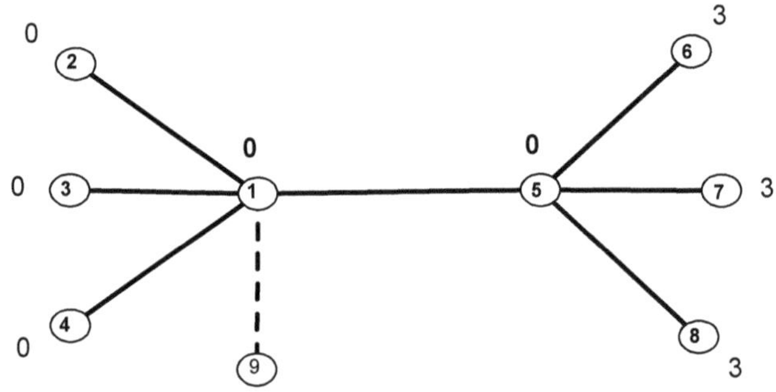

Fig. 1 Example 1

$$\widetilde{A} = \begin{bmatrix} 0 & 1 & 1 & 1 & 1 & 0 & 0 & 0 & \sqrt{6}e^{i\theta} \\ 1 & 0 & 0 & 0 & 0 & 0 & 0 & 0 & 0 \\ 1 & 0 & 0 & 0 & 0 & 0 & 0 & 0 & 0 \\ 1 & 0 & 0 & 0 & 0 & 0 & 0 & 0 & 0 \\ 1 & 0 & 0 & 0 & 0 & 1 & 1 & 1 & 0 \\ 0 & 0 & 0 & 0 & 1 & 3 & 0 & 0 & 0 \\ 0 & 0 & 0 & 0 & 1 & 0 & 3 & 0 & 0 \\ 0 & 0 & 0 & 0 & 1 & 0 & 0 & 3 & 0 \\ \sqrt{6}e^{-i\theta} & 0 & 0 & 0 & 0 & 0 & 0 & 0 & 0 \end{bmatrix}$$

Example 2 Let B be an Hermitian matrix as below,

$$B = \begin{bmatrix} 0 & 1 & 1 & 1 & 1 & 0 & 0 & 0 & 0 & 0 & 0 \\ 1 & 0 & 0 & 0 & 0 & 0 & 0 & 0 & 0 & 0 & 0 \\ 1 & 0 & 0 & 0 & 0 & 0 & 0 & 0 & 0 & 0 & 0 \\ 1 & 0 & 0 & 0 & 0 & 0 & 0 & 0 & 0 & 0 & 0 \\ 1 & 0 & 0 & 0 & 0 & 1 & 1 & 1 & 0 & 0 & 0 \\ 0 & 0 & 0 & 0 & 1 & 2 & 0 & 0 & 1 & 0 & 0 \\ 0 & 0 & 0 & 0 & 1 & 0 & 2 & 0 & 0 & 1 & 0 \\ 0 & 0 & 0 & 0 & 1 & 0 & 0 & 2 & 0 & 0 & 1 \\ 0 & 0 & 0 & 0 & 0 & 1 & 0 & 0 & 2 & 0 & 0 \\ 0 & 0 & 0 & 0 & 0 & 0 & 1 & 0 & 0 & 2 & 0 \\ 0 & 0 & 0 & 0 & 0 & 0 & 0 & 1 & 0 & 0 & 2 \end{bmatrix}$$

The graph of B is represented in Fig. 2. The values assigned to vertices are placed outside the circles. B has eigenvalues 1 and 3 with multiplicity 2 each. And $B(1)$ also has eigenvalues 1 and 3 with multiplicity 2 respectively. So, vertex 1 is neutral for both eigenvalues 1 and 3. In this example, we show that the multiplicities of 1 and 3 increase simultaneously by adding one pendent vertex to a vertex in T that is neutral vertex for the two eigenvalues.

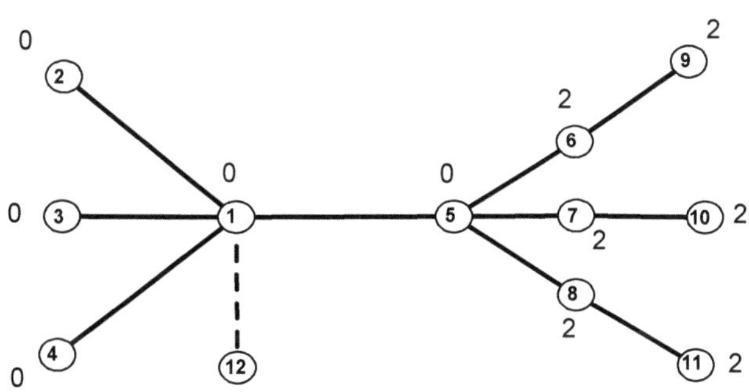

Fig. 2 Example 2

To make the multiplicity of 1 and 3 increase simultaneously, the next equations must hold, in which α is the value assigned to the pendent vertex,

$$\alpha = 3 - |\tilde{b}_{1,12}|^2 f(3) = 1 - |\tilde{b}_{1,12}|^2 f(1),$$

in which $f(x) = \frac{p_{B(1)}(x)}{p_B(x)}$. We have $f(x) = \frac{p_{B(1)}(x)}{p_B(x)} = \frac{x(x^3-4x^2+6)}{(x-2)(x^4-2x^3-8x^2+6x+9)}$, then $f(3) = 0.5$, $f(1) = -0.5$. Then $\tilde{b}_{1,12} = \sqrt{2}e^{i\theta}$. Therefore, $\widetilde{B}$ is as follows with $\alpha = 2$ and $\tilde{b}_{1,12} = \tilde{b}_{12,1} = \sqrt{2}$, then the multiplicity of each eigenvalue 1 and 3 simultaneously goes up to multiplicity 3.

$$\widetilde{B} = \begin{bmatrix} 0&1&1&1&1&0&0&0&0&0&0&\sqrt{2} \\ 1&0&0&0&0&0&0&0&0&0&0&0 \\ 1&0&0&0&0&0&0&0&0&0&0&0 \\ 1&0&0&0&0&0&0&0&0&0&0&0 \\ 1&0&0&0&0&1&1&1&0&0&0&0 \\ 0&0&0&0&1&2&0&0&1&0&0&0 \\ 0&0&0&0&1&0&2&0&0&1&0&0 \\ 0&0&0&0&1&0&0&2&0&0&1&0 \\ 0&0&0&0&0&1&0&0&2&0&0&0 \\ 0&0&0&0&0&0&1&0&0&2&0&0 \\ 0&0&0&0&0&0&0&1&0&0&2&0 \\ \sqrt{2}&0&0&0&0&0&0&0&0&0&0&2 \end{bmatrix}$$

References

1. Horn, R., Johnson, C.R.: Matrix Analysis, 2nd edn. Cambridge University Press, New York (2013)
2. Johnson, C.R.: Leal Duarte, A.: The maximum multiplicity of an eigenvalue in a matrix whose graph is a tree. Linear Multilinear Algebra **46**, 139–144 (1999)
3. Johnson, C.R.: Leal Duarte, A.: On the possible multiplicities of the eigenvalues of a Hermitian matrix whose graph is a tree. Linear Algebra Appl. **348**, 7–21 (2002)
4. Johnson, C.R., Leal Duarte, A., Saiago, C.M.: The Parter-Wiener theorem: refinement and generalization. SIAM J. Matrix Anal. Appl. **25**(2), 352–361 (2003)
5. Johnson, C.R., Leal Duarte, A., Saiago, C.M.: The change in eigenvalue multiplicity associated with perturbation of a diagonal entry. Linear and Multilinear Algebra **60**(5), 525–532 (2012)
6. Johnson, C.R., Li Andrew, A., Walker Andrew, J.: Ordered multiplicity lists for eigenvalues of symmetric matrices whose graph is a linear tree. Discret. Math. **333**, 39–55 (2014)
7. Leal Duarte, A., Johnson, C.R.: On the minimum number of distinct eigenvalues for a symmetric matrix whose graph is a given tree. Math. Inequal. Appl. **5**(2), 175–180 (2002)
8. Parter, S.: On the eigenvalues and eigenvectors of a class of matrices. J. Soc. Indust. Appl. Math. **8**, 376–388 (1960)
9. Wiener, G.: Spectral multiplicity and splitting results for a class of qualitative matrices. Linear Algebra Appl. **61**, 15–29 (1984)

Nonlinear Local Invertibility Preservers

M. Bendaoud, M. Jabbar and M. Sarih

Abstract Let $\mathscr{L}(X)$ be the algebra of all bounded linear operators on a complex Banach space X. Complete descriptions are given of the nonlinear maps of $\mathscr{L}(X)$ preserving local invertibility of $T * S$ for different kinds of binary operations $*$ on operators such as the sum $T + S$, the difference $T - S$, and the product TS. Extensions of these results to the case of different Banach spaces are also established. As application, mappings from $\mathscr{L}(X)$ onto itself that preserve the inner local spectral radius zero of such binary operations on operators are described.

Keywords Local spectrum · Local (inner) spectral radius · Single-valued extension property · Nonlinear preservers

1 Introduction

Throughout this paper, X and Y will denote complex Banach spaces and $\mathscr{L}(X, Y)$ will denote the space of all bounded linear operators from X into Y. As usual, when $X = Y$ we simply write $\mathscr{L}(X)$ for the algebra of all bounded linear operators on X with identity operator I. The local resolvent set of an operator $T \in \mathscr{L}(X)$ at a vector $x \in X$, $\rho_T(x)$, is the set of all λ in the complex field $\mathbb{C}$ for which there exists an open neighborhood U_λ of λ in $\mathbb{C}$ and an X-valued analytic function $f : U_\lambda \to X$ such that $(\mu - T)f(\mu) = x$ for all $\mu \in U_\lambda$. The local spectrum of T at x, denoted by $\sigma_T(x)$, is defined by

M. Bendaoud (✉) · M. Jabbar
Moulay Ismail University, ENSAM, Marjane II, B.P. 1529, Meknès,
Al Mansour, Morocco
e-mail: m.bendaoud@ensam-umi.ac.ma

M. Jabbar
e-mail: m.jabbar@ensam-umi.ac.ma

M. Sarih
Faculty of Sciences, BP 11201 Zitoune, Meknès, Morocco
e-mail: m.sarih@fs-umi.ac.ma

N. Bebiano (ed.), *Applied and Computational Matrix Analysis*,
Springer Proceedings in Mathematics & Statistics 192,
DOI 10.1007/978-3-319-49984-0_9

$$\sigma_T(x) := \mathbb{C} \setminus \rho_T(x),$$

and is a compact (possibly empty) subset of the usual spectrum $\sigma(T)$ of T. The local spectral radius of T at x is given by the formula

$$r_T(x) := \limsup_{n \to +\infty} \|T^n x\|^{\frac{1}{n}}.$$

Its counterpart the so-called inner local spectral radius of T at x is defined by

$$\iota_T(x) := \sup\{\varepsilon \geq 0 : x \in \mathscr{X}_T(\mathbb{C} \setminus D_\varepsilon)\},$$

where D_ε denotes the open disc of radius ε centered at 0 and $\mathscr{X}_T(\mathbb{C} \setminus D_\varepsilon)$ is the glocal spectral subspace of T associated with $\mathbb{C} \setminus D_\varepsilon$, that is, the set of all $x \in X$ for which there is an X-valued analytic function f on D_ε such that $(\lambda - T)f(\lambda) = x$ for all $\lambda \in D_\varepsilon$. The local (resp. inner local) spectral radius of T at x coincides with the maximum (resp. minimum) modulus of $\sigma_T(x)$ provided that T has the single-valued extension property. Recall that T is said to have the single-valued extension property (or SVEP, for short) if for every open subset U of $\mathbb{C}$, the equation $(\mu - T)f(\mu) = 0$, $(\mu \in U)$, has no nontrivial X-valued analytic solution f on U. Clearly, every operator $T \in \mathscr{L}(X)$ for which the interior of the set of its eigenvalues is empty enjoys this property.

Local spectra are a useful tool for analyzing operators, furnishing information well beyond that provided by classical spectral analysis. They play a very natural role in automatic continuity and in harmonic analysis, for instance in connection with the Wiener-Pitt phenomenon. For further details on the local spectral theory, as well as investigations and applications in numerous fields, we refer to the books [1, 25, 28].

The problem of characterizing linear or additive maps on matrix or operator algebras that leave invariant a given subset, function or relation defined on the underlying algebras represents one of the most active research areas. Plenty of deep and interesting results have been obtained by now and these results often reveal the algebraic or the merely ring structure of these algebras. Recently, a more challenging approach, attracting a lot of attention of researchers in the fields, consider the general preserver problems with respect to various algebraic operations on $\mathscr{M}_n$, the algebra of $n \times n$ complex matrices, or on operator algebras; see for instance [13, 17, 19–21, 24, 27, 29] and the references therein.

On the problem of describing mappings leaving invariant the local spectra, we mention: [22], where linear maps on $\mathscr{M}_n$ preserving the local spectrum at a fixed nonzero vector are characterized, [15] concerned with the infinite dimensional case, and in [9, 10] preserver problems that have to do with locally spectrally bounded linear maps or additive local spectrum compressors on the matrix spaces and on $\mathscr{L}(X)$ are considered. While, non-linear preserver problems on the subject were studied in [3–5, 7] where complete descriptions are given of the nonlinear transformations of $\mathscr{M}_n$ or of $\mathscr{L}(X)$ leaving invariant the local spectra of different kinds of binary operations on matrices or on operators such as the sum, the difference, the product, and the Jordan triple product. The corresponding problem for the local invertibility has been initiated by Bendaoud et al. in [6]. Fixing a Banach space X of dimension at least two, they proved that the only additive map ϕ from $\mathscr{L}(X)$ onto itself satisfying

$$0 \in \sigma_{\phi(T)}(x) \Longleftrightarrow 0 \in \sigma_T(x) \qquad (T \in \mathscr{L}(X), x \in X) \tag{1}$$

is the identity up a nonzero scalar. It is interesting to relax the additivity assumption and to know what kind of nonlinear transformations ϕ on $\mathscr{L}(X)$ will leave invariant the local invertibility property. Clearly, if one just assume (1) on ϕ, the structure of ϕ can be quite arbitrary. So, it is reasonable to impose a more restrictive condition on such transformations relating the local spectra of a pair of operators.

In this note, by strengthening the preservability condition, we consider the nonlinear preservers of local invertibility on $\mathscr{L}(X)$, and we obtain characterizations for mappings with less smoothness assumptions on them. In the next section, we consider maps on $\mathscr{L}(X)$ that preserve the local invertibility of the product of operators. It is shown that such maps are the identity up a scalar functions, and investigation of several extensions of these results to the case of different Banach spaces were obtained. While, in Sect. 3 we describe nonlinear transformations on $\mathscr{L}(X)$ that preserve the local invertibility of the sum (difference) of operators. As application, we describe in the last section mappings from $\mathscr{L}(X)$ onto itself that preserve the inner local spectral radius zero of operators.

2 Preservers of Local Invertibility of Operator Products

We first fix some notation. The duality between the Banach spaces X and its dual X^* will be denoted by $\langle ., . \rangle$. For $x \in X$ and $f \in X^*$, as usual we denote by $x \otimes f$ the rank at most one operator on X given by $z \mapsto \langle z, f \rangle x$. For $T \in \mathscr{L}(X)$ we will denote by $\ker(T)$, T^*, $\sigma(T)$,

$$\sigma_{su}(T) := \{\lambda \in \mathbb{C} : \lambda - T \text{ is not surjective}\},$$

and $r(T)$, the null space, the adjoint, the spectrum, the surjectivity spectrum, and the spectral radius of T; respectively.

Before stating the main results of this section, we provide some elementary lemmas needed in the sequel. The first one relies the SVEP and the local spectrum, see for instance [1, Theorems 2.20 and 2.22].

Lemma 1 *For an operator $T \in \mathscr{L}(X)$, the following statements hold.*

(i) For every $\lambda \in \mathbb{C}$ and every nonzero vector x in $\ker(\lambda - T)$ *we have $\sigma_T(x) \subseteq \{\lambda\}$.*
(ii) T has the SVEP if and only if for every $\lambda \in \mathbb{C}$ and every nonzero vector x in $\ker(\lambda - T)$ *we have $\sigma_T(x) = \{\lambda\}$.*

The second lemma is a simple consequence of [25, Proposition 1.2.16] and [1, Theorem 2.22], and its proof is therefore omitted here.

Lemma 2 *Let e be a fixed nonzero vector in X and let $R = x \otimes f$ be a non-nilpotent rank one operator. Then $0 \in \sigma_R(e)$ if and only if $\langle e, f\rangle = 0$ or e and x are linearly independent.*

The third lemma, established in [12, Proposition 3.1], gives some common local spectral properties shared by the operators TS and ST.

Lemma 3 *Let $T, S \in \mathscr{L}(X)$ and let x be a nonzero vector in X. Then $\sigma_{TS}(Tx) \subseteq \sigma_{ST}(x) \subseteq \sigma_{TS}(Tx) \cup \{0\}$. If moreover T is one-to-one, then $\sigma_{TS}(Tx) = \sigma_{ST}(x)$.*

The next lemma is quoted from [23, Theorem 1.1].

Lemma 4 *If ϕ is a surjective map on $\mathscr{M}_n$ satisfying*

$$\phi(T) - \phi(S) \text{ is invertible} \Longleftrightarrow T - S \text{ is invertible} \qquad (T, S \in \mathscr{M}_n), \tag{2}$$

then ϕ is additive.

We will say that a map ϕ on $\mathscr{L}(X)$ preserves the local invertibility of operators in both directions if for every $x \in X$ and $T \in \mathscr{L}(X)$ we have $0 \in \sigma_{\phi(T)}(x)$ if and only if $0 \in \sigma_T(x)$.

The following is one of the main results of this section. It characterizes nonlinear maps on $\mathscr{L}(X)$ that preserve local invertibility of operator products and extends the above mentioned result [6, Theorem 1.1] to the following more general scope.

Theorem 1 *A map ϕ from $\mathscr{L}(X)$ into itself satisfies*

$$0 \in \sigma_{\phi(T)\phi(S)}(x) \Longleftrightarrow 0 \in \sigma_{TS}(x) \qquad (T \in \mathscr{L}(X), x \in X) \tag{3}$$

if and only if there exists a map $\eta : \mathscr{L}(X) \to \mathbb{C}$ such that $\eta(T) \neq 0$ for every nonzero operator T and $\phi(T) = \eta(T)T$ for all $T \in \mathscr{L}(X)$.

As variant theorems, in the case of two different Banach spaces, the followings give similar results but at the price of the additional assumption that ϕ is surjective.

Theorem 2 *Let $\phi : \mathscr{L}(X) \to \mathscr{L}(Y)$ be a surjective map for which there exists $B \in \mathscr{L}(Y, X)$ such that for every $y \in Y$ we have*

$$0 \in \sigma_{\phi(T)\phi(S)}(y) \Longleftrightarrow 0 \in \sigma_{TS}(By) \qquad (T, S \in \mathscr{L}(X)). \tag{4}$$

Then B is invertible and there exists a map $\eta : \mathscr{L}(X) \to \mathbb{C}$ such that $\eta(T) \neq 0$ for every nonzero operator T and $\phi(T) = \eta(T)B^{-1}TB$ for all $T \in \mathscr{L}(X)$.

Theorem 3 *Let $\phi : \mathscr{L}(X) \to \mathscr{L}(Y)$ be a surjective map for which there exists $A \in \mathscr{L}(X, Y)$ such that for every $x \in X$ we have*

$$0 \in \sigma_{\phi(T)\phi(S)}(Ax) \Longleftrightarrow 0 \in \sigma_{TS}(x) \qquad (T, S \in \mathscr{L}(X)). \tag{5}$$

Then A is invertible and there exists a map $\eta : \mathscr{L}(X) \to \mathbb{C}$ *such that* $\eta(T) \neq 0$ *for every nonzero operator T and* $\phi(T) = \eta(T)ATA^{-1}$ *for all* $T \in \mathscr{L}(X)$.

The following examples shows that the assumption "ϕ is surjective" in Theorems 2 and 3 cannot be removed.

Example 1 Let $E \in \mathscr{L}(Y)$ be an arbitrary invertible operator, and let $\phi : \mathscr{L}(X) \to \mathscr{L}(Y)$ be defined by $\phi(T) := E \ \ (T \in \mathscr{L}(X))$. Let $B \in \mathscr{L}(Y, X)$ be given by $By := 0$ $(y \in Y)$. For any $T, S \in \mathscr{L}(X)$ and $y \in Y$, we have

$$\sigma_{TS}(By) = \emptyset \ \text{ and } \ \sigma_{\phi(T)\phi(S)}(y) \subseteq \sigma(E^2) \subseteq \mathbb{C} \setminus \{0\},$$

and so (4) is satisfied. However, B is not invertible.

Example 2 Let $A \in \mathscr{L}(X, X \oplus X)$ be given by $Ax := x \oplus x$ for every $x \in X$, and set $\phi(T) := T \oplus T$ for all $T \in \mathscr{L}(X)$. The map ϕ satisfies (5), but A is not invertible.

Proof of Theorem 2. Assume that ϕ satisfies

$$0 \in \sigma_{\phi(T)\phi(S)}(y) \Longleftrightarrow 0 \in \sigma_{TS}(By)$$

for any $T, S \in \mathscr{L}(X)$ and $y \in Y$.

We first claim that B is injective. If $By = 0$, then $\sigma_{TS}(By) = \emptyset$ and $0 \notin \sigma_{\phi(T)\phi(S)}(y)$ for any $T, S \in \mathscr{L}(X)$. This together with the surjectivity of ϕ entail that $0 \notin \sigma_{T'}(y)$ for each $T' \in \mathscr{L}(Y)$. Therefore $y = 0$, as claimed.

Next, let us prove that the operators $B\phi(T)$ and TB are linearly dependent for every operator $T \in \mathscr{L}(X)$. Let A be a fixed operator in $\mathscr{L}(X)$. Observe that for every $y \in Y$, the vectors $B\phi(T)y$ and TBy are linearly dependent. Indeed, assume for a contradiction that there exists $y \in Y$ such that $B\phi(T)y$ and TBy are linearly independent. Let $f \in X^*$ be a linear functional such that $f(B\phi(T)y) = 0$ and $f(TBy) = 1$, and set $R := By \otimes f$. Note that, the operators TR and RT are of rank one and have the SVEP as well as R. So, by Lemmas 1 and 3, we have

$$\begin{aligned} 0 \in \sigma_{TR}(B\phi(T)y) &\Longleftrightarrow 0 \in \sigma_{\phi(T)\phi(R)}(\phi(T)y) \\ &\Longrightarrow 0 \in \sigma_{\phi(R)\phi(T)}(y) \\ &\Longrightarrow 0 \in \sigma_{RT}(By); \end{aligned}$$

which contradicts the fact that $\sigma_{RT}(By) = \{1\}$. Its follows that for every $y \in Y$ the vector $B\phi(T)y$ belong to the linear span of TBy. By [16, Theorem 2.3], either $B\phi(T)$ and TB are linearly dependent, or they are both of rank one with the same image. In the first case we are done, while in the second case we have $B\phi(T) = u \otimes f$ and $TB = u \otimes g$ for some nonzero $u \in X$ and some nonzero $f, g \in Y^*$. We must prove that f and g are linearly dependent. Assume the contrary. Then we can find $y \in Y$ such that $f(y) = 0$ and $g(y) = 1$. The fact that B is injective implies that the operator

$\phi(T)$ is of rank one and $\phi(T)y = 0$. From this together with Lemma 1 and the fact that $\phi(S)\phi(T)$ has the SVEP, we have

$$\sigma_{\phi(S)\phi(T)}(y) = \{0\}$$

for all $S \in \mathcal{L}(X)$. Choose $S \in \mathcal{L}(X)$ with $STBy = By$. For such S we have $\sigma_{ST}(y) \subseteq \{1\}$. This contradicts (4) and shows that $B\phi(T)$ and TB are linearly dependent in this case, too.

Thus, for every nonzero operator $T \in \mathcal{L}(X)$, there exists a scalar λ_T such that $B\phi(T) = \lambda_T TB$.

Now, we assert that B is surjective. Assume on the contrary that B is not surjective, and let x be a nonzero vector in $X \setminus \text{range}\,(B)$. Pick an arbitrary non zero vector y in Y, and note that $By \neq 0$. Choose a linear functional f in X^* and $T \in \mathcal{L}(X)$ such that $Tx = By$ and $\langle By, f\rangle = 1$, and set $R = x \otimes f$. Firstly assume that $B\phi(R)y \neq 0$. From Lemmas 1, 2 and 3 together with the fact that $RT = x \otimes f \circ T$ and x and $B\phi(R)y$ are linearly independent, we have

$$\begin{aligned} 0 \in \sigma_{RT}(B\phi(R)y) &\Longleftrightarrow 0 \in \sigma_{\phi(R)\phi(T)}(\phi(R)y) \\ &\Longrightarrow 0 \in \sigma_{\phi(T)\phi(R)}(y) \\ &\Longrightarrow 0 \in \sigma_{TR}(By) = \{1\}, \end{aligned}$$

arriving to a contradiction.

In the remainder case when $B\phi(R)y = 0$, we have $\lambda_R RBy = 0$. From this we infer that $\lambda_R = 0$, and so $B\phi(R) = 0$. Consequently, $\phi(R) = 0$ since B is bijective. In particular, $0 \in \sigma_{\phi(T)\phi(R)}(y) = \{0\}$, and therefore $0 \in \sigma_{TR}(By) = \{1\}$; which leads to a contradiction in this case, too.

The contradictions obtained in all cases imply that B is surjective, as asserted.

Our next step is the prove, $\lambda_T \neq 0$ for all nonzero operator $T \in \mathcal{L}(X)$. Suppose by way of contradiction that there exists a nonzero operator $T \in \mathcal{L}(X)$ such that $\lambda_T = 0$, and let $x \in X$ be a nonzero vector such that $Tx \neq 0$. By the surjectivity of B, we can find a nonzero vector $y \in Y$ such that $By = Tx$. Choose a linear functional $f \in X^*$ such that $\langle By, f\rangle = 1$, and set $R := x \otimes f$. Note that $\phi(T) = 0$, and so $\sigma_{\phi(T)\phi(R)}(y) = \{0\}$ contradicting the fact that $\sigma_{TR}(By) = \{1\}$.

In order to complete the proof, let us observe that $\phi(0) = 0$ since otherwise we can find a nonzero vector $y \in Y$ such that $\phi(0)y \neq 0$. Let $x \in X$ such that $B^{-1}x = y$, and let $f \in Y^*$ be a linear functional such that $\langle B\phi(0)B^{-1}x, f\rangle = 1$. Then the nonzero operator $T := x \otimes f$ satisfies

$$0 \in \sigma_{\phi(T)\phi(0)}(y) = \sigma_{\lambda_T B^{-1}TB\phi(0)}(y) = \{\lambda_T\},$$

a contradiction. The proof is therefore complete. □

Proof of Theorem 3. Assume that ϕ satisfies

$$0 \in \sigma_{\phi(T)\phi(S)}(Ax) \Longleftrightarrow 0 \in \sigma_{TS}(x)$$

for any $T, S \in \mathscr{L}(X)$ and $x \in X$. We first assert that A is injective. If $Ax = 0$, then $0 \notin \sigma_{\phi(T)\phi(I)}(Ax)$ for every $T \in \mathscr{L}(X)$. So, (5) gives $0 \notin \sigma_T(x)$ for each $T \in \mathscr{L}(X)$, and consequently $x = 0$.

Next, we claim that A is surjective. Assume by the way of contradiction that A is not surjective, and let y be a nonzero vector in $Y \setminus \text{range}(A)$. Let $g \in Y^*$ be an arbitrary linear functional, and set $R' := y \otimes g$. We will show that $\phi(0) = y \otimes g$. The surjectivity of ϕ implies that there exists $R \in \mathscr{L}(X)$ such that $\phi(R) = R'$. For every nonzero vector $x \in X$ and $S \in \mathscr{L}(X)$, Lemma 2 tell us that

$$0 \in \sigma_{y\otimes g\phi(S)}(Ax) = \sigma_{\phi(R)\phi(S)}(Ax)$$

since y and Ax are linearly independent; implying that $0 \in \sigma_{RS}(x)$. From this we infer that $R = 0$ since otherwise we can find $x \in X$ and $S \in \mathscr{L}(X)$ such that $Rx \neq 0$ and $SRx = x$. This shows that $0 \in \sigma_{RS}(Rx) \subseteq \sigma_{SR}(x)$, and contradicts the fact that $\sigma_{SR}(x) \subseteq \{1\}$ since $SRx = x$; see Lemma 1. Hence, $R = 0$ and $\phi(0) = y \otimes g$. The arbitrariness of g give a contradiction, and shows that A is surjective, as claimed.

Thus, A is bijective and ϕ satisfies

$$0 \in \sigma_{\phi(T)\phi(S)}(y) \Longleftrightarrow 0 \in \sigma_{TS}(A^{-1}y)$$

for any $T, S \in \mathscr{L}(X)$ and $y \in Y$. The desired conclusion follows from Theorem 2; which achieves the proof. □

Remark 1 By inspecting the proof of Theorems 2 and 3, with no extra efforts, one can see that Theorem 2 (resp. Theorem 3) remains valid when the assumption "ϕ is surjective" is replaced by "B is surjective (resp. A is surjective)".

Proof of Theorem 1. The sufficiency condition is easily verified, and the necessity is a consequence of Theorem 3 and the above remark. □

3 Preservers of Local Invertibility of Operator Sums

In this section, we describe mappings ϕ from $\mathscr{L}(X)$ onto itself that preserve the local invertibility of operator sums. The following is one the purposes of this section. It generalizes [6, Theorem 1.1] and gives a partial response to [6, Problem].

Theorem 4 *A surjective map ϕ from $\mathscr{L}(X)$ into itself satisfies*

$$0 \in \sigma_{\phi(T)-\phi(S)}(x) \Longleftrightarrow 0 \in \sigma_{T-S}(x) \qquad (T \in \mathscr{L}(X), x \in X) \tag{6}$$

if and only if there exist $R \in \mathscr{L}(X)$ and a map $\eta : \mathscr{L}(X) \to \mathbb{C}$ such that $\eta(T) \neq 0$ for every nonzero operator T and $\phi(T) = \eta(T)T + R$ for all $T \in \mathscr{L}(X)$.

Proof Checking the "if" part is straightforward, so we will only deal with the "only if" part. So assume that (6) holds. Replacing ϕ by the mapping $T \mapsto \phi(T) - \phi(0)$, we may assume that $\phi(0) = 0$.

From the fact that

$$\sigma_{su}(T) = \bigcup_{x \in X} \sigma_T(x) \tag{7}$$

for every $T \in \mathscr{L}(X)$ (see [25, Lemma 2.3]), we have

$$\begin{aligned} T - S \text{ is not surjective} &\iff \exists x \in X : 0 \in \sigma_{T-S}(x) \\ &\iff \exists x \in X : 0 \in \sigma_{\phi(T)-\phi(S)}(x) \\ &\iff \phi(T) - \phi(S) \text{ is not surjective} \end{aligned}$$

for all $T \in \mathscr{L}(X)$. So, if X is an finite dimensional Banach space, then from Lemma 4 together with the fact that, in this case, an operator T is surjective if and only if it is invertible one can see that ϕ is additive. In the case when X is an infinite dimensional Banach space, the map ϕ is also additive; see [14, Theorem 4.2]. Thus, the desired conclusion follows from [6, Theorem 1.1], and the proof is complete. □

We obtain similar conclusion when using sums in (6) instead of subtractions.

Theorem 5 *A surjective map ϕ from $\mathscr{L}(X)$ into itself satisfies*

$$0 \in \sigma_{\phi(T)+\phi(S)}(x) \iff 0 \in \sigma_{T+S}(x) \qquad (T \in \mathscr{L}(X), x \in X)$$

if and only if there exists a map $\eta : \mathscr{L}(X) \to \mathbb{C}$ such that $\eta(T) \neq 0$ for every nonzero operator T and $\phi(T) = \eta(T)T$ for all $T \in \mathscr{L}(X)$.

Proof The sufficiency condition is easily verified. To prove the necessity, assume that

$$0 \in \sigma_{\phi(T)+\phi(S)}(x) \iff 0 \in \sigma_{T+S}(x)$$

for any $T, S \in \mathscr{L}(X)$ and $x \in X$. We first claim that $\phi(0) = 0$. To do so, let $A \in \mathscr{L}(X)$ such that $\phi(A) = 0$, and note that for every $T \in \mathscr{L}(X)$, we have

$$\begin{aligned} \exists x \in X : 0 \in \sigma_{T+A}(x) &\iff \exists x \in X : 0 \in \sigma_{T+A}(x) \\ &\iff \exists x \in X : 0 \in \sigma_{\phi(T)}(x) \\ &\iff \exists x \in X : 0 \in \sigma_{2\phi(T)}(x) \\ &\iff \exists x \in X : 0 \in \sigma_{2T}(x) \\ &\iff \exists x \in X : 0 \in \sigma_T(x). \end{aligned}$$

From this together with the equality (7), we infer that

$$T + A \text{ is not surjective} \iff T \text{ is not surjective}$$

for every $T \in \mathscr{L}(X)$. Upon replacing T by $T - \lambda$, we deduce that

$$\sigma_{su}(T + A) = \sigma_{su}(T)$$

for all $T \in \mathscr{L}(X)$. As the surjectivity spectrum contains the boundary of the spectrum, we conclude that $r(T + A) = r(T)$ for all $T \in \mathscr{L}(X)$. Thus, by the Zemánek's spectral characterization of the radical, [2, Theorem 5.3.1], $A = 0$ as desired.

Next, we assert that ϕ is additive. Similar argument as above allows to get that

$$T + S \text{ is surjective } \iff \phi(T) + \phi(S) \text{ is surjective}$$

for any $T, S \in \mathscr{L}(X)$. So, if X is a finite dimensional Banach space, then from the fact that Lemma 4 remains valid when using sums in (2) instead of subtractions and $\phi(0) = 0$, we deduce that the map ϕ is additive. In the case when X is an infinite dimensional Banach space, by [14, Theorem 5.1], ϕ is also additive, as asserted.

Thus, the map ϕ satisfies (6), and the desired conclusion follows from Theorem 4; which concludes the proof. $\square$

4 Preservers of the Inner Local Spectral Radius Zero

This section is devoted to deriving some consequences of the above obtained results of this paper. These consequences describe maps from $\mathscr{L}(X)$ onto itself that preserve the inner local spectral radius zero of operators. A map ϕ from $\mathscr{L}(X)$ into itself is said to preserve the inner local spectral radius zero if

$$\iota_{\phi(T)}(x) = 0 \iff \iota_T(x) = 0$$

for all $T \in \mathscr{L}(X)$ and $x \in X$.

The first consequence, extending [6, Theorem 1.6], describes nonlinear mappings that preserve the inner local spectral radius zero of operator products.

Theorem 6 *A map ϕ from $\mathscr{L}(X)$ into itself satisfies*

$$\iota_{\phi(T)\phi(S)}(x) = 0 \iff \iota_{TS}(x) = 0 \qquad (T \in \mathscr{L}(X), x \in X)$$

if and only if there exists a map $\eta : \mathscr{L}(X) \to \mathbb{C}$ such that $\eta(T) \neq 0$ for every nonzero operator T and $\phi(T) = \eta(T)T$ for all $T \in \mathscr{L}(X)$.

The second consequence extends the main results of [8, 11].

Theorem 7 *Let X be a complex Banach space of dimension at least two. A surjective map ϕ from $\mathscr{L}(X)$ into itself satisfies*

$$\iota_{\phi(T)-\phi(S)}(x) = 0 \iff \iota_{T-S}(x) = 0 \qquad (T \in \mathscr{L}(X)).$$

if and only if there exist $R \in \mathscr{L}(X)$ and a map $\eta : \mathscr{L}(X) \to \mathbb{C}$ such that $\eta(T) \neq 0$ for every nonzero operator T and $\phi(T) = \eta(T)T + R$ for all $T \in \mathscr{L}(X)$.

Proof of Theorems 6 *and* 7. As the notion of local invertibility encompasses inner spectral radius zero: for any $x \in X$ and $T \in \mathscr{L}(X)$ we have

$$0 \in \sigma_T(x) \Longleftrightarrow \iota_T(x) = 0$$

(see [26]), Theorems 1 and 4 remain valid when the hypothesis "$0 \in \sigma_.(.)$" is replaced by "$\iota_.(.) = 0$"; which yield the desired conclusions in Theorems 6 and 7. □

From the above comment, Theorems 2, 3, and 5 also remain valid when the assumption "$0 \in \sigma_.(.)$" is replaced by "$\iota_.(.) = 0$", and the obtained results in these theorems and in Theorems 6 and 7 lead to the nonlinear inner local spectral radius versions of the main results of [18] which describe surjective linear maps on $\mathscr{L}(X)$ that are local spectral radius zero-preserving.

Acknowledgements The first author thank the hospitality of the organizers of MatTriad'2015, Coimbra, Portugal, September 6–11, 2015, where the main results of this chapter were announced. This work was partially supported by a grant from MIU-SRA, Morocco.

References

1. Aiena, P.: Fredholm and Local Spectral Theory with Applications to Multipliers. Kluwer, Boston (2004)
2. Aupetit, B.: A Primer on Spectral Theory. Springer, New York (1991)
3. Bendaoud, M.: Preservers of local spectra of matrix sums. Linear Algebra Appl. **438**, 2500–2507 (2013)
4. Bendaoud, M.: Preservers of local spectrum of matrix Jordan triple products. Linear Algebra Appl. **471**, 604–614 (2015)
5. Bendaoud, M., Douimi, M., Sarih, M.: Maps on matrices preserving local spectra. Linear Multilinear Algebra **61**, 871–880 (2013)
6. Bendaoud, M., Jabbar, M., Sarih, M.: Additive local invertibility preservers. Publ. Math. Debr. **85**, 467–480 (2014)
7. Bendaoud, M., Jabbar, M., Sarih, M.: Preservers of local spectra of operator products. Linear Multilinear Algebra **63**, 806–819 (2015)
8. Bendaoud, M., Sarih, M.: Surjective linear maps preserving certain spectral radii. Rocky Mountain J. Math. **41**, 727–735 (2011)
9. Bendaoud, M., Sarih, M.: Locally spectrally bounded linear maps. Math. Bohem. **136**, 81–89 (2011)
10. Bendaoud, M., Sarih, M.: Additive local spectrum compressors. Linear Algebra Appl. **435**, 1473–1478 (2011)
11. Bendaoud, M., Sarih, M.: Additive maps preserving the inner local spectral radius. Oper. Theory Adv. Appl. **236**, 95–102 (2014)
12. Benhida, C., Zerouali, E.H.: Local spectral theory of linear operators *RS* and *SR*. Integr. Equ. Oper. Theory **73**, 1–8 (2006)
13. Bhatia, R., Šemrl, P., Sourour, A.R.: Maps on matrices that preserve the spectral radius distance. Studia Math. **134**, 99–110 (1999)
14. Bourhim, A., Mashreghi, J., Stepanyan, A.: Nonlinear maps preserving the minimum and surjectivity moduli. Linear Algebra Appl. **463**, 171–189 (2014)

15. Bračič, J., Müller, V.: Local spectrum and local spectral radius at a fixed vector. Studia Math. **194**, 155–162 (2009)
16. Brešar, M., Šemrl, P.: On locally linearly dependent perators and derivations. Trans. Amer. Math. Soc. **351**, 1257–1275 (1999)
17. Chan, J.T., Li, C.K., Sze, N.S.: Mappings preserving spectra of products of matrices. Proc. Amer. Math. Soc. **135**, 977–986 (2007)
18. Costara, C.: Linear maps preserving operators of local spectral radius zero. Integr. Equ. Oper. Theory **73**, 7–16 (2012)
19. Cui, J., Li, C.-K.: Maps preserving peripheral spectrum of Jordan products of operators. Oper. Matrices **6**, 129–146 (2012)
20. Dolinar, G., Houb, J., Kuzma, B., Qi, X.: Spectrum nonincreasing maps on matrices. Linear Algebra Appl. **438**, 3504–3510 (2013)
21. Gao, H.: $*$-Jordan-triple multiplicative surjective maps on B(H). J. Math. Anal. Appl. **401**, 397–403 (2013)
22. González, M., Mbekhta, M.: Linear maps on $M_n(\mathbb{C})$ preserving the local spectrum. Linear Algebra Appl. **427**(2–3), 176–182 (2007)
23. Havlicek, H., Šemrl, P.: From geometry to invertibility preservers. Studia Math. **174**, 99–109 (2006)
24. Hou, J.C., Li, C.-K., Wong, N.C.: Maps preserving the spectrum of generalized Jordan product of operators. Linear Algebra Appl. **432**, 1049–1069 (2010)
25. Laursen, K.B., Neumann, M.M.: An Introduction to Local Spectral Theory. Oxford University Press, New York (2000)
26. Miller, T.L., Miller, V.G., Neumann, M.: Local spectral properties of weighted shifts. J. Operator Theory **51**, 71–88 (2004)
27. Molnár, L.: Some characterizations of the automorphisms of $B(H)$ and $C(H)$. Proc. Amer. Math. Soc. **130**, 111–120 (2001)
28. Müller, V.: Spectral Theory of Linear Operators and Spectral Systems in Banach Algebras. Operator Theory: Advances and Applications, vol. 39. Springer, Berlin (2007)
29. Zhang, W., Hou, J.: Maps preserving peripheral spectrum of Jordan semi-triple products of operators. Linear Algebra Appl. **435**, 1326–1335 (2011)

More on the Hankel Pencil Conjecture—News on the Root Conjecture

Alexander Kovačec

Abstract The Hankel pencil conjecture concerns certain pencils of $n \times n$ Hankel matrices and has a control theoretic origin; see [4]. For each specific n it was abbreviated in [2] as HPnC and reduced to a conjecture RnC about roots of pairs of certain polynomials of degree $n-2$. To be solved, each conjecture RnC would be laboriously translated into a system of equations for the elementary symmetric polynomials and solved by Gröbner basis methods (we stopped at $n = 8$). In this paper we present conjecturally a parametrized system of equations in the symmetric polynomials which permits to prove specific cases of the root conjecture and hence of the Hankel pencil conjecture by much lighter computation. Other formulations of the root conjecture are also given.

Keywords Matrix pencils · Control theory · Root conjecture · Systems of algebraic equations

1 Introduction

The Hankel pencil conjecture is a deceptively simple looking conjecture on a certain family of Hankel or equivalently Toeplitz matrices. It was published by Schmale and Sharma who showed in [4] that its solution would significantly advance a 1981 conjecture by Bumby, Sontag, Sussmann, and Vasconcelos in control theory.

With x an indeterminate, and $c_i \in \mathbb{C}^* = \mathbb{C} \setminus \{0\}$, define the $n \times n$ Hankel matrix

A. Kovačec (✉)
Department of Mathematics, University of Coimbra, Coimbra, Portugal
e-mail: kovacec@uc.pt

N. Bebiano (ed.), *Applied and Computational Matrix Analysis*,
Springer Proceedings in Mathematics & Statistics 192,
DOI 10.1007/978-3-319-49984-0_10

$$H_n(x) = H_n(x; c_1, \ldots, c_{n-1}) = \begin{bmatrix} & & & x & c_1 & c_2 \\ & & x & c_1 & c_2 & c_3 \\ & \cdot^{\cdot^{\cdot}} & \cdot^{\cdot^{\cdot}} & & \vdots & \vdots \\ x & c_1 & \ldots & \ldots & c_{n-2} & c_{n-1} \\ c_1 & c_2 & \ldots & \ldots & c_{n-1} & c_n \\ c_2 & c_3 & \ldots & \ldots & c_n & c_{n+1} \end{bmatrix}.$$

A formal definition of this matrix is $H_n(x) = (c_{i+j-n+1})$, where $c_0 = x$ is an indeterminate, $c_l = 0$ for $l < 0$, and $c_l \in \mathbb{C}^*$ for $l \geq 1$.

Conjecture (Hankel Pencil Conjecture HPnC) If $\det H_n(x) \equiv 0$, then the last two columns are dependent, i.e. there exists a λ such that for all i $c_i = \lambda^{i-1} c_1$.

We begin with a short outline of how we reduced in the paper [2] the Hankel pencil conjecture to another conjecture which we called "root conjecture" and how we proved this latter and hence the former conjecture for various special cases. We then report on a twist we introduced in the root conjecture which led to an almost purely combinatorial conjecture and show why its proof could mean significant progress in the Hankel pencil conjecture.

The following facts were shown.

- Reference [2, Corollary 2.5] *If HPnC is true for the subclass of admissible matrices for which $c_1 = c_2 = 1$, then HPnC is true in general.*
  Thus $c_1 = c_2 = 1$ together with $\det H_n(x) = 0$ should imply $c_3 = \cdots = c_n = 1$.
- Sylvester's identity implies that there are polynomials $m_{ij}(x)$, $i, j \in \{n-1, n\}$, such that there holds the formula:

$$m_{nn}(x) \cdot m_{n-1,n-1}(x) - m_{n-1,n}^2(x) = \delta_{n-1} x^{n-2} \cdot \det H_n(x), \quad (\delta_n = (-1)^{\lfloor (n-1)/2 \rfloor}).$$

Combinatorial reasoning allowed us to determine the polynomials explicitly and for modified reciprocals of these polynomials, defined via $\hat{m}_{ij}(x) = \delta_n x^{n-2} m_{ij}(1/x)$, we found the following formulae (and a similar one for $\hat{m}_{n-1,n-1}(x)$ which we do not need here).

$$\hat{m}_{nn}(x) = (-1)^n \sum_{j=0}^{n-2} \left(\sum_i c_{i_1} \cdots c_{i_j} c_{i_{j+1}} \right) (-x)^j;$$

$$\hat{m}_{n-1,n}(x) = (-1)^n \sum_{j=0}^{n-2} \left(\sum_i c_{i_1} \cdots c_{i_j} c_{1+i_{j+1}} \right) (-x)^j,$$

where the inner sums $\sum_i, \ldots$ changing with j, are always over all $i = (i_1, \ldots, i_{j+1}) \in \mathbb{Z}_{\geq 1}^{j+1}$ for which $|i| = i_1 + \cdots + i_{j+1} = n-1$, a set of indices we designate I_{j+1}, if n is clear.

It is easy to see that if $c_1 = c_2 = 1$, then all the polynomials $\hat{m}_{ij}$, $i, j \in \{n-1, n\}$ are monic and this is the reason why we preferred working with the $\hat{m}_{ij}$ rather than with the m_{ij}.

• The above relation between the m_{ij} and det $H_n(x)$ and the fact that the hypothesis of HPnC requires det $H_n(x) = 0$ then led to the following

Reference [2, Proposition 5.1] HPnC, $n \geq 3$, *is equivalent to the following assertion for modified reciprocal polynomials.*

$$\hat{m}_{nn}(x) \cdot \hat{m}_{n-1,n-1}(x) = \hat{m}^2_{n-1,n}(x) \ \& \ c_1 = c_2 = 1 \text{ implies } c_3 = \cdots = c_{n+1} = 1.$$

Looking at the equation in the premisses of this implication one sees that every root of $\hat{m}_{nn}(x)$ must be a root of $\hat{m}^2_{n-1,n}(x)$. This observation finally led to conjecture that this apparently weaker hypothesis already implies the conclusion and herewith the Hankel pencil conjecture. That is, we formulated [2], Conjecture 5.3. (Root conjecture RnC)

$$If \, \text{roots}(\hat{m}_{n,n}) \subseteq \text{roots}(\hat{m}_{n-1,n}) \ \& \ c_1 = c_2 = 1, \ then \, \text{roots}(\hat{m}_{n-1,n}) = \{1\}.$$

It is easy to show that the conclusion roots$(\hat{m}_{n-1,n}) = \{1\}$ is equivalent to $c_1 = c_2 = \cdots = c_n = 1$. The discussion then shows

- Reference [2, Proposition 5.4] *For every $n \geq 3$, RnC implies HPnC.*
- We finally proceeded to show RnC and hence HPnC for all $n \leq 8$. We give an example how we did this. For $n = 5$ and from now on always assuming $c_1 = c_2 = 1$, one finds

$$\begin{aligned} \hat{m}_{55} &= -c_4 + (1 + 2c_3)x - 3x^2 + x^3, \\ \hat{m}_{45} &= -c_5 + (2c_3 + c_4)x - (2 + c_3)x^2 + x^3. \end{aligned}$$

These polynomials each have three not necessarily distinct roots. Let roots$(\hat{m}_{45}) = \{a, b, g\}$. The hypothesis of the root conjecture is roots$(\hat{m}_{55}) \subseteq$ roots$(\hat{m}_{45}) = \{a, b, g\}$. Of course, if we have here equality in the sense of multisets, then $\hat{m}_{55} = \hat{m}_{45}$, since the polynomials are monic. In this case we can do a direct comparison of coefficients and get $c_3 = c_4 = c_5 = 1$. Then the polynomials are equal to $(-1 + x)^3$ hence only admitting the root 1. If $\hat{m}_{55}$ has only one root (of multiplicity 3), say a, then Viéte's rules say $3 = 3a$, so $a = 1$. Now assume $\hat{m}_{55}$ has roots we can write roots$(\hat{m}_{55}) = \{a, a, b\}$. Then Viéte's rules allow us to write this system of equations as

$$\begin{aligned} 3 &\overset{1}{=} 2a + b & 2 + c_3 &\overset{2}{=} a + b + g \\ 1 + 2c_3 &\overset{1'}{=} a^2 + 2ab, & 2c_3 + c_4 &\overset{2'}{=} ab + ag + bg \\ c_4 &\overset{1''}{=} a^2 b & (c_5 &\overset{2''}{=} abg). \end{aligned}$$

Using "$\overset{1}{=}$" one has $b = 3 - 2a$. Then "$\overset{1'}{=}$" yields $c_3 = \frac{1}{2}(-3a^2 + 6a - 1)$, and then by "$\overset{2}{=}$", $g = -\frac{3}{2}a^2 + 4a - \frac{3}{2}$, while "$\overset{1''}{=}$" gives $c_4 = -2a^3 + 3a^2$. Substituting these expressions in a in "$\overset{2'}{=}$," yields $0 = \frac{7}{2}(a-1)^3$. Hence $a = 1$. Thus $b = 1$, and $g = 1$, showing roots$(\hat{m}_{45}) = \{1\}$, hence R5C, thus HP5C.

Before solving, one may opt, alternatively, to eliminate the c_i altogether and obtain a homogeneous polynomial system in only the roots of $m_{n-1,n}$.

For illustration let us continue with R5C. Write $\hat{e}_j = e_j(a,a,b)$, and $e_j = e_j(a,b,g)$, where $e_j(\ldots)$ signifies the j-th elementary symmetric function; here of three variables. Then we could alternatively substitute the right hand side of the left system by $\hat{e}_1, \hat{e}_2, \hat{e}_3$, respectively, and the right hand side of the right system by e_1, e_2, e_3 respectively. Having done this, we can eliminate c_3, c_4, c_5 and obtain the system

$$\begin{aligned} 0 &= \hat{e}_1 - 3 \\ 0 &= \hat{e}_2 - 2e_1 + 3 \\ 0 &= \hat{e}_3 - e_2 + 2e_1 - 4, \end{aligned}$$

which is a system solely in a, b, g. In [2] we used this technique to eliminate the c_i similarly for the cases $n = 6, 7, 8$. Each of these cases requires to treat a number of subcases which correspond to the various possibilities in which roots$(\hat{m}_{n,n}) \subseteq$ roots$(\hat{m}_{n-1,n})$ can happen.

For example in the case $n = 6$ supposing roots$(\hat{m}_{n-1,n}) = \{a, b, g, h\}$, one has to examine the subcases in which roots$(\hat{m}_{n,n})$ is equal to $\{a, a, a, a\}$, $\{a, a, a, b\}$, $\{a, a, b, b\}$, $\{a, a, a, b\}$, $\{a, a, b, g\}$, or $\{a, b, g, h\}$, respectively. In the first and the last case it is easy to show that the system of equations obtained admits only the solution $a = b = g = h = 1$, but for the other cases we solved the system computing the solution via Gröbner bases.

One of the difficulties we did not know how to overcome at the time is that the system in the e_i and $\hat{e}_i$ had to be computed for every n anew and we did not see any pattern by which these systems evolve. The principal news of the present paper was obtained by formulating the root conjecture not for the polynomials $\hat{m}_{nn}(x)$ and $\hat{m}_{n-1,n}(x)$, but rather the polynomials $\hat{m}_{nn}(1+x)$ and $\hat{m}_{n-1,n}(1+x)$. Using these polynomials we are now able to conjecture a pattern according to which the system in the $\hat{e}_j$ and e_j develops; these now defined w.r.t. the new polynomials analogously as before $\hat{e}_j$ and e_j were defined w.r.t $\hat{m}_{nn}(x)$ and $\hat{m}_{n-1,n}(x)$. Section 3 reports these developments. Furthermore we prove in Sect. 4 that *if* this conjecture is correct, then it is equivalent to a homogeneous polynomial system which has as many equations as it has unknowns. The Hankel pencil conjecture then follows if this latter system of equations has only the trivial solution. The fact that in a certain sense "almost all" systems of homogeneous equations of the referred type have only the trivial solution, see [1, p. 80], earns the Hankel pencil conjecture well founded credibility.

Before we launch into those sections and in order to whet a reader's appetite to work on RnC, we present in Sect. 2 (without proofs) alternative formulations of the root conjecture. Although we have not yet used these formulations for progress in RnC, they merit mention since they permit to present the root conjecture from scratch in a succinct way.

2 Alternative Formulations of the Root Conjecture

Miguel R. Moreira [3], a medalist of the International Mathematical Olympiads, showed that the polynomials $\hat{m}_{n-1,n}(x)$ and $\hat{m}_{nn}(x)$ stand in close relationship with a simple inductively defined sequence of polynomials.

Given a sequence $c_1 = c_2 = 1, c_3, \ldots$ of nonzero complex numbers, define the polynomials (P_n), $n = 1, 2, 3, \ldots$ by the rules

$$P_1(x) = 1, \quad P_2(x) = 1 + x, \quad P_n(x) = c_n + x\left(\sum_{i=1}^{n-1} c_i P_{n-i}(x)\right).$$

One then can prove the following lemma

Lemma 1 *There hold the relations*

i. $(-1)^n \hat{m}_{nn}(-x) = P_{n-1}(x)$.
ii. $(-1)^n \hat{m}_{n,n-1}(-x) = P_n(x) - c_1 x P_{n-1}(x)$.

Thus it is easy to see the following conjecture as being (equivalent to) RnC.

Conjecture 1 (RnC). $\text{roots}(P_{n-1}) \subseteq \text{roots}(P_n)$ implies $\text{roots}(P_{n-1}) = \{-1\}$.

It is in certain contexts reasonable to define "simplicity" as "having as many zeros as possible". From this point of view the following further formulation of RnC may appeal to the reader. By a simple variable transformation one can introduce polynomials of which one expects they have 0 as the only root. Since we also expect then all c_i will have value 1, we also put $c_i' = 1 + c_i$. If one defines now polynomials $Q_j(x) = P_j(-1 + x)$, one gets an inductively defined sequence given by

$$\begin{aligned}
Q_1 &= 1,\\
Q_2(x) &= x,\\
Q_n(x) &= (1 + c_n') + (-1 + x)(Q_{n-1} + Q_{n-2} + (1 + c_3')Q_{n-3} + \cdots + (1 + c_{n-1}')Q_1).
\end{aligned}$$

Then the first few polynomials read

$$\begin{aligned}
Q_3 &= c_3' + x^2,\\
Q_4 &= (-2c_3' + c_4') + 2c_3' x + x^3,\\
Q_5 &= (c_3' - 2c_4' + c_5') + (-4c_3' + 2c_4')x + 3c_3' x^2 + x^4.
\end{aligned}$$

This way one gets:

Conjecture 2 (RnC, version Q) $\text{roots}(Q_{n-1}) \subseteq \text{roots}(Q_n)$ implies $c_3' = c_4' = \cdots = c_n' = 0$, or equivalently, $\text{roots}(Q_n) = \{0\}$.

3 A Parametrized System of Equations for Elementary Symmetric Functions

Instead of formulating the root conjecture for polynomials $\hat{m}_{n-1,n}(x)$ and $\hat{m}_{nn}(x)$, one can, of course, use the polynomials $\hat{m}_{n-1,n}(1+x)$ and $\hat{m}_{nn}(1+x)$ and formulate RnC this way reminiscent of Conjecture 2 above.

RnC: If $c_1 = c_2$ and the inclusion roots$(\hat{m}_{n,n}(1+x)) \subseteq$ roots$(\hat{m}_{n-1,n}(1+x))$ holds in set theoretic sense, then roots$(\hat{m}_{n-1,n}(1+x)) = \{0\}$.

One can now invoke a simple lemma relating the coefficients of polynomials $f(x)$ and $f(1+x)$.

Lemma 2 *Let $f(x) = f_0 + f_1x + \cdots + f_nx^n$ be a polynomial and let $g(x) = f(1+x) = g_0 + g_1x + \cdots + g_nx^n$. Then, for $l = 0, 1, \ldots, n$, there holds the relation $g_l = \sum_{j=l}^{n} \binom{j}{l} f_j$. In particular if f is monic, then g is.*

Now we use the formulae in Sect. 1 and name the roots of $\hat{m}_{n-1,n}(1+x)$ by $z_1, \ldots, z_{n-2}$; and similarly the roots of $\hat{m}_{n,n}(1+x)$ by $z'_1, \ldots, z'_n$. We use Viète, and get for $l = 0, 1, \ldots, n-2$:

$$\hat{e}_l := e_l(z'_1, \ldots, z'_{n-2}) = \sum_{j=n-2-l}^{n-2} (-1)^{j+n+l} \binom{j}{n-2-l} \sum_{i \in I_{j+1}} c_{i_1} \cdots c_{i_j} c_{i_{j+1}}.$$

For $e_l = e_l(z_1, \ldots, z_{n-2})$ use the same formula, but replace $c_{i_{j+1}}$ by $c_{1+i_{j+1}}$.

The first few of these formulae are the following. Again these could be simplified somewhat introducing $c'_i = -1 + c_i$, but the result in the e_i and $\hat{e}_i$ after elimination of c_i or c'_i would be the same.

$$\begin{aligned}
e_1 &= -1 + c_3,\\
e_2 &= 4 + c_4 + c_3(-5+n) - n,\\
e_3 &= -9 + c_5 + c_3(19-4n) + c_4(-6+n) + c_3^2(-5+n) + 2n,\\
e_4 &= c_6 + c_4(23-4n) + c_5(-7+n) + 2c_3c_4(-6+n) + c_3(-69+18n-n^2) +\\
&\quad (c_3^2(76-19n+n^2))/2 + (52-15n+n^2)/2,
\end{aligned}$$

$$\begin{aligned}
\hat{e}_1 &= 0\\
\hat{e}_2 &= 3 + c_3(-3+n) - n,\\
\hat{e}_3 &= -4 - 2c_3(-4+n) + c_4(-4+n) + n,\\
\hat{e}_4 &= -2c_4(-5+n) + c_5(-5+n) - c_3(-7+n)(-5+n) + ((-6+n)(-5+n))/2\\
&\quad +(c_3^2(-6+n)(-5+n))/2,\\
\hat{e}_5 &= -2c_5(-6+n) + c_6(-6+n) - c_4(-8+n)(-6+n) + c_3c_4(-7+n)(-6+n)\\
&\quad -(-6+n)^2 - c_3^2(-6+n)(-13+2n) + c_3(-6+n)(-19+3n).
\end{aligned}$$

Curiously, although these formulas are more complicated than the corresponding ones in [2, Lemma 6.1] the elimination of the c_i yields relations between the e_i and $\hat{e}_i$ that are simpler and allow explicit parametrization. We explain the conjecture which we came up with after a number of computer experiments.

It is not hard to show that the polynomials $\hat{e}_i$ as symbolic expressions in n, $c_3, \ldots, c_{i+1}$ are divisible by $n - i - 1$ in the realm of integer coefficient polynomials; that is defining

$$p_i = \begin{cases} \hat{e}_i/(n-i-1) & \text{if } i \leq n-2 \\ 0 & \text{if } i > n-2, \end{cases},$$

the p_i are polynomials in $\mathbb{Z}[n, c_3, \ldots, c_{i+1}]$.

More generally, define the abbreviations $p_{ij} = p_i p_j$, $p_{ijk} = p_i p_j p_k$, etc.

Recall that a finite sequence of positive integers, $\lambda = (\lambda_1, \lambda_2, \ldots, \lambda_k)$ is a *partition* of an integer n, written $\lambda \vdash n$, if $\lambda_1 \geq \lambda_2 \geq \lambda_3 \geq \cdots \geq \lambda_k$, and $\sum_i \lambda_i = n$. The length of λ, $\lg(\lambda) = k$. Further we define the *typenumber* of λ as the product of the factorials of the multiplicities with which the positive components of λ occur. Thus for example, for $\lambda = (3, 3, 2, 2, 2, 2, 2, 1, 1, 1)$, we have $\lambda \vdash 19$, $\lg(\lambda) = 10$, and $\mathrm{typenb}(\lambda) = 2!5!3!$.

The mentioned conjecture is the following.

Conjecture 3 Considering the e_j and $\hat{e}_j$ as polynomials in $\mathbb{Q}[n, c_3, c_4, \ldots]$, for every $j = 0, 1, 2, \ldots$ define

$$q_{1+j} = p_{1+j} + (n - j - 1) \times \sum_{\substack{\lambda \vdash j+1 \\ \lg \lambda \geq 2}} \frac{(-1)^{\lg \lambda} \prod_{j=1}^{\lg \lambda - 2} (jn - j - 1)}{\mathrm{typenb}(\lambda)} p_\lambda.$$

Then there hold the relations $0 = e_j - \hat{e}_j - q_{1+j}$, where $j = 0, 1, 2, \ldots$.

We show the first few equations. Note that the first equation says that $p_1 = 0$, so that all λ in the sum which have a component 1 can be suppressed.

$$\begin{aligned}
&0.\ 0 = \hat{e}_1 \\
&1.\ 0 = e_1 - \hat{e}_1 - p_2 \\
&2.\ 0 = e_2 - \hat{e}_2 - p_3 \\
&3.\ 0 = e_3 - \hat{e}_3 - p_4 - (n-4)(p_{22}/2) \\
&4.\ 0 = e_4 - \hat{e}_4 - p_5 - (n-5)p_{32} \\
&5.\ 0 = e_5 - \hat{e}_5 - p_6 - (n-6)\left(p_{42} + \tfrac{1}{2}p_{33} - \tfrac{n-2}{6}p_{222}\right) \\
&\quad\vdots
\end{aligned}$$

This conjecture which was tested up to $j = 9$ gives a relation between the e_j and $\hat{e}_j$ in a parametrized form as desired. It is not overly difficult to obtain this conjecture - one tries to write for sufficiently many j e_j as a linear combinations of the p_i, and its products $i = 1, 2, .., j, j + 1$ and finds after a number of observations the above pattern. Unfortunately even very special cases of the conjecture, for example the one arising when wishing to prove that the c_i-free 'constant' coefficients of both sides are equal, are hard to prove and the author has not yet succeeded in this endeavour.

However the system of equations in the elementary symmetric functions this way obtained is considerably simpler than the previous one. While for example the sixth equation for the case $n = 8$ in [2], p. 1523, involves symmetric polynomials from degrees zero to six - it is $0 = 6392 - 2740e_1 + 314e_1^2 - 6e_1^3 + 210e_2 - 36e_1e_2 + e_2^2 - 18e_3 + 2e_1e_3 + 2e_4 - e_5 + \hat{e}_6$, - the corresponding equation number 5 above, involves only degrees 5 and 6 - it is (up to multiples)

$$0 = 3000(e_5 - \hat{e}_5 - \hat{e}_6) - 800\hat{e}_4\hat{e}_2 - 375\hat{e}_3^2 + 96\hat{e}_2^3.$$

We see in the next section how the equations so obtained permit, by much lighter and more insightful calculations than previously was possible, confirmation of the root conjectures.

4 Transforming the Quasi-homogeneous Systems into Homogeneous Ones and Solving Them

We begin by recalling a simple lemma for elementary symmetric polynomials. It is convenient to introduce the convention to let $e_j(\ldots)$ stand for the elementary symmetric polynomial of the variables indicated in a specific case and to assume $e_j(\ldots) = 0$ if $j < 0$ or $j >$number of variables. So for example $e_2(x_1, x_2, x_3) = x_1x_2 + x_1x_3 + x_2x_3$ but $e_5(x_1, x_2, x_3) = 0$. Furthermore $x_{i:j}$ means in case $1 \le i \le j$ the $(j - i + 1)-$tuple $(x_i, \ldots, x_j)$.

Lemma 3 *There holds for any integers j and $1 \le k \le n$ the identity*

$$e_j(x_{1:n}) = \sum_\nu e_\nu(x_{1:k})e_{j-\nu}(x_{1+k:n}),$$

where the sum is over the integers.

We will use this lemma in a moment only in cases in which n is replaced with $n - 2$ and the $x$es by zs, names for the solutions of the enumerated system of equations above. Keep in mind that the root conjecture says that *whatever* choice $z_1', \ldots, z_{n-2}' \in \{z_1, \ldots, z_{n-2}\}$, we suppose, the particular system of equations in $z_1, \ldots, z_{n-2}$ which arises from such a choice, will admit only the trivial solution.

The following explanations are exemplified in the example after the proposition below.

Assume we make a choice in which certain k of the z_i are not contained in the left set. Then, by symmetry, $\hat{e}_j = e_j(z_1', \ldots, z_{n-2}')$ can be thought of as being $\hat{e}_j = e_j(u_{1:k}, z_{1+k:n-2})$ where $\{u_1, \ldots, u_k\} \subseteq \{z_{1+k}, \ldots, z_{n-2}\} \subseteq \{z_1, \ldots, z_{n-2}\}$, while $e_j = e_j(z_1, \ldots, z_{n-2})$.

Now apply the previous lemma as well to $e_j(z_1, \dots, z_{n-2}) = e_j(z_{1:k}, z_{1+k:n-2})$, as to $e_j(u_{1:k}, z_{1+k:n-2})$ and define the shorthands

$$\dot{e}_j = e_j(z_{1+k:n-2}), \quad E_j = e_j(z_{1:k}) - e_j(u_{1:k}).$$

Note that for $j \geq 1+k$, $E_j = 0$. We see that the subsystem of the equations numbered $0, 1, \dots, n-3$ in Sect. 3 gains the following aspect:

$$\begin{aligned}
0.\ 0 &= \hat{e}_1 \\
1.\ 0 &= E_1 - q_2 \\
2.\ 0 &= E_1\dot{e}_1 + E_2 - q_3 \\
3.\ 0 &= E_1\dot{e}_2 + E_2\dot{e}_1 + E_3 - q_4 \\
&\vdots \\
k-1.\ 0 &= E_1\dot{e}_{k-2} + E_2\dot{e}_{k-3} + E_3\dot{e}_{k-4} + \dots + E_{k-1} - q_k \\
k.\ 0 &= E_1\dot{e}_{k-1} + E_2\dot{e}_{k-2} + E_3\dot{e}_{k-3} + \dots + E_{k-1}\dot{e}_1 + E_k - q_{k+1} \\
k+1.\ 0 &= E_1\dot{e}_k + E_2\dot{e}_{k-1} + E_3\dot{e}_{k-2} + \dots + E_{k-1}\dot{e}_2 + E_k\dot{e}_1 - q_{k+2} \\
&\vdots \\
n-3.\ 0 &= E_1\dot{e}_{n-4} + E_2\dot{e}_{n-5} + E_3\dot{e}_{n-6} + \dots + E_{k-1}\dot{e}_{n-k-2} + E_k\dot{e}_{n-k-3} - q_{n-2}
\end{aligned}$$

This system has furthermore the following features:

- Each $\dot{e}_j, \hat{e}_j, E_j, q_j$ is a homogeneous polynomial of degree j (or possibly zero).
- $\dot{e}_j, \hat{e}_j$, and q_j depend only on $z_{1+k}, \dots, z_{n-2}$, i.e., $\dot{e}_j = \dot{e}_j(z_{1+k:n-2})$; $q_j = q_j(z_{1+k:n-2})$.

(• E_j may depend on all variables.)

Proposition 1 *A system of polynomial equations of this form and with these features has a system of $n-2-k$ homogeneous equations of respective degrees* $1; k+2, k+3, \dots, n-2$ *in the $n-2-k$ variables $z_{1+k}, \dots, z_{n-2}$ as a consequence.*

Proof Equation 0 can be written as $0 = e_1(u_{1:k}) + e_1(z_{1+k:n-2})$. This equation is homogeneous of degree 1 and as $u_1, \dots, u_k \in \{z_{1+k}, \dots, z_{2+n}\}$, it is an equation in $z_{1+k}, \dots, z_{n-2}$. We show now that the remaining equations $1, \dots, n-3$ have as a consequence a system of $n-3-k$ homogeneous equations of respective degrees $k+2, k+3, \dots, n-2$ in variables $z_{1+k}, \dots, z_{n-2}$.

To see this note that equation 1 justifies to substitute q_2 for E_1; that is to do $E_1 \to q_2$ in all the following equations. Next equation 2 justifies the substitution $E_2 \to q_3 - E_1\dot{e}_1$, that is $E_2 \to q_3 - q_2\dot{e}_1$, in equations 3, 4, …Next we do $E_3 \to q_4 - q_2\dot{e}_2 - (q_3 - q_2\dot{e}_1)\dot{e}_1$ in the equations $4, 5, \dots$ We see by this process that, having substituted $E_1, E_2, \dots, E_{j-1}$, the first $j-1$ terms of equation j turn into terms of degree $j+1$. In particular, when we use equation $k-1$ to do a substitution $E_{k-1} \to \dots$ in equations $k, 1+k, \dots, n-3$, the first $k-1$ terms in these equations turn into terms of degrees $1 +$ number of equation. Once more doing this, using now equation k to substitute E_k we see that equation $k+1$ turns into a homogeneous equation of degree $k+2$, and in general equation $l \geq k+1$ into a homogeneous equation of degree $l+1$. In particular equation $n-3$ will turn into a homogeneous equation of degree $n-2$. So we have at the end $(n-3)-(k+1)+1 = n-3-k$ homogeneous equations. Finally observe that we have replaced the E_j, $j = 1, 2, \ldots, k$ by polynomials in $\dot{e}_1, \dot{e}_2, \ldots, \dot{e}_{k-1}, q_2, \ldots, q_{k+1}$. These polynomials as well as $q_{k+2}, \ldots, q_{n-2}$ are polynomials in $z_{1+k}, \ldots, z_{n-2}$. The proposition follows. □

A system of m homogeneous polynomial equations in m variables has typically only the trivial solution. Let us assume that the homogeneous system obtained by the process of the proof of the proposition is "typical". Then we get $x_{1+k} = \cdots = x_{n-2} = 0$. This implies also that all q_j are 0 and that $u_{1:k} = 0$. Then from the system we see $0 = E_j = e_j(x_{1:k}) - e_j(u_{1:k}) = e_j(x_{1:k})$, $j = 1, 2, \ldots, k$. Since the map $\mathbb{C}^k \ni x_{1:k} \mapsto (e_1(x_{1:k}), \ldots, e_k(x_{1:k})) \in \mathbb{C}^n$ defines (by the fundamental theorem of algebra and by Viéte's rules) a bijection from $\mathbb{C}^k$ to $\mathbb{C}^k$, we find $x_{1:k} = 0$.

Example 1 If $n = 7$, then we speak of variables z_1, z_2, z_3, z_4, z_5. Assume for $\hat{e}_j$ roots $\{z_4, z_5, z_5, z_4, z_5\}$. Then the system of equations in explicit form is found to be

$$
\begin{aligned}
0 &= 2z_4 + 3z_5,\\
0 &= z_1 + z_2 + z_3 - z_4 - z_4^2/4 - 2z_5 - (3z_4z_5)/2 - (3z_5^2)/4,\\
0 &= z_1z_2 + z_1z_3 + z_2z_3 + z_1z_4 + z_2z_4 + z_3z_4 - z_4^2 + z_1z_5 + z_2z_5 + z_3z_5 - 5z_4z_5 - z_4^2z_5\\
&\quad -3z_5^2 - 2z_4z_5^2 - z_5^3/3,\\
0 &= z_1z_2z_3 + z_1z_2z_4 + z_1z_3z_4 + z_2z_3z_4 - (3z_4^4)/32 + z_1z_2z_5 + z_1z_3z_5 + z_2z_3z_5\\
&\quad +z_1z_4z_5 + z_2z_4z_5 + z_3z_4z_5 - 3z_4^2z_5 - (9z_4^3z_5)/8 - 6z_4z_5^2 - (87z_4^2z_5^2)/16 - z_5^3\\
&\quad -(35z_4z_5^3)/8 - (27z_5^4)/32,\\
0 &= z_1z_2z_3z_4 + z_1z_2z_3z_5 + z_1z_2z_4z_5 + z_1z_3z_4z_5 + z_2z_3z_4z_5 - z_4^4z_5 - 3z_4^2z_5^2 - 8z_4^3z_5^2\\
&\quad -2z_4z_5^3 - (49z_4^2z_5^3)/3 - 8z_4z_5^4 - z_5^5.
\end{aligned}
$$

In practical work it is not necessary to write this system down explicitly. In fact it would be sufficient to use equations 0 and 4 of the last of the following four blocks below.

We now treat the system according to the proof of the proposition. According to the above, $u_{1:3} = (z_4, z_5, z_5)$, and so $k = 3$, $n - 3 = 4$. In the form of Sect. 3 the system takes the form as shown at the left.

0. $0 = \hat{e}_1$
1. $0 = E_1 - q_2$
2. $0 = E_1\dot{e}_1 + E_2 - q_3$
3. $0 = E_1\dot{e}_2 + E_2\dot{e}_1 + E_3 - q_4$
4. $0 = E_1\dot{e}_3 + E_2\dot{e}_2 + E_3\dot{e}_1 - q_5$

Substituting $E_1 \to q_2$ the system takes the form

0. $0 = \hat{e}_1$
1. $0 = 0$
2. $0 = q_2\dot{e}_1 + E_2 - q_3$
3. $0 = q_2\dot{e}_2 + E_2\dot{e}_1 + E_3 - q_4$
4. $0 = q_2\dot{e}_3 + E_2\dot{e}_2 + E_3\dot{e}_1 - q_5$

Next substituting $E_2 \to q_3 - q_2\dot{e}_1$, the system becomes

$$\begin{aligned}
0.\ 0 &= \hat{e}_1\\
1.\ 0 &= 0\\
2.\ 0 &= 0\\
3.\ 0 &= q_2\dot{e}_2 + (q_3 - q_2\dot{e}_1)\dot{e}_1 + E_3 - q_4\\
&= q_2\dot{e}_2 + q_3\dot{e}_1 - q_2\dot{e_1}^2 + E_3 - q_4.\\
4.\ 0 &= q_2\dot{e}_3 + (q_3 - q_2\dot{e}_1)\dot{e}_2 + E_3\dot{e}_1 - q_5
\end{aligned}$$

Finally substituting $E_3 \to q_4 - q_2\dot{e}_2 - q_3\dot{e}_1 + q_2\dot{e_1}^2$ one gets

$$\begin{aligned}
0.\ 0 &= \hat{e}_1\\
1.\ 0 &= 0\\
2.\ 0 &= 0\\
3.\ 0 &= 0\\
4.\ 0 &= q_2\dot{e}_3 + q_3\dot{e}_2 - q_2\dot{e}_1\dot{e}_2\\
&\quad +(q_4 - q_2\dot{e}_2 - q_3\dot{e}_1 + q_2\dot{e_1}^2)\dot{e}_1 - q_5\\
&= q_2\dot{e}_3 + q_3\dot{e}_2 - q_2\dot{e}_1\dot{e}_2 + q_4\dot{e}_1\\
&\quad -q_2\dot{e}_2\dot{e}_1 - q_3\dot{e_1}^2 + q_2\dot{e_1}^3 - q_5
\end{aligned}$$

In the case at hand $\dot{e}_j = e_j(z_4, z_5)$, so $\dot{e}_1 = z_4 + z_5$, $\dot{e}_2 = z_4z_5$, and for $j \geq 3$, $\dot{e}_j = 0$; furthermore one finds

$$\begin{aligned}
q_2 &= (z_4^2 + 6z_4z_5 + 3z_5^2)/4;\\
q_3 &= (3z_4^2z_5 + 6z_4z_5^2 + z_5^3)/3;\\
q_4 &= (3z_4^4 + 36z_4^3z_5 + 174z_4^2z_5^2 + 140z_4z_5^3 + 27z_5^4)/32;\\
q_5 &= (3z_4^4z_5 + 24z_4^3z_5^2 + 52z_4^2z_5^3 + 24z_4z_5^4 + 3z_5^5)/6;
\end{aligned}$$

This then leads to these equations (after multiplying equation 4 with 96):

$$\begin{aligned}
0.\ 0 &= 2z_4 + 3z_5\\
4.\ 0 &= 33z_4^5 + 141z_4^4z_5 + 198z_4^3z_5^2 + 30z_4^2z_5^3 + 109z_4z_5^4 + 73z_5^5,
\end{aligned}$$

which yields quite easily $z_4 = z_5 = 0$.

Once we knows this, we infer $q_2 = q_3 = q_4 = q_5 = 0$, and hence from the original equations 1, 2, 3, $E_1 = E_2 = E_3 = 0$. We also find $u_{1:3} = (0, 0, 0)$. Now $E_j = e_j(z_{1:3}) - e_j(u_{1:3})$. Thus $e_j(z_{1:3}) = 0$ for $j = 1, 2, 3$, and so $z_1 = z_2 = z_3 = 0$.

This case was treated in [2] (as case $n = 7$, subcase 32) by solving a system of equations obtained from a Gröbner basis with 6 polynomials of lengths 3, 6, 6, 13, 13 and large coefficients. Thus while the new methods are still not as light as desirable, we see that the case $n = 7$ can be still be done by hand, if necessary. This was completely out of question previously.

Note added in proof. By a variation of the reasoning above we recently established a conjecture analogous to Conjecture 3 but directly claiming a fully homogeneous system. This result would make Proposition 1 superfluous.

Acknowledgements The author received support from Centro de Matemática da Universidade de Coimbra – UID/MAT/00324/2013, funded by the Portuguese Government through FCT/MEC and co-funded by the European Regional Development Fund through the Partnership Agreement PT2020. He also thanks his wife for lending him a computer which digested the documentclass used for this article.

References

1. Cox, D., Little, J., O'Shea, D.: Using Algebraic Geometry. Springer, Heidelberg (1998)
2. Kovačec, A., Gouveia, M.C.: The Hankel pencil conjecture. Linear Algebra Appl. **431**, 1509–1525 (2009)
3. Moreira, M.: Private communication
4. Schmale, W., Sharma, P.K.: Cyclizable matrix pairs over $\mathbb{C}[x]$ and a conjecture on Toeplitz pencils. Linear Algebra Appl. **389**, 33–42 (2004)

Componentwise Products of Totally Non-Negative Matrices Generated by Functions in the Laguerre–Pólya Class

Prashant Batra

Abstract In connection with the characterisation of real polynomials which have exclusively negative zeros Holtz and Tyaglov exposed in 2012 a new, totally non-negative, infinite matrix. This matrix resembles the matrices considered in the stability problem, and was called a matrix of "Hurwitz-type". No precise connection to the Hurwitz matrices of the stability problem or structural properties could be established. We identify those matrices as limits of Hurwitz matrices generated by Hurwitz-stable polynomials. This allows to give a new and concise proof of the Holtz–Tyaglov characterisation as we connect it here to the classical theorem of Aissen, Edrei, Schoenberg and Whitney. Our approach naturally extends to entire functions in the Laguerre–Pólya class which have exclusively non-negative Taylor coefficients. Results on Hurwitz-stable polynomials are employed to show that certain positive pairs of real functions in the Laguerre–Pólya class generate totally non-negative matrices. Finally, we give the first composition result on the structured, infinite matrices considered: We show that the componentwise product of any of the considered infinite matrices is totally non-negative.

Keywords Schur-Hadamard product · Infinite matrices · Aperiodic polynomials · Positive pairs · Hurwitz-stability · Totally positive matrices

1 Introduction

It was shown by Holtz and Tyaglov [9] that the total non-negativity of all minors of the infinite matrix

P. Batra (✉)
Institute for Reliable Computing, Hamburg University of Technology,
21071 Hamburg, Germany
e-mail: batra@tuhh.de

N. Bebiano (ed.), *Applied and Computational Matrix Analysis*,
Springer Proceedings in Mathematics & Statistics 192,
DOI 10.1007/978-3-319-49984-0_11

$$E(f) := \begin{pmatrix} a_0 & a_1 & a_2 & a_3 & a_4 & a_5 & a_6 & \cdots \\ 0 & a_1 & 2a_2 & 3a_3 & 4a_4 & 5a_5 & 6a_6 & \cdots \\ 0 & a_0 & a_1 & a_2 & a_3 & a_4 & a_5 & \cdots \\ 0 & 0 & a_1 & 2a_2 & 3a_3 & 4a_4 & 5a_5 & \cdots \\ 0 & 0 & a_0 & a_1 & a_2 & a_3 & a_4 & \cdots \\ 0 & 0 & 0 & a_1 & 2a_2 & 3a_3 & 4a_4 & \cdots \\ \vdots & \vdots & \vdots & \vdots & \vdots & \vdots & \vdots & \ddots \end{pmatrix} \tag{1}$$

is characteristic for $f(x) = \sum_{k=0}^{d} a_k x^k \in \mathbb{R}[x]$ with $a_d > 0$, $a_0 \neq 0$, to have all roots exclusively on the negative real axis. Thus, a new class of totally non-negative matrices was found in these "infinite matrices of Hurwitz-type", (*cf.* [9], Definitions 1.40 and 1.42, p. 455*f.*). These matrices were considered as somehow related, but not identical with, the classical Hurwitz matrices as considered in connection with the stability problem [12].

We will show in the following that the matrix (1) is the limit of matrices possessing the classical Hurwitz structure (properly defined below *cf.* Definition 2), and which matrices are generated by Hurwitz-stable polynomials or functions. This approach allows us to extend the above characterisation of root-location to real entire functions of low order with sufficiently separated zeros lying exclusively in the open left half-plane, see Theorem 4. The connection to classical results via the Hurwitz matrix-structure facilitates the independent, concise proof of the generalisation Theorem 3 as well as of the result by Holtz and Tyaglov. The interested reader finds a different proof of this generalisation in [5], together with the most general results. Focussing on the Laguerre–Polyá class, we were able to use only simple polynomial tools besides the classical canon of results.

Moreover, our interpretation allows to use a result of Garloff and Wagner [8] on the Hadamard product of real Hurwitz-stable polynomials, and we thus show that the Schur–Hadamard product of matrices $E(f) \circ E(g)$, generated by certain real entire functions f and g with exclusively negative zeros, is totally non-negative, see Proposition 3. We extend the mentioned results to generalised positive pairs of polynomials and their uniform limits in the Laguerre–Pólya class, see Proposition 4 and Theorem 6.

Paper outline: In the following subsection, we collect a number of definitions and facts related to Hurwitz matrices and Hurwitz-stability. In Sect. 3 we use the important result by Aissen *et al.* to prove our Theorem 3 generalising the characterisation of exclusively negative roots via total non-negativity of (1) to entire functions. In Sect. 4 we show that this approach naturally leads to the fact that the Schur–Hadamard product of matrices of the form (1) is totally non-negative. Moreover, we extend these results in Theorem 6 and Proposition 4 to matrices generated from generalised positive pairs and their uniform limits. To achieve this, we use in Sect. 4 classical results on entire functions with zeros exclusively in the upper half-plane $\{z \in \mathbb{C} : \Im z > 0\}$ (*cf.* [13]) in reformulations suitable for entire functions which are the uniform limits of Hurwitz-stable polynomials.

Terminology: By $\mathbb{R}_{>0}$ we denote the set of positive real numbers. An entire function is a complex function analytic everywhere in $\mathbb{C}$. Such functions can be classified using order and genus of the function and the genus of its zeros, for these notions *cf.*, e.g., [4]. Polynomials are entire functions of order and genus zero.

For two entire functions f and g, with Taylor expansions $f(x) = \sum_{i=0}^{\infty} a_i x^i$ and $g(x) = \sum_{i=0}^{\infty} b_i x^i$, we denote by $(f \circ g)(x)$ the power series $\sum_{i=0}^{\infty}(a_i \cdot b_i)x^i$. This power series is everywhere convergent in the complex plane (as a computation of the radius of convergence by the Cauchy–Hadamard formula [1] shows), and we denote the corresponding function by $f \circ g$. The function $f \circ g$ is called the *Hadamard product* of f and g.

For two matrices A and B of identical dimensions, with entries a_{ij} and b_{ij} respectively, we denote by $A \circ B$ the matrix with entries $a_{ij} \cdot b_{ij}$. We call $A \circ B$ the *Schur–Hadamard product* of A and B.

2 Hurwitz-Stability, Hurwitz Matrices and Total Non-Negativity

In this paper, the following definitions regarding stability will be used.

Definition 1 An entire function f is said to be *Hurwitz-stable* if all solutions of $f(z) = 0$ lie in $\{z \in \mathbb{C} : \Re z < 0\}$.

We call here *quasi-stable* any entire function f for which all solutions of $f(z) = 0$ lie in the closed left half-plane $\{z \in \mathbb{C} : \Re z \leq 0\}$.

Many authors have discussed the related questions of root-location on the real axis, real roots of a single definite sign or the question of Hurwitz-stability using expansions at Infinity. (The surveys [9, 12, 16] are no exception and contain references to many more examples.) Especially, normalisations of a polynomial

$$\sum_{i=0}^{d} q_i x^{d-i}$$

would occur for the leading term $q_0 x^d$. But to extend a result naturally from polynomials with only non-negative Taylor coefficients to transcendental entire functions it is more convenient to consider expansions at Zero, not at Infinity. So we choose the (in this context uncommon) expansion near the origin. (And we are but little surprised that this is exactly the type of expansion Hurwitz had used to derive his determinant results *viz.* [10, p. 281 *ff.*]). Moreover, we phrase here Hurwitz' classical stability result for polynomials in terms of an infinite matrix.

Theorem 1 (Hurwitz) *Given a real polynomial p of degree $d \in \mathbb{N}$, positive at the origin ($p(0) > 0$), with even-odd decomposition $p(x) = h(x^2) + xg(x^2)$ into polynomials $h, g \in \mathbb{R}[x]$, and with Taylor expansion*

$$p(x) = h(x^2) + xg(x^2) = \sum_{i=0}^{d} p_i x^i = p_0 + p_1 x + p_2 x^2 + \cdots + p_d x^d \in \mathbb{R}[x].$$

The polynomial p is Hurwitz-stable if and only if the first d consecutive initial principal minors of the infinite matrix

$$H(p) := H(g,h) := \begin{pmatrix} p_1 & p_3 & p_5 & p_7 & \cdots \\ p_0 & p_2 & p_4 & p_6 & \cdots \\ 0 & p_1 & p_3 & p_5 & \cdots \\ 0 & p_0 & p_2 & p_4 & \cdots \\ \vdots & \vdots & \vdots & \vdots & \ddots \end{pmatrix} \tag{2}$$

are positive.

While the preceding formulation of Hurwitz' result might seem unconventional if not unnecessary, the changed set-up allows for a smooth transition to entire functions and totally non-negative infinite matrices as detailed further below.

The following definition of "Hurwitz matrix" may be found in [11, p. 331] or [14, Sect. 4.8, p. 117].

Definition 2 We call *Hurwitz matrix* any finite or infinite matrix $M = (m_{i,j})_{i=1;j=1}^{\omega}$ with entries $m_{i,j}$ generated from a single fixed, finite or infinite ($1 \leq \omega \leq +\infty$) sequence

$$(m_\nu)_{\nu=0}^{\mu},$$

indexed by $\nu \in \mathbb{N}_0$ with $0 \leq \mu \leq +\infty$, such that we have $m_{i,j} = m_{2j-i}$ whenever $0 \leq 2j - i \leq \mu$, and $m_{i,j} = 0$ otherwise.

Thus, the matrix considered in (2) is a Hurwitz matrix. Let us additionally point out the hitherto overlooked fact that the matrix $E(f) = (e_{i,j})$ defined in (1) is a Hurwitz matrix: The matrix can be described as $E(f) = H(f, id \cdot f')$ (where $(id \cdot f')(z) = id(z)f'(z) = zf'(z)$). But the matrix $E(f)$ is not generated by a Hurwitz-stable polynomial. Thus, the following important non-negativity result derived independently first by Asner, and afterwards Kemperman [3, 11] does not apply.

Proposition 1 *A real polynomial p with expansion*

$$p(x) = h(x^2) + xg(x^2) = \sum_{i=0}^{d} p_i x^i, \text{ where } h, g \in \mathbb{R}[x],$$

which is positive at the origin, and which is Hurwitz-stable or quasi-stable, yields a totally non-negative, infinite Hurwitz matrix $H(g, h)$.

Although the preceding proposition cannot be applied directly to the matrix $E(f)$, the matrix $E(f)$ is totally non-negative by the mentioned result of Holtz and Tyaglov [9, Theorem 4.29, p. 503]. We will show in the following that there is a connection between the Hurwitz matrices in (2) and (1), and point out the common source of the total non-negativity of both structures. To this end let us consider in the following pairs of polynomials with exclusively real, interlacing zeros.

2.1 Hurwitz-Stability and Positive Pairs

Hurwitz-stability of a real polynomial $p(x) = h(x^2) + xg(x^2)$ hinges on the following inter-connecting properties of h and g (as we re-call in Proposition 2) *viz.* [7, Sect. 16.14].

Definition 3 Two real, non-zero polynomials h and g constitute a *positive pair* (h, g), if

(i) $\deg(h) \geq \deg(g)$,
(ii) $sign(h^{(\deg h)}(0)) = sign(g^{(\deg g)}(0))$,
(iii) h and g both have exclusively simple, negative roots, denoted by λ_i and γ_i respectively, indexed in decreasing order and which alternate (interlace each other) on the negative real axis beginning with the largest root of h:

$$0 > \lambda_1 > \gamma_1 > \lambda_2 > \gamma_2 > \cdots .$$

A tuple (h, g) of real, non-zero polynomials with exclusively real, non-positive roots λ_i and γ_i which satisfies (i) and (ii) in Definition 3, is called here (comp. [8, p. 799/800]) a *generalised positive pair* if the weak version of (iii) holds true, i.e., for which instead of (iii) above it holds true with root-indexing such that $\lambda_1 \geq \lambda_2 \geq \ldots$, and $\gamma_1 \geq \gamma_2 \geq \ldots$, that

$$0 \geq \lambda_1 \geq \gamma_1 \geq \lambda_2 \geq \gamma_2 \geq \cdots .$$

We have the following connection of (generalised) positive pairs to Hurwitz-stability *cf.* [7], and to quasi-stability *cf.* [8].

Proposition 2 *Two real, non-zero polynomials (h, g) generate a Hurwitz-stable polynomial $p(x) = h(x^2) + xg(x^2)$ if and only if (h, g) constitute a positive pair.*

Two real, non-zero polynomials (h, g) generate a quasi-stable polynomial $p(x) = h(x^2) + xg(x^2)$ if and only if (h, g) constitute a generalised positive pair.

The sign of the Taylor coefficients of a positive pair is not necessarily positive, but with a suitable normalisation the combination of Proposition 2 with Proposition 1 yields the following.

Corollary 1 *A positive, or generalised positive, pair of real polynomials (h, g) such that $h(0) > 0$ generates a totally non-negative infinite Hurwitz matrix $H(g, h)$.*

The polynomial tuple $(f, id \cdot f') \hat{=} (f(x), xf'(x)) \in \mathbb{R}[x] \times \mathbb{R}[x]$ which generates the matrix $E(f) = H(f, id \cdot f')$ is not a generalised positive pair as $f(0) \neq 0$. Thus, the preceding two results cannot directly yield the total non-negativity of the matrix $H(f, id \cdot f') = E(f)$. The total non-negativity of $E(f)$ will be shown to be a consequence of a limiting process. In the next section, total non-negativity of $E(f)$ turns out to be characteristic due to a related characterisation of exclusively negative zeros involving upper triangular Toeplitz matrices.

3 Characterising Exclusively Negative Zeros

Let us change our perspective on the characterisation by Holtz and Tyaglov: Perceiving it as a result on root-location of (rational) entire functions rather than one on total non-negativity of structured matrices, we see that it complements the well-known theorem for meromorphic functions by Aissen, Edrei, Schoenberg and Whitney (for reference, *cf.* [2, p. 306, Theorem 5] or, e.g., [6]) in our polynomial case. Let us spell out the restriction to entire functions of the latter theorem.

Theorem 2 (Aissen et al.) *Let f be an entire function with Taylor expansion $f(x) = \sum_{k=0}^{\infty} a_k x^k$ such that $a_0 > 0$. The function f has exclusively negative zeros and is of the form*

$$f(x) = g(x) \cdot e^{\beta \cdot x}, \ \beta \geq 0, \quad \text{where } g \text{ is a real entire function of genus } 0, \tag{3}$$

if and only if the upper triangular Toeplitz matrix

$$AESW(f) := \begin{pmatrix} a_0 & a_1 & a_2 & a_3 & a_4 & a_5 & a_6 & \cdots \\ 0 & a_0 & a_1 & a_2 & a_3 & a_4 & a_5 & \cdots \\ 0 & 0 & a_0 & a_1 & a_2 & a_3 & a_4 & \cdots \\ \vdots & \vdots & \vdots & \ddots & \ddots & \ddots & \ddots & \ddots \end{pmatrix} = (a_{i,j})_{i,j=0}^{\infty}, \tag{4a}$$

$$\text{where } a_{i,j} := 0 \text{ if } j - i < 0, \text{ and } a_{i,j} := a_{j-i} \text{ otherwise}, \tag{4b}$$

is totally non-negative, i.e., every minor is non-negative.

Viewing the matrix $AESW(f)$ as a sub-matrix of $E(f)$ is now crucial to properly identify the nature of $E(f)$, and to extend the Holtz–Tyaglov result on the total non-negativity of $E(f), f \in \mathbb{R}_{>0}[x]$, to entire functions. Our extension covers naturally those real functions f of the form (3), positive at the origin, which have exclusively negative zeros. These functions make the essential part of the Laguerre–Pólya class $\mathscr{L}\text{-}\mathscr{P}^{+}$ (*cf.* [15]) of real, entire functions with a product expansion of the form

$$cx^m e^{\beta x} \prod_{i=1}^{\infty}(1 + xx_i) \quad \text{with} \quad c, \beta \geq 0, x_i \geq 0, \sum_{i=1}^{\infty} x_i < \infty; \quad m \in \mathbb{N}_0. \tag{5}$$

Regarding this class, we establish here the following characterisation.

Theorem 3 *An entire real function* f *of the form* (3), *with Taylor expansion* $f(x) = \sum_{k=0}^{\infty} a_k x^k$, *and such that* $f(0) = a_0 > 0$, *has exclusively negative zeros if and only if all minors of the matrix* (1), *repeated here as*

$$E(f) = H(f, id \cdot f') = \begin{pmatrix} a_0 & a_1 & a_2 & a_3 & a_4 & a_5 & a_6 & \cdots \\ 0 & a_1 & 2a_2 & 3a_3 & 4a_4 & 5a_5 & 6a_6 & \cdots \\ 0 & a_0 & a_1 & a_2 & a_3 & a_4 & a_5 & \cdots \\ 0 & 0 & a_1 & 2a_2 & 3a_3 & 4a_4 & 5a_5 & \cdots \\ 0 & 0 & a_0 & a_1 & a_2 & a_3 & a_4 & \cdots \\ 0 & 0 & 0 & a_1 & 2a_2 & 3a_3 & 4a_4 & \cdots \\ \vdots & \vdots & \vdots & \vdots & \vdots & \vdots & \vdots & \vdots \end{pmatrix}, \tag{6}$$

are non-negative.

Remark 1 The original result by Holtz/Tyaglov (*cf.* [9], Theorem 4.29, p. 503, as well as p. 423) characterising exclusively negative roots of real polynomials of degree d had the normalisation conditions: $f^{(d)}(0) > 0, f(0) \neq 0$. The normalisation assumptions of our Theorem 3, $f^{(d)}(0) \neq 0, f(0) > 0$, are equivalent to the former conditions in either case of the theorem.

Proof of Theorem 3.
"$\Leftarrow$": If the real, entire function f with $f(0) > 0$ generates a totally non-negative matrix (6), we take the infinite submatrix of (6) consisting of the first, third, fifth etc. rows, and the first, second, third etc. columns. This totally non-negative submatrix is actually the matrix $AESW(f)$ defined in (4a), hence Theorem 2 implies that f has exclusively negative zeros.
"$\Rightarrow$": If f is a positive constant, the claim is trivial. Let f be a real polynomial of degree $d \in \mathbb{N}$ such that $f(0) > 0$, and with exclusively negative zeros, say $\zeta_i, i = 1, \ldots, d$. Then f and f' have positive non-trivial Taylor coefficients. Let us assume first that f has exclusively *simple* zeros. Then the ordered tuple (f, f') is a *positive pair* as the leading coefficients are of the same sign, and by Rolle's theorem the roots of f and f' are negative as well as simple, and interlace each other - beginning with the largest root $\max_{i=1,\ldots,d} \zeta_i = -\min_{i=1,\ldots,d} |\zeta_i|$ of f, the pair's first member.

For $\varepsilon > 0$ chosen such that $\varepsilon < \min\{\min_{i=1,\ldots,d} |\zeta_i|, \min_{i=1,\ldots,d} 1/|\zeta_i|\}$, we define

$$F_\varepsilon(x) := f'(x) \cdot (x + \varepsilon) \cdot (\varepsilon x + 1), \text{ with Taylor expansion, say, } F_\varepsilon(x) = \sum_{i=0}^{d+2} \beta_i x^i.$$

The choice of ε yields that (F_ε, f) is a positive pair. The positive pair (F_ε, f) generates the Hurwitz matrix $H(f, F_\varepsilon)$ which we write as

$$H(f, F_\varepsilon) = \begin{pmatrix} a_0 & a_1 & a_2 & a_3 & a_4 & a_5 & a_6 & \cdots \\ \beta_0 & \beta_1 & \beta_2 & \beta_3 & \beta_4 & \beta_5 & \beta_6 & \cdots \\ 0 & a_0 & a_1 & a_2 & a_3 & a_4 & a_5 & \cdots \\ 0 & \beta_0 & \beta_1 & \beta_2 & \beta_3 & \beta_4 & \beta_5 & \cdots \\ 0 & 0 & a_0 & a_1 & a_2 & a_3 & a_4 & \cdots \\ 0 & 0 & \beta_0 & \beta_1 & \beta_2 & \beta_3 & \beta_4 & \cdots \\ \vdots & \vdots & \vdots & \vdots & \vdots & \vdots & \vdots & \vdots \end{pmatrix}.$$

The matrix $H(f, F_\varepsilon)$ is totally non-negative, by Corollary 1, for all sufficiently small, positive ε, and hence by continuity also for $\varepsilon = 0$. With $\varepsilon \to 0+$, the Taylor coefficients of $F_\varepsilon(x)$ tend to those of $xf'(x)$. Explicitly, we have that

$$\beta_0 = \varepsilon a_1 \to 0,\ \beta_1 = (1+\varepsilon^2)a_1 + \varepsilon 2a_2 \to a_1,$$
$$\beta_k = \varepsilon(k-1)a_{k-1} + (1+\varepsilon^2)ka_k + \varepsilon 2(k+1)a_{k+1} \to ka_k \quad \text{for } k = 2, \ldots, d-1, \text{ and}$$
$$\beta_d = (1+\varepsilon^2)da_d + \varepsilon(d-1)a_{d-1} \to da_d,\ \beta_{d+1} = \varepsilon da_d \to 0.$$

Hence, letting $\varepsilon \to 0$ we obtain $H(f, F_\varepsilon) \to H(f, id \cdot f') = E(f)$, and our claim is proved in this case.

If f is a real polynomial of degree $d \geq 1$, with leading coefficient α, and exclusively negative, pairwise different zeros $\zeta_k, k = 1, \ldots, v$, of arbitrary multiplicity $\mu_k \in$, we consider $\tilde{f}_n(x) := \alpha \prod_{k=1}^{v} \prod_{j=1}^{\mu_k} (x - \zeta_k(1 + \frac{\mu_k - 1}{j \cdot n}))$. The sequence $(\tilde{f}_n)_{n \in \mathbb{N}}$ approximates f (with $f(0) > 0$) uniformly on the unit disc. From the above, we have that the matrices $E(\tilde{f}_n)$ generated by $\tilde{f}_n$ are totally non-negative, and this remains true, by continuity, for $E(f)$.

If f is a transcendental entire, real function of the form (3) with exclusively negative zeros, and such that $f(0) > 0$, then the function f may be written as $f(x) = ce^{\beta x} \prod_{i=1}^{\infty}(1 + xx_i)$ with $\beta \geq 0, c, x_i \geq 0,\ \sum_{i=1}^{\infty} x_i < \infty$, and hence has only positive Taylor coefficients. We have to show that any minor of $E(f)$ is non-negative. Let us consider an arbitrary, but fixed minor which is the determinant of l rows, indexed by $r_1, \ldots, r_l$, and l columns, indexed by $c_1, \ldots, c_l$. Let us denote the thus specified minor by μ. To approximate the minor μ let us define for entire functions g with $g(0) \neq 0$ the minors $M(g)$, determinant of the $l \times l$ submatrix of $E(g)$, composed from the rows $r_1, \ldots, r_l$ and the columns $c_1, \ldots, c_l$. Suppose the minor $\mu = M(f)$ under consideration contains only rows and columns of the first $k+1$ of $E(f)$ (i.e., $\max_{i=1,\ldots,n} \max\{r_i, c_i\} = k+1$). This implies especially that only the first $k+1$ coefficients of $f(x)$ and $xf'(x)$ are involved in the determinant. These coefficients are less in modulus than $m := \max_{i=0,\ldots,k} |(i+1) \cdot a_i|$. We will show for all sufficiently small $\varepsilon > 0$ that $\mu = M(f) \geq -\varepsilon$.

The transcendental function f can be obtained as the uniform limit of the polynomials $f_{2n}(x) = c(1 + \beta x/n)^n \prod_{i=1}^{n}(1 + xx_i) = \sum_{i=0}^{2n} a_i^{(2n)} x^i$ with only positive coefficients (*cf.* [15], p. 96). Let $b := (2^k \max\{1; m^k\})^{-1}$. As $f(0) > 0$, and f is entire, there exists $\varepsilon_0 > 0$ such that there are no negative zeros of f smaller than ε_0 in modulus. Let us now take $\varepsilon > 0$ such that

$$\varepsilon < b \cdot \varepsilon_0.$$

There exists $n = n(m, k, \varepsilon)$ such that the Taylor coefficients $a_i^{(2n)}, i = 0, \ldots, k$, of the functions f_{2n} approximate the corresponding coefficients a_i of f as follows:

$$|a_i - a_i^{(2n)}| \leq \varepsilon/2^{k+1}, \text{ for } i = 0, \ldots, k.$$

Thus, $|M(f) - M(f_{2n})| \leq \varepsilon/2$. As f_{2n} is a polynomial with exclusively negative roots it can be uniformly approximated by polynomials $\tilde{f}_N$ having exclusively simple, negative roots for which $M(\tilde{f}_N) \geq 0$. Choosing $N = N(n, m, k, \varepsilon)$ suitably large, the estimate $|M(f_{2n}) - M(\tilde{f}_N)| \leq \varepsilon/2$ must eventually hold. Thus, we have for all suitably small $\varepsilon > 0$ that $M(f) \geq -\varepsilon$. This yields $M(f) \geq 0$, i.e., non-negativity of the arbitrarily chosen minor $\mu = M(f)$. □

We may use the above technique for positive pairs other than (f, f'), and their *polynomial* limits. We easily obtain the following via our preceding arguments.

Theorem 4 *For a positive pair* $(h, g) \in {}_{>0}[x] \times {}_{>0}[x]$ *the infinite Hurwitz matrix* $H(h, id \cdot g)$ *is totally nonnegative.*

Our new approach allows to discuss in the following section the entrywise product of the nonnegative matrices $E(f_i) = H(f_i, id \cdot f_i'), i = 1, 2,$ and $H(h_i, id \cdot g_i), i = 1, 2,$ considered above.

4 The Entrywise Product of the Considered Matrices

The Cauchy–Binet determinant formula (*cf.*, e.g., [7]) implies that the matrix product of totally non-negative (TNN) matrices is again totally non-negative. For certain classes of structured TNN matrices, we even know that the componentwise product in the class is again a TNN matrix (*cf.*, e.g., [14]).

4.1 The Polynomial Case

In the class of Hurwitz matrices generated by Hurwitz-stable polynomials $p(x) = h(x^2) + xg(x^2) \in {}_{>0}[x]$ we may deduce total non-negativity of the Schur–Hadamard product from a fundamental result of Garloff and Wagner [8, Theorem 3.b].

Theorem 5 (Garloff/Wagner) *For two positive pairs (resp. generalised positive pairs) of polynomials,* (h_1, g_1) *and* (h_2, g_2)*, the componentwise Hadamard product* $(h_1 \circ h_2, g_1 \circ g_2)$ *is a positive pair (resp. generalised positive pair).*

This implies that the Schur–Hadamard product of two totally non-negative Hurwitz matrices $H(g_i, h_i), i = 1, 2,$ generated by generalised positive pairs (h_i, g_i) is itself a totally non-negative Hurwitz matrix generated by a generalised positive pair. From our approximation approach leading to Theorems 3 and 4 we obtain from Theorem 5 the following for the matrices $H(h_i, id \cdot g_i)$.

Proposition 3 *For two quasi-stable real polynomials* f_1 *and* f_2*, both positive at the origin and with even-odd polynomial decomposition* $f_i(x) = h_i(x^2) + xg_i(x^2), i = 1, 2,$ *the matrix*

$$H(h_1, id \cdot g_1) \circ H(h_2, id \cdot g_2) = H(h_1 \circ h_2, id \cdot (g_1 \circ g_2))$$

is totally non-negative. Especially, we obtain for two polynomials $f_i \in \; _{>0}[x], i = 1, 2,$ *with exclusively negative zeros that*

$$E(f_1) \circ E(f_2) = H(f_1, id \cdot f_1') \circ H(f_2, id \cdot f_2') = H(f_1 \circ f_2, id \cdot (f_1' \circ f_2'))$$

is totally non-negative.

Before we begin a discussion of Hurwitz matrices $H(h, id \cdot g)$ generated by transcendental entire functions f, let us gather information on the limit of sequences of Hurwitz-stable polynomials in the following sub-section.

4.2 Function-Theoretic Description and Characterisation of the Limits of Hurwitz-Stable Polynomials

What happens to the matrices H and E discussed in Sect. 3 if we consider a sequence of Hurwitz-stable polynomials f_n with uniform limit $f \not\equiv 0$? Let us first think of the nature of the limit function f. In view of $(1 + x\beta/n)^n \to e^{\beta x}$ for $n \to \infty$, factors like $e^{\beta x}$, $\beta \geq 0$, may appear in the limit function. Thus, the uniform limits of real Hurwitz-stable polynomials with positive coefficients contain the Laguerre–Pólya class $\mathcal{L}\text{-}\mathcal{P}^+$ of entire functions as described above in (5). Discussion of Hurwitz-stable polynomial sequences with non-real root pairs tending towards the imaginary axis shows that a term $e^{\gamma x^2}$, $\gamma \geq 0$, cannot generally be avoided in the limit function. The following description is essentially well-known (but usually formulated for functions with zeros exclusively in $\Im z \geq 0$).

Fact 1 (Comp. [13, Theorem 3, p. 331], and [13, (7.23), p. 318)])
If a sequence (f_n) *of* real *Hurwitz-stable polynomials converges uniformly (on some open, non-empty neighbourhood of the origin) to a function* $f \not\equiv 0$*, it converges uniformly on every bounded domain, and the function* f *is a real entire function of the form*

$$f(x) = cx^q e^{\gamma \cdot x^2 + \beta \cdot x} \prod_{k=1}^{\infty} (1 - xa_k)e^{xa_k}, \tag{7}$$

where $c \in \;, q \in \; _0; \beta, \gamma \geq 0; a_k \in \;, \sum_{k=1}^{\infty} |a_k|^2 < \infty.$

For sake of brevity, we introduce the following definition (compare [13, p. 334]) for entire functions f of order and genus at most $m + 1$.

Definition 4 We say here that an entire function f is of *lifted genus at most m*, if $f(x) = k(x) \cdot e^{\alpha \cdot x^{m+1}}$, $\alpha \geq 0$, where k is an entire function of genus *at most m*.

The Hurwitz-stable entire functions of the form (7), i.e., the uniform limits of Hurwitz-stable polynomials thus are functions of lifted genus at most one. Not all of the latter functions can be obtained as those uniform limits, e.g., e^{-x}. We have the following characterisation of those uniform limits.

Fact 2 (Compare [13, Theorem 4, p. 334*f.* and p. 313])
An entire function f is the uniform limit of Hurwitz-stable polynomials if and only if it is of lifted genus at most 1, with no roots in the right open half-plane and such that

$$|f(z)| \leq |\overline{f(-\overline{z})}| \quad \text{for all } z \text{ with } \Re z < 0 \tag{8}$$

holds true.

For *real* entire functions with expansion (7) and roots lying exclusively in the left half-plane the condition (8) obviously holds true.

4.3 Schur–Hadamard Matrix Product Arising from Hurwitz-Stable Transcendental Functions

If the real entire function $f \not\equiv 0$ is the uniform limit (on every bounded domain) of a sequence of real Hurwitz-stable polynomials f_n with positive Taylor coefficients and even-odd polynomial decomposition $f_n(x) = h_n(x^2) + xg_n(x^2)$, we have the uniform approximations

$$h_n(x^2) \to \frac{f(x) + f(-x)}{2} =: f^e(x^2), \text{ and}$$
$$g_n(x^2) \to \frac{f(x) - f(-x)}{2x} =: f^o(x^2).$$

As the (h_n, g_n) are positive pairs with exclusively positive non-trivial Taylor coefficients, the Hurwitz matrices $H(h_n, id \cdot g_n)$ are totally non-negative by Theorem 4. Considering individual minors as in the proof of Theorem 3 we see that this property transfers minor-wise to the matrix generated by the limit functions.

Proposition 4 *For a real, Hurwitz-stable entire function f, positive at the origin, with product representation (7) and even-odd decomposition $f(x) = f^e(x^2) + xf^o(x^2)$, where f^e and f^o are real entire functions, the Hurwitz matrix*

$$H(f^e, id \cdot f^o) \tag{9}$$

is totally non-negative.

The Schur–Hadamard product of Hurwitz matrices $H(f_i^e, id \cdot f_i^o)$, $i = 1, 2$, as considered in (9), generated by Hurwitz-stable f_i of the form (7), essentially inherits totally non-negativity from the positive pairs uniformly approximating f_i as we will see next.

Theorem 6 *Given two real, Hurwitz-stable entire functions f_1, f_2 such that $f_1(0) > 0$ and $f_2(0) > 0$, of the form (7), and with even-odd decomposition $f_i(x) = f_i^e(x^2) + xf_i^o(x^2)$, where f_i^e, f_i^o are real entire functions.*

The matrices $H(f_1^e, id \cdot f_1^o)$ and $H(f_2^e, id \cdot f_2^o)$ as well as their product $H(f_1^e, id \cdot f_1^o) \circ H(f_2^e, id \cdot f_2^o) = H(f_1^e \circ f_2^e, id \cdot (f_1^o \circ f_2^o))$ are totally non-negative.

Proof The real functions f_i are Hurwitz-stable, of lifted genus at most one, and of the form (7). Hence, condition (8) holds true. Thus, the functions f_i can by approximated uniformly by Hurwitz-stable polynomials $f_n^{[i]}$, $n \in$, with even-odd polynomial decomposition given by $f_n^{[i]}(x) = h_n^{[i]}(x^2) + xg_n^{[i]}(x^2)$.

By Theorem 4, a positive pair (h_n, g_n) with non-negative, real coefficients generates a totally non-negative, infinite Hurwitz matrix $H(h_n, id \cdot g_n)$. Thus, by Theorem 4 the matrices $H_n^{[i]} := H(h_n^{[i]}, id \cdot g_n^{[i]})$ are totally non-negative. By Proposition 3 and Theorem 5, the product $\pi_n := H(h_n^{[1]}, id \cdot g_n^{[1]}) \circ H(h_n^{[2]}, id \cdot g_n^{[2]}) = H(h_n^{[1]} \circ h_n^{[2]}, id \cdot (g_n^{[1]} \circ g_n^{[2]}))$ is a Hurwitz matrix generated from the limit of positive pairs, and hence totally non-negative.

The total non-negativity of the minors of the matrices $H_n^{[i]}$, $i = 1, 2$, and of the minors of their Schur–Hadamard product π_n transfers to their limits as in the proof of Theorem 3. □

Regarding function pairs with common zeros we consider here merely pairs (f, f'), where $f \in \mathscr{L}\text{-}\mathscr{P}^+$ is positive at the origin, i.e., f is a real entire function of lifted genus at most 0 with exclusively negative zeros and exclusively positive (non-trivial) Taylor coefficients. From the proof of Theorem 3 we obtain *mutatis mutandis* the following.

Theorem 7 *Given two real entire functions $f_1, f_2 \in \mathscr{L}\text{-}\mathscr{P}^+$ which are both positive at the origin, the Schur–Hadamard product $E(f_1) \circ E(f_2)$ of the two matrices $E(f_1) = H(f_1, id \cdot f_1')$, $E(f_2) = H(f_2, id \cdot f_2')$ (generated according to* (1)) *is totally non-negative, and we have*

$$E(f_1) \circ E(f_2) = H(f_1 \circ f_2, id \cdot (f_1' \circ f_2')).$$

References

1. Ahlfors, L.V.: Complex Analysis, 3rd edn. McGraw-Hill Inc., New York (1979)
2. Aissen, M., Edrei, A., Schoenberg, I.J., Whitney, A.: On the generating functions of totally positive sequences. Proc. Natl. Acad. Sci. U. S. A. **37**, 303–307 (1951)
3. Asner, B.A.: On the total nonnegativity of the Hurwitz matrix. SIAM J. Appl. Math. **18**, 407–414 (1970)
4. Boas, R.P.: Entire Functions. Academic Press Inc., New York (1954)
5. Dyachenko, A.: Total nonnegativity of infinite Hurwitz matrices of entire and meromorphic functions. Complex Anal. Oper. Theory **8**(5), 1097–1127 (2014)
6. Fallat, S.M., Johnson, C.R.: Totally Nonnegative Matrices. Princeton University Press, Princeton (2011)
7. Gantmacher, F.: Matrizentheorie. Springer, Berlin (1986)
8. Garloff, J., Wagner, D.G.: Hadamard products of stable polynomials are stable. J. Math. Anal. Appl. **202**, 797–809 (1996)
9. Holtz, O., Tyaglov, M.: Structured matrices, continued fractions, and root localization of polynomials. SIAM Rev. **54**(3), 421–509 (2012)
10. Hurwitz, A.: Über die Bedingungen, unter welchen eine Gleichung nur Wurzeln mit negativen reellen Theilen besitzt. Math. Ann. **46**, 273–284 (1895)
11. Kemperman, J.H.B.: A Hurwitz matrix is totally positive. SIAM J. Math. Anal. **13**, 331–341 (1982)
12. Krein, M.G., Naimark, M.A.: The method of symmetric and hermitian forms in the theory of the separation of the roots of algebraic equations. Linear Multilinear Algebra **10**, 265–308 (1981)
13. Levin, B.J.: Distribution of Zeros of Entire Functions, Translation of Mathematical Monographs, vol. 5 (AMS, Providence, Rhode Island, 1980), 2nd, revised edition

14. Pinkus, A.: Totally Positive Matrices, 2nd edn. AMS, Rhode Island (1980)
15. Pólya, G., Schur, I.: Über zwei Arten von Faktorenfolgen in der Theorie der algebraischen Gleichungen. J. Reine Angew. Math. **144**, 89–113 (1914)
16. Rahman, Q.I., Schmeisser, G.: Analytic Theory of Polynomials. Oxford University Press, Oxford (2002)

Fields of Values of Linear Pencils and Spectral Inclusion Regions

Natália Bebiano, João da Providência, Ana Nata and João P. da Providência

Abstract We propose efficient methods for the numerical approximation of the field of values of the linear pencil $A - \lambda B$, when one of the matrix coefficients A or B is Hermitian and $\lambda \in \mathbb{C}$. Our approach builds on the fact that the field of values can be reduced under compressions to the bidimensional case, for which these sets can be exactly determined. The presented algorithms hold for matrices both of small and large size. Furthermore, we investigate spectral inclusion regions for the pencil based on certain fields of values. The results are illustrated by numerical examples. We point out that the given procedures complement the known ones in the literature.

Keywords Field of values · Linear pencil · Selfadjoint linear pencil

1 Introduction

Consider the linear pencil $A - \lambda B$, where A and B are $n \times n$ complex matrices and $\lambda \in \mathbb{C}$. The study of linear pencils has a rich and long history that goes back to Weierstrass and Kronecker in the nineteenth century, usually in the context of their spectral analysis. A complex number λ is said to be an eigenvalue of the pencil if there exists a nonzero $x \in \mathbb{C}^n$ such that

N. Bebiano (✉)
Department of Mathematics, CMUC, University of Coimbra, 3001-454 Coimbra, Portugal
e-mail: bebiano@mat.uc.pt

J. da Providência
Department of Physics, CFisUC, University of Coimbra, 3004-516 Coimbra, Portugal
e-mail: providencia@teor.fis.uc.pt

A. Nata
Department of Mathematics, CMUC, Polytechnic Institute of Tomar, 2300-313 Tomar, Portugal
e-mail: anata@ipt.pt

J.P. da Providência
Department of Physics, University of Beira Interior, 6201-001 Covilhã, Portugal
e-mail: joaodaprovidencia@daad-alumni.de

N. Bebiano (ed.), *Applied and Computational Matrix Analysis*,
Springer Proceedings in Mathematics & Statistics 192,
DOI 10.1007/978-3-319-49984-0_12

$$Ax = \lambda Bx. \tag{1}$$

The vector x is called an eigenvector of the pencil corresponding to the eigenvalue λ. The set of all eigenvalues is known as the spectrum of $A - \lambda B$ and denoted by $\sigma(A, B)$.

In the present work we are particularly interested in the numerical computation of certain fields of values, that are spectral inclusion regions for linear pencils. Motivations to investigate this problem come from stability theory and from the study of certain over-damped vibration systems, e.g. see [6].

The field of values of a linear pencil is denoted and defined as

$$W(A, B) = \{\lambda \in \mathbb{C} : x^*(A - \lambda B)x = 0, \quad \text{for some } 0 \neq x \in \mathbb{C}^n\}, \tag{2}$$

(cf. [8, 10, 12]). The set (2) does not contain the point at infinity. If B is singular, then $\sigma(A, B)$ may have an infinite eigenvalue. Therefore, from the above definition, $W(A, B)$ is not necessarily a spectral inclusion region for the generalized eigenvalue problem (1). So, we consider a slightly modified definition: if A, B have a common null space, then $W(A, B) = \mathbb{C} \cup \{\infty\}$; otherwise

$$W(A, B) = \left\{ \frac{x^*Ax}{x^*Bx} : 0 \neq x \in \mathbb{C}^n \right\}, \tag{3}$$

where $1/0$ is understood as the point at infinity. When $B = I$, (3) reduces to the classical field of values of the $n \times n$ matrix A,

$$W(A) = \{x^*Ax : x \in \mathbb{C}^n, \ \|x\| = 1\},$$

where $\|x\| = \langle x, x\rangle^{1/2} = (x^*x)^{1/2}$ is the usual Euclidean norm in $\mathbb{C}^n$. This concept has been extensively investigated; see, for instance, [5, 7].

Psarrakos [12] investigated the problem of the numerical computation of $W(A, B)$, when one of the coefficients A or B is Hermitian. His approach uses the algorithm of Li and Rodman [9] to compute boundary points (u, v, w) of the so-called joint numerical range

$$JNR(B, H, S) = \{(x^*Bx, x^*Hx, x^*Sx) : x \in \mathbb{C}^n, \ \text{with } x^*x = 1\},$$

where $A = H + iS$ and H and S are Hermitian. Given a point (u, v, w) of $JNR(B, H, S)$ the solutions of the equations $u\lambda + v + iw = 0$ $(u \neq 0)$ are points of $W(A, B)$. Psarrakos method performs specially well for matrices of small size. So, for large matrices, there is place for improvement and this is one of our main concerns. Our second goal is to obtain eigenvalue inclusion regions for matrix linear pencils, based on fields of values.

If B is Hermitian positive definite, we clearly have $W(A, B) = W\left(B^{-1/2}AB^{-1/2}\right)$ and due to the convexity of the classical field of values (stated by the Toeplitz-Hausdorff Theorem [4]), $W(A, B)$ is a convex set. However, $W(A, B)$ is not always convex and not even bounded or connected [8]. If $0 \in W(B)$, then $W(A, B)$ is unbounded and consequently this set is not an informative spectral inclusion region for the pencil. This motivated the investigation of other inclusion regions of field of values type. If B is nonsingular, the spectrum of $B^{-1}A$ coincides with that of the pencil $A - \lambda B$. Henceforth, $W(B^{-1}A)$ and $W(AB^{-1})$ are inclusion regions for the eigenvalues of (1). Interchanging the roles of A and B and considering the generalized eigenvalue problem $Bx = \lambda^{-1}Ax$, the sets $1/W(A^{-1}B)$ and $1/W(BA^{-1})$, for nonsingular A, are also inclusion regions for (1). Division is interpreted elementwise.

The paper is organized as follows. In Sect. 2 we characterize the field of values of selfadjoint linear pencils, i.e., with Hermitian matrices as coefficients. In Sect. 3, auxiliary background is presented. In Sect. 4 we give a method to approximate $W(A, B)$ for Hermitian positive semi-definite B. In Sect. 5, a procedure to numerically approximate $W(A, B)$ for indefinite invertible B is presented, based on the connection of this set with the Krein space field of values. Finally, in Sect. 6, some conclusions are included. A few illustrative examples are provided. All images were computed numerically using MATLAB.

The key idea behind the algorithms here proposed is the following: we use subspace projection methods, a line of attack exploited by Hochstenbach in [6], stressing the fact that the field of values is often well approximated from a low dimension Krylov space. Our attempts are in this vein, and in summary, their advantages over the existing ones are that we perform projections on bidimensional spaces, in which case the fields of values are easily and exactly determined.

2 Selfadjoint Linear Pencils

In the sequel, M_n denotes the algebra of $n \times n$ complex matrices. If the matrices A and B have a common nonzero isotropic vector, i.e., $x^*Ax = 0$ and $x^*Bx = 0$, then $W(A, B) = \mathbb{C}$. To avoid this situation, we assume that A and B do not have a common isotropic vector and so $W(A, B) \neq \mathbb{C}$. For A and B Hermitian, we define

$$\sigma^+(A, B) = \{\lambda \in \mathbb{C} : Au - \lambda Bu = 0,\ 0 \neq u \in \mathbb{C}^n,\ u^*Bu > 0\},$$

$$\sigma^-(A, B) = \{\lambda \in \mathbb{C} : Au - \lambda Bu = 0,\ 0 \neq u \in \mathbb{C}^n,\ u^*Bu < 0\}.$$

The shape of $W(A, B)$ when A and B are Hermitian is described in Theorem 4.1 of [8]. The statement of this theorem is not correct, and is incorrectly reproduced in [12, Theorem 9]. Next we present the proper result and proof.

Theorem 1 *Let $A - \lambda B$ be a $n \times n$ self-adjoint pencil with $W(A, B) \neq \mathbb{C}$.*

(a) If B is positive or negative definite, then $W(A, B)$ is a closed interval in $\mathbb{R}$.
(b) If B is positive (or negative) semi-definite, then $W(A, B)$ is an unbounded interval of the form $[a, +\infty[$ or $]-\infty, a]$.

(c) If B is indefinite and A is positive (negative) definite, then $W(A, B)$ is the union of 2 disjoint unbounded intervals and $0 \notin W(A, B)$.

(d) If B is indefinite and A is semi-definite positive (or negative), then one of the following holds

(1) $W(A, B) =]-\infty, a] \cup [0, +\infty[$ with $a < 0$,

(2) $W(A, B) =]-\infty, 0] \cup [b, +\infty[$ with $0 < b$.

(e) If both B and A are indefinite, then two possibilities may occur:

(1) $W(A, B) =]-\infty, a] \cup [b, +\infty[$, with $0 \in W(A, B)$ and $a < b$.

(2) $W(A, B) = \mathbb{R}$.

In all cases, the endpoints of the intervals are eigenvalues of the pencil.

Proof (a) Let $B = \text{diag}\,(\beta_1, \ldots, \beta_n)$ with all β's positive and let $\sigma(A, B) = \{\alpha_1, \ldots \alpha_n\}$, $\alpha_1 \geq \cdots \geq \alpha_n$. There exists a non-singular matrix T such that

$$T^*AT = \text{diag}\,(\alpha_1\beta_1, \ldots, \alpha_n\beta_n), \quad T^*BT = B.$$

Let $v = \sum_{i=1}^n \gamma_i e_i$, where $\gamma_i \in \mathbb{C}$ and e_i is the column vector with 1 in place i and 0 everywhere else. Then, we have

$$v^*T^*ATv = \sum_{i=1}^n |\gamma_i|^2 \alpha_i \beta_i, \quad v^*Bv = \sum_{i=1}^n |\gamma_i|^2 \beta_i,$$

and so

$$\frac{v^*T^*ATv}{v^*Bv} = \frac{\sum_{i=1}^n |\gamma_i|^2 \alpha_i \beta_i}{\sum_{i=1}^n |\gamma_i|^2 \beta_i}.$$

Hence,

$$\alpha_1 \geq \frac{v^*T^*ATv}{v^*Bv} \geq \alpha_n,$$

and consequently

$$W(A, B) = [\alpha_n, \alpha_1].$$

(b) If B is positive semi-definite with rank r, we can take $B = \text{diag}\,(\beta_1, \ldots, \beta_r, 0, \ldots, 0)$. Let $\sigma(A, B) = \{\alpha_1, \ldots \alpha_r\}$, $\alpha_1, \ldots \alpha_r \in \mathbb{R}$. There exists a non-singular matrix T such that

$$T^*AT = \text{diag}\,(\alpha_1\beta_1, \ldots, \alpha_r\beta_r, \alpha_{r+1}, \ldots, \alpha_n), \quad \alpha_{r+1}, \ldots, \alpha_n \in \mathbb{R}, \quad T^*BT = B.$$

Since $W(A, B) \neq \mathbb{C}$, the eigenvalues $\alpha_{r+1}, \ldots, \alpha_n$ are non-vanishing and have all the same sign. In fact, suppose $\alpha_{r+1} > 0$, $\alpha_{r+2} < 0$. We may choose $v = \gamma e_{r+1} + \delta e_{r+2}$, $\gamma, \delta \in \mathbb{C}$, such that

$$v^*T^*ATv = |\gamma|^2\alpha_{r+1} + |\delta|^2\alpha_{r+2} = 0,$$

which is impossible because, by hypothesis, A and B do not have common isotropic eigenvectors. Let $v = \sum_{i=1}^{n} \gamma_i e_i$, $\gamma_i \in \mathbb{C}$. Then, assuming that $\alpha_{r+1} \geq \cdots \geq \alpha_n > 0$, we get

$$v^*T^*ATv = \sum_{i=1}^{r} |\gamma_i|^2\alpha_i\beta_i + \sum_{i=r+1}^{n} |\gamma_i|^2\alpha_i \geq \sum_{i=1}^{r} |\gamma_i|^2\alpha_i\beta_i, \quad v^*Bv = \sum_{i=1}^{r} |\gamma_i|^2\beta_i,$$

and so we obtain

$$\frac{v^*T^*ATv}{v^*Bv} \geq \frac{\sum_{i=1}^{r} |\gamma_i|^2\alpha_i\beta_i}{\sum_{i=1}^{r} |\gamma_i|^2\beta_i} \geq \alpha_r.$$

On the other hand, if $0 > \alpha_{r+1} \geq \cdots \geq \alpha_n$, we find

$$\frac{v^*T^*ATv}{v^*Bv} \leq \alpha_1.$$

(c) Let B be indefinite with inertia $(r, n-r)$. Let $1/\alpha > 0 > 1/\beta$ be the largest and smallest eigenvalue of the pencil (B, A), so that $W(B, A) = [1/\beta, 1/\alpha]$. Since $W(A, B) = 1/W(B, A)$, we have

$$W(A, B) =]-\infty, \beta] \cup [\alpha, +\infty[,$$

and $0 \notin W(A, B)$.

(d) Similar to (c).

(e) Let A be indefinite and let B have inertia $(r, n-r)$. As before, B may be taken of the form

$$B = \operatorname{diag}(\beta_1, \ldots, \beta_n), \quad \beta_1 \geq \cdots \geq \beta_r > 0 > \beta_{r+1} \geq \cdots \geq \beta_n.$$

According to hypothesis $W(A, B) \neq \mathbb{C}$, the eigenvalues of the pencil (A, B) are all real and the associated eigenvectors are non-isotropic. Let us consider the matrix $T = (u_1, \ldots, u_n)$ where $u_1, \ldots, u_n$ are B-orthogonal eigenvectors of the pencil. Clearly, the number of columns with positive B-norm is r and the number of columns with negative B-norm is $n-r$. This matrix is non-singular and may be chosen so that

$$T^*AT = \operatorname{diag}(\alpha_1\beta_1, \ldots, \alpha_n\beta_n), \quad T^*BT = B.$$

We further assume that $\sigma^+(A, B) = \{\alpha_1, \ldots, \alpha_r\}$, $\alpha_1 \geq \cdots \geq \alpha_r$, and $\sigma^-(A, B) = \{\alpha_{r+1}, \ldots, \alpha_n\}$, $\alpha_{r+1} \geq \cdots \geq \alpha_n$. For $v = \sum_{i=1}^{n} \gamma_i e_i$, with $\gamma_i \in \mathbb{C}$, we find

$$v^*T^*ATv = \sum_{i=1}^{n} |\gamma_i|^2\alpha_i\beta_i, \quad v^*Bv = \sum_{i=1}^{n} |\gamma_i|^2\beta_i.$$

Let $z = v^*T^*ATv/v^*Bv \in W(A, B)$, so we may write

$$z = \frac{ap - bq}{p - q},$$

where

$$a = \frac{\sum_{i=1}^{r} |\gamma_i|^2 \alpha_i \beta_i}{\sum_{i=1}^{r} |\gamma_i|^2 \beta_i}, \quad b = \frac{\sum_{i=r+1}^{n} |\gamma_i|^2 \alpha_i \beta_i}{\sum_{i=r+1}^{s} |\gamma_i|^2 \beta_i}, \quad p = \sum_{i=1}^{r} |\gamma_i|^2 \beta_i, \quad q = -\sum_{i=r+1}^{s} |\gamma_i|^2 \beta_i.$$

Thus, $a \in [\alpha_r, \alpha_1]$ and $b \in [\alpha_{r+1}, \alpha_n]$. Moreover, $z \in]-\infty, a] \cup [b, +\infty[$ if $a < b$ and $z \in]-\infty, b] \cup [a, +\infty[$ if $a > b$. If $\alpha_r > \alpha_{r+1}$ it follows that $]-\infty, b] \cup [a, +\infty] \subseteq]-\infty, \alpha_{r+1}] \cup [\alpha_r, +\infty[$, so that α_r is the lowest value which z may assume if $p > q$, while α_{r+1} is the highest value which z may assume if $p < q$. Thus,

$$W(A, B) =]-\infty, \alpha_{r+1}] \cup [\alpha_r, +\infty[.$$

It may be seen that $\alpha_r \alpha_{r+1} > 0$ so that $0 \in W(A, B)$.

If $\alpha_n > \alpha_1$ it follows that $]-\infty, a] \cup [b, +\infty[\subseteq]-\infty, \alpha_1] \cup [\alpha_n, +\infty[$, and so

$$W(A, B) =]-\infty, \alpha_1] \cup [\alpha_n, +\infty[\,,$$

with $\alpha_n \alpha_1 > 0$ and so $0 \in W(A, B)$.

If $\alpha_n < \alpha_1$ and $\alpha_r < \alpha_{r+1}$ it follows that $W(A, B) = \mathbb{R}$.

Next, let A be indefinite and let B have inertia $(r, s - r, n - s)$. We may consider

$$B = \operatorname{diag}(\beta_1, \ldots, \beta_s, 0, \ldots, 0), \quad \beta_1 \geq \cdots \geq \beta_r > 0 > \beta_{r+1} \geq \cdots \geq \beta_s.$$

Since by hypothesis $W(A, B) \neq \mathbb{C}$, the eigenvalues of the pencil (A, B) are all real and the associated eigenvectors are non-isotropic. We may assume that $\sigma^+(A, B) = \{\alpha_1, \ldots, \alpha_r\}$, $\alpha_1 \geq \cdots \geq \alpha_r$, $\sigma^-(A, B) = \{\alpha_{r+1}, \ldots, \alpha_s\}$, $\alpha_{r+1} \geq \cdots \geq \alpha_n$, so that there exists a non-singular T such that

$$T^*AT = \operatorname{diag}(\alpha_1 \beta_1, \ldots, \alpha_s \beta_s, \alpha_{s+1}, \ldots, \alpha_n), \quad \alpha_{s+1}, \ldots, \alpha_n \in \mathbb{R}, \quad T^*BT = B.$$

Indeed, we consider the matrix $T = (u_1, \ldots, u_s, u_{s+1}, \ldots, u_n)$ where $u_1, \ldots, u_s$ are B-orthogonal eigenvectors of the pencil and $u_{s+1}, \ldots, u_n$ are eigenvectors of the projection of A to the eigenspace of B associated with the eigenvalue 0. Moreover, the eigenvalues $\alpha_{s+1}, \ldots, \alpha_n$ are non-vanishing, since $W(A, B) \neq \mathbb{C}$, and all them have the same sign.

For $v = \sum_{i=1}^{n} \gamma_i e_i$, $\gamma_i \in \mathbb{C}$, we find

$$v^*T^*ATv = \sum_{i=1}^{s} |\gamma_i|^2 \alpha_i \beta_i + \sum_{i=s+1}^{n} |\gamma_i|^2 \alpha_i, \quad v^*Bv = \sum_{i=1}^{n} |\gamma_i|^2 \beta_i.$$

Let $z = v^*T^*ATv/v^*Bv \in W(A, B)$. We may write

$$z = \frac{ap - bq}{p - q} + t,$$

where

$$a = \frac{\sum_{i=1}^{r} |\gamma_i|^2 \alpha_i \beta_i}{\sum_{i=1}^{r} |\gamma_i|^2 \beta_i}, \quad b = \frac{\sum_{i=r+1}^{n} |\gamma_i|^2 \alpha_i \beta_i}{\sum_{i=r+1}^{s} |\gamma_i|^2 \beta_i}, \quad p = \sum_{i=1}^{r} |\gamma_i|^2 \beta_i, \quad q = -\sum_{i=r+1}^{s} |\gamma_i|^2 \beta_i,$$

$$t = \sum_{i=s+1}^{n} \frac{|\gamma_i|^2}{p - q} \alpha_i.$$

Thus, $a \in [\alpha_r, \alpha_1]$ and $b \in [\alpha_{r+1}, \alpha_n]$. Furthermore, $z \in] - \infty, a] \cup [b, +\infty[+t$ if $a < b$, while $z \in] - \infty, b] \cup [a, +\infty[+t$ if $a > b$.

Let $\alpha_r > \alpha_{r+1}$ and $\alpha_i > 0,\ i = s + 1, \ldots, n$. Then, $a > b$ and $t > 0$ if $p > q$, while $t < 0$ if $p < q$. Since $(ap - bq)/(p - q) \in] - \infty, b] \cup [a, +\infty[$, $a \geq \alpha_r$ and $b \leq \alpha_{r+1}$, it follows that

$$z \in] - \infty, b] \cup [a, +\infty[\subseteq] - \infty, \alpha_{r+1}] \cup [\alpha_r, +\infty[,$$

so that α_r is the lowest value which z may assume if $p > q$. Indeed, $z = \alpha_r$ if and only if all the γ_i vanish except γ_r. On the other hand, α_{r+1} is the highest value which z may assume if $p < q$ and $z = \alpha_{r+1}$ if and only if all the γ_i vanish except γ_{r+1}. Thus,

$$W(A, B) =] - \infty, \alpha_{r+1}] \cup [\alpha_r, +\infty[.$$

As $\alpha_r\alpha_{r+1} \geq 0$, $\alpha_r \neq \alpha_{r+1}$, we have $0 \in W(A, B)$.

Similarly, if $\alpha_n > \alpha_1$ and $\alpha_i < 0,\ i = s + 1, \ldots, n$, we get

$$z \in] - \infty, a] \cup [b, +\infty] \subseteq] - \infty, \alpha_1] \cup [\alpha_n, +\infty[,$$

so that

$$W(A, B) =] - \infty, \alpha_1] \cup [\alpha_n, +\infty[,$$

with $0 \in W(A, B)$.

If $\alpha_n < \alpha_1$ and $\alpha_r < \alpha_{r+1}$ it follows that $W(A, B) = \mathbb{R}$.

If $\alpha_r > \alpha_{r+1}$ and $\alpha_i < 0,\ i = s + 1, \ldots, n$, or if $\alpha_n > \alpha_1$ and $\alpha_i > 0,\ i = s + 1, \ldots, n$, we may also conclude that $W(A, B) = \mathbb{R}$. □

3 Background

3.1 Compression of $W(A,B)$

The classical field of values may be characterized as a union of elliptical disks. This result is many times referred as the Marcus–Pesce Theorem [11], although it was already known long before. In the following, we recall the standard compression of $W(A,B)$ into fields of values of 2×2 pencils [3].

Theorem 2 (Chien and Nakazato) *For any $A, B \in M_n$,*

$$W(A,B) = \bigcup_{u,v} W(A_{uv}, B_{uv}),$$

where u and v vary over all pairs of orthonormal vectors in $\mathbb{C}^n$ and

$$A_{uv} = \begin{bmatrix} \langle Au, u\rangle & \langle Av, u\rangle \\ \langle Au, v\rangle & \langle Av, v\rangle \end{bmatrix}, \quad B_{uv} = \begin{bmatrix} \langle Bu, u\rangle & \langle Bv, u\rangle \\ \langle Bu, v\rangle & \langle Bv, v\rangle \end{bmatrix}. \tag{4}$$

When B is Hermitian positive definite, then also B_{uv} is Hermitian positive definite, because it is a principal submatrix of a positive definite matrix. If B is indefinite, B_{uv} may be definite or indefinite. The field of values $W(A,B)$ in the 2 by 2 case, can be easily drawn from the entries of the matrices according to the Elliptical Range Theorem, the Hyperbolical Range Theorem, and the Parabolical Range Theorem (cf. [2, Sect. 2]).

3.2 Connection of $W(A,B)$ with the Krein Space Field of Values for B Indefinite

There is an interesting relation of $W(A,B)$ when B is indefinite Hermitian, with the Krein space field of values [1]. Indeed, suppose that B is an $n \times n$ indefinite Hermitian matrix with inertia $(r, n-r)$. Consider $\mathbb{C}^n$ endowed with indefinite inner product $[x,y] = y^*Bx,\ x,y \in \mathbb{C}^n$. The Krein space field of values of $A \in M_n$ is defined by

$$W_B(A) = \left\{ \frac{[Aw,w]}{[w,w]} : w \in \mathbb{C}^n,\ [w,w] \neq 0 \right\}.$$

We easily find the connection of $W_B(A)$ with the field of values of the pencil $A - \lambda B$. Indeed, we easily get

$$W_B(A) = W(BA, B) = \left\{ \frac{\langle BAw, w\rangle}{\langle Bw, w\rangle} : w \in \mathbb{C}^n,\ \langle Bw, w\rangle \neq 0 \right\},$$

and so $W(A,B) = W_B(B^{-1}A)$.

4 Approximation of $W(A, B)$ for B Positive Semidefinite

Throughout, for $A \in M_n$, we consider the Cartesian decomposition $A = H(A) + iK(A)$ where $H(A) = (A + A^*)/2$ and $K(A) = (A - A^*)/(2i)$ are Hermitian.

4.1 Algorithm 1

Input: A matrix $A \in M_n$, a Hermitian positive semidefinite matrix B and m angles.
Output: An approximation for $W(A, B)$.

1. Set $\theta_k = (k - 1)\pi/m$, $k = 1, \ldots, m + 1$ for some positive integer $m \geq 3$.
2. Starting with $k = 1$ and up to $k = m$, take the following steps:

- (i) Compute an eigenvector u_k associated to $\lambda_{\min}(H\left(\mathrm{e}^{-i\theta_k}A\right) - \lambda B)$, if $W(H\left(\mathrm{e}^{-i\theta_k}A\right), B) = [a, +\infty[$ (to $\lambda_{\max}(H\left(\mathrm{e}^{-i\theta_k}A\right) - \lambda B)$, if $W(H\left(\mathrm{e}^{-i\theta_k}A\right), B) =]-\infty, a]$).
- (ii) Compute the compressions of A and B to $\mathrm{span}\{u_k, u_{k+1}\}$, denoted by $A_{\tilde{u}_k\tilde{u}_{k+1}}$ and $B_{\tilde{u}_k\tilde{u}_{k+1}}$.
- (iii) Compute and draw the boundary of $W(A_{\tilde{u}_k\tilde{u}_{k+1}}, B_{\tilde{u}_k\tilde{u}_{k+1}})$ denoted by Γ_k.
- (iv) If $k < m$, take next k value and return to (i). Otherwise, continue.

3. Take the convex-hull of the collection of curves $\Gamma_1, \ldots, \Gamma_m$, as an approximation for $W(A, B)$.

According to the Elliptical and the Parabolical Range Theorems, the collection of curves in Step 3 of Algorithm 1 is constituted by ellipses and parabolas.

4.2 Approximation of $W(A, B)$ for B Positive Definite

Algorithm 1 may be applied when B is positive definite with the following replacements of Sub-steps (i), (ii), (iii) of Step 2:

- (i) Compute eigenvectors u_k and v_k associated, respectively, to $\lambda_{\min}(H\left(\mathrm{e}^{-i\theta_k}A\right) - \lambda B)$ and $\lambda_{\max}(H\left(\mathrm{e}^{-i\theta_k}A\right) - \lambda B)$.
- (ii) Compute the compressions of A to $\mathrm{span}\{u_k, u_{k+1}\}$ and $\mathrm{span}\{v_k, v_{k-1}\}$, denoted by $A_{\tilde{u}_k\tilde{u}_{k+1}}$ and $A_{\tilde{v}_k\tilde{v}_{k-1}}$, and do the same for B, notation: $B_{\tilde{u}_k\tilde{u}_{k+1}}$ and $B_{\tilde{v}_k\tilde{v}_{k-1}}$.
- (iii) Compute and draw the boundary of $W(A_{\tilde{u}_k\tilde{u}_{k+1}}, B_{\tilde{u}_k\tilde{u}_{k+1}})$ denoted by Γ_k and the boundary of $W(A_{\tilde{v}_k\tilde{v}_{k-1}}, B_{\tilde{v}_k\tilde{v}_{k-1}})$ denoted by Λ_k,

and the following replacement of Step 3:
3. Take the convex-hull of the collection of curves $\Gamma_1, \ldots, \Gamma_m, \Lambda_1, \ldots, \Lambda_m$ as an approximation for $W(A, B)$.

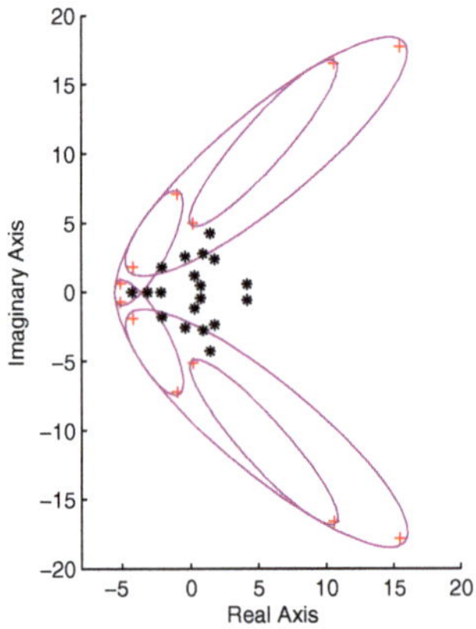

Fig. 1 Eigenvalues of $A - \lambda B$ (*asterisks*) and part of the boundary of $W(A, B)$. Here, B is PSD. (Example 1)

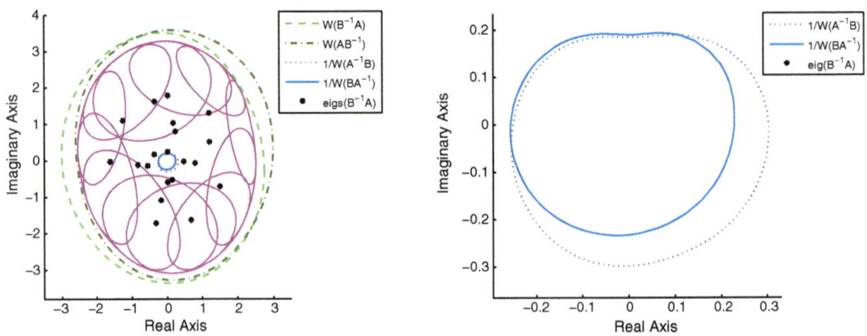

(a) $A = C_1 \oplus \mathrm{i}C_2$, $B = I_{20} + 0.1DD^*$. (b) Zoom-in of (a)

Fig. 2 **a** Eigenvalues of $A - \lambda B$ (*asterisks*) and the boundaries of $W(A, B)$, $W(B^{-1}A)$, $W(AB^{-1})$, $1/W(AB^{-1})$, $1/W(B^{-1}A)$, for Example 2, $m = 6$. **b** Exclusion regions for the eigenvalues of the pencil, $1/W(AB^{-1})$, $1/W(B^{-1}A)$

Example 1 We take the matrix $A = \mathrm{randn}(20)$, and the positive semidefinite matrix $B = I_{19} \oplus 0_1$. We carry out Algorithm 1 with $m = 6$. Considered as a spectral inclusion region, $W(A, B)$ has drawbacks since it is unbounded. See Fig. 1.

Example 2 We take the matrix $A = C_1 \oplus \mathrm{i}\, C_2$ and the positive definite matrix $B = I_{20} + 0.1DD^*$, with $C_1 = \mathrm{randn}(10)$, $C_2 = \mathrm{randn}(10)$, $D = \mathrm{randn}(20)$ with $m = 6$. See Fig. 2. The Zoom shows that the bounded complements of $1/W(A^{-1}B)$ and $1/W(BA^{-1})$ are spectral exclusion regions for the eigenvalues of the pencil.

5 Approximation of $W(A, B)$ for B Indefinite

To avoid trivial situations, assume that the matrices A and B have no common nonzero isotropic vector. Let us define

$$W_+(A, B) = \{\lambda \in \mathbb{C} : u^*Au - \lambda u^*Bu = 0,\ u \in \mathbb{C}^n,\ u^*Bu > 0\},$$

$$W_-(A, B) = \{\lambda \in \mathbb{C} : u^*Au - \lambda u^*Bu = 0,\ u \in \mathbb{C}^n,\ u^*Bu < 0\},$$

and so $W(A, B) = W_+(A, B) \cup W_-(A, B)$. To avoid trivial cases of degeneracy of $W(A, B)$, we shall be specially concerned with the class of matrices in M_n, for which there exists a real interval $[\theta_1, \theta_2]$, with $0 < \theta_2 - \theta_1 < \pi$, such that for θ ranging over that interval, the Hermitian pencil

$$H(\mathrm{e}^{-i\theta}A) - \lambda B, \tag{5}$$

has real eigenvalues satisfying simultaneously the following conditions:

(i) $\lambda_1(H(\mathrm{e}^{-i\theta}A), B) \geq \cdots \geq \lambda_r(H(\mathrm{e}^{-i\theta}A), B) \in \sigma^+(H(\mathrm{e}^{-i\theta}A), B)$;
(ii) $\lambda_{r+1}(H(\mathrm{e}^{-i\theta}A), B) \geq \cdots \geq \lambda_n(H(\mathrm{e}^{-i\theta}A), B) \in \sigma^-(H(\mathrm{e}^{-i\theta}A), B)$;
(iii) $\lambda_r(H(\mathrm{e}^{-i\theta}A), B) > \lambda_{r+1}(H(\mathrm{e}^{-i\theta}A), B)$.

For the pencils of this class, $W(H(\mathrm{e}^{-i\theta}A), B)$ is non-degenerate, that is, it is not a singleton, a whole line (possibly without a point), or the whole complex plane (possibly without a line). This class of pencils is called *class* $\mathscr{N}\mathscr{D}$, the acronym for *non-degenerate*.

When B is indefinite Hermitian nonsingular, B_{uv} may be indefinite or definite. If B_{uv} is indefinite, $\partial W(A_{uv}, B_{uv})$, the boundary of $W(A_{uv}, B_{uv})$, is the union of two hyperbolical arcs, one in $W_+(A_{uv}, B_{uv})$ and the other one in $W_-(A_{uv}, B_{uv})$. If B_{uv} is definite, $\partial W(A_{uv}, B_{uv})$ may be in $W_+(A, B)$ or in $W_-(A, B)$. Let the curves $C_1^+, C_2^+, \ldots, C_r^+$ ($C_1^-, C_2^-, \ldots, C_s^-$) denote the arcs of $\partial W(A_{uv}, B_{uv})$ in $W_+(A, B)$ ($W_-(A, B)$). Let $K^+ = \operatorname{conv}(C_1^+, C_2^+, \ldots, C_r^+)$ and $K^- = \operatorname{conv}(C_1^-, C_2^-, \ldots, C_s^-)$. The *pseudo-convex hull* of $C_1^+, C_2^+, \ldots, C_r^+, C_1^-, C_2^-, \ldots, C_s^-$, denoted $\operatorname{pconv}(C_1^+, C_2^+, \ldots, C_r^+, C_1^-, C_2^-, \ldots, C_s^-)$, is the union of all half-rays of the lines passing through $z^+ \in K^+$, $z^- \in K^-$ with endpoint in z^+ not containing z^-, or with endpoint in z^- not containing z^+.

As a preliminary stage to Algorithm 2, we start by searching an *admissible* angle θ. If the matrix is complex, we test the angle $-\pi/2$ for this property. If the answer is positive, we go to Step 0. If not, we test the admissibility of $\theta = 0$. In the affirmative case, we proceed to Step 0. Otherwise, we test the admissibility of the angles

$$\theta_{\ell,k} = -2^{k-1}\pi/2^k + (2\ell - 1)\pi/2^k,\ \ell = 0, 1, \ldots, 2^{k-1}, \quad k = 1, 2, 3, \ldots$$

until an admissible angle is found, and then we proceed to Step 0. It is worth noticing that replacing the matrix A by $\mathrm{e}^{-i\theta_{\ell,k}}A$, where $\theta_{\ell,k}$ is admissible for the pencil $H(A) -$

λB, then $\theta = 0$ is admissible for the rotated pencil

$$H(\mathrm{e}^{-i\theta_{\ell,k}}A) - \lambda B$$

Step 0. Choice of $[\boldsymbol{\theta}_{\mathbf{min}}, \boldsymbol{\theta}_{\mathbf{max}}]$ Fix a tolerance $tol = \pi/2^N$, $N \geq 4$ and let $\theta = 0$ be an admissible angle. Starting with $\theta_0 = 0$, construct a set of admissible angles, as follows. Bisect successively the interval $[0, \pi/2]$ until we find an admissible angle $\theta_1 = \pi/2^{\nu_1}$, the integer ν_1 being such that the angle $\theta_1 + \pi/2^{\nu_1}$ is non-admissible. Proceed in this way until we find a new admissible angle $\theta_2 = \pi/2^{\nu_1} + \pi/2^{\nu_1+\nu_2}$, the integer ν_2 being such that the angle $\theta_2 + \pi/2^{\nu_1+\nu_2}$ is non-admissible, and so on, until we reach the admissible angle $\theta_k = \pi/2^{\nu_1} + \pi/2^{\nu_1+\nu_2} + \cdots + \pi/2^{\nu_1+\nu_2+\ldots+\nu_k}$, such that $\theta_k + \pi/2^{\nu_1+\nu_2+\cdots+\nu_k}$ is non-admissible, being $\nu_1 + \nu_2 + \cdots + \nu_k \leq N$. The admissible angles $\bar{\theta}_1 = -\pi/2^{\bar{\nu}_1}$, $\bar{\theta}_2 = -\pi/2^{\bar{\nu}_1} - \pi/2^{\bar{\nu}_1+\bar{\nu}_2}, \ldots, \bar{\theta}_\ell = -\pi/2^{\bar{\nu}_1} - \pi/2^{\bar{\nu}_1+\bar{\nu}_2} - \cdots - \pi/2^{\bar{\nu}_1+\bar{\nu}_2+\cdots+\bar{\nu}_\ell}$ are analogously obtained. If the matrix is real, we obviously have $\bar{\theta}_j = -\theta_j$, $j = 1, \ldots k$. The interval of admissible angles is $[\theta_{min}, \theta_{max}] = [\bar{\theta}_\ell, \theta_k]$.

5.1 Algorithm 2

Input: A matrix $A \in M_n$, an indefinite nonsingular matrix B and m angles.
Output: An approximation for $W(A, B)$.

1. Set $\theta_k = \theta_{\min} + \frac{k-1}{m}(\theta_{\max} - \theta_{\min})$, $k = 1, \ldots, m+1$ for some positive integer $m \geq 3$.
2. Starting with $k = 1$ and up to $k = m$, take the following steps:
   (i) Compute eigenvectors u_k and v_k associated, respectively, to
   $$\lambda_{\max}(H(\mathrm{e}^{-i\theta}A), B) \in \sigma^-(H(\mathrm{e}^{-i\theta_k}A), B)$$
   and
   $$\lambda_{\min}(H(\mathrm{e}^{-i\theta}A), B) \in \sigma^+(H(\mathrm{e}^{-i\theta_k}A), B).$$
   (ii) Compute the compressions of A and B to $\mathrm{span}\{u_k, u_{k+1}\}$ and $\mathrm{span}\{v_k, v_{k+1}\}$, $A_{\tilde{u}_k\tilde{u}_{k+1}}, A_{\tilde{v}_k\tilde{v}_{k+1}}, B_{\tilde{u}_k\tilde{u}_{k+1}}$ and $B_{\tilde{v}_k\tilde{v}_{k+1}}$, respectively.
   (iii) Compute and draw $\partial W(A_{\tilde{u}_k\tilde{u}_{k+1}}, B_{\tilde{u}_k\tilde{u}_{k+1}})$ and $\partial W(A_{\tilde{v}_k\tilde{v}_{k+1}}, B_{\tilde{v}_k\tilde{v}_{k+1}})$, denoted by Γ_k and Λ_k, respectively.
   (iv) If $k < m$, take next k value and return to (i). Otherwise, continue.
3. Take the pseudo-convex hull of the collection of curves $\Gamma_1, \ldots, \Gamma_m, \Lambda_1, \ldots, \Lambda_m$ as an approximation for $W(A, B)$.

We now present an illustrative example.

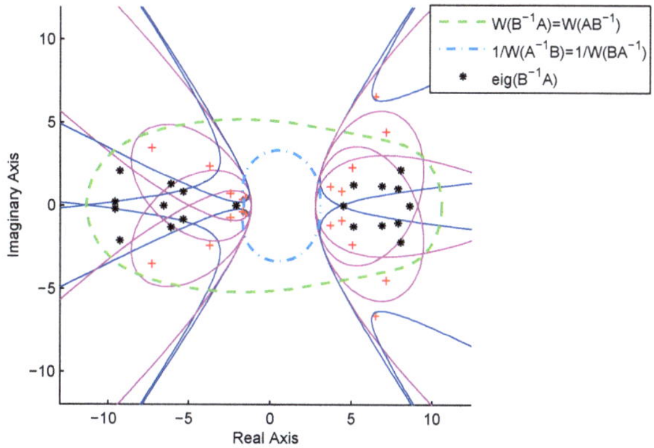

Fig. 3 Field of values $W(A, B)$, eigenvalues of the pencil (*asterisks*), boundaries of $W(B^{-1}A)$ (*green*) and of $1/W(B^{-1}A)$ (*blue*) for $A = \text{randn}(20) + 7I_{20}$, $B = I_{10} \oplus -I_{10}$, $m = 6$ (Example 3)

Table 1 Performance of Algorithm 2 and Psarrakos Algorithm [12], for the matrix of Example 3. The computed area is the one of the domain bounded by the approximation of $\partial W(A, B)$ and by the vertical lines $x = -8$ and $x = 8$

	m	Eigenanalyses	Area	Acc. digits	Seconds
Algorithm 2	6	24	160.7854	2	0.121777
	12	30	161.5071	3	0.230439
	24	42	161.6953	3	0.322045
	48	66	161.7327	5	0.494298
	96	114	161.7378	5	1.082117
	192	210	161.7391	6	2.783337
Psarrakos algorithm	6	602	159.2174	1	0.387542
	12	2354	160.6860	2	0.731580
	24	9314	161.5101	3	1.967806
	48	37058	161.6773	3	7.043684
	96	147842	161.7255	4	27.077536
	192	590594	161.7368	5	111.270553

Example 3 The fields of values $W(A, B)$ and $W(B^{-1}A)$, where $A = \text{randn}(20) + 7I_{20}$ and $B = I_{10} \oplus -I_{10}$, have been obtained using Algorithm 2 and are plotted in Fig. 3. We have used $\theta_{\max} = -\theta_{\min} = 0.5915413$ and $m = 6$. To compare, in accuracy, Algorithm 2 with Psarrakos Algorithm, we have computed the area of the domain bounded by the obtained approximation of $\partial W(A, B)$ and by the lines parallel to the imaginary axis with abscissas $x = 8$ and $x = -8$. We have also considered higher values of m in order to improve the accuracy. As Table 1 shows, Algorithm 2 requires much fewer eigenanalyses and reaches faster a given number of accurate digits.

6 Conclusions

We have given procedures to numerically approximate $W(A, B)$, of which at least one of the two matrices is Hermitian. Several matrices in [13] have been tested. In our approach we used the key fact that the field of values of a linear pencil is efficiently approximated by the compression into bidimensional linear pencils. Our algorithms compute the extreme eigenvalues of a small number of rotated pencils $H(e^{-i\theta_j}A) - \lambda B$ together with the respective eigenvectors u_j. In a second stage compression matrices of size 2 for the span$\{u_j, u_{j+1}\}$ for each $j = 2, \ldots, m$ are constructed. Elliptical and hyperbolical arcs generated from the compression matrices provide a quick and quite accurate approximation of the searched boundaries. Evaluating eigenvalues and eigenvectors involves $O(n^3)$ operations for n sized matrices. Performing 2-by-2 compressions is an $O(n^2)$ process and determining ellipses, parabolas or hyperbolas by using the Elliptical, Parabolical and Hyperbolical Range Theorems takes almost no time. Variations in relative speed and accuracy occur for varying dimensions, varying matrices and obviously changing the prescribed degree of accuracy. The preliminary stages for Algorithm 2 take negligible time. We stress that the proposed algorithms hold for both matrices of small and large dimensions. Psarrakos method [12] can be used for pairs of matrices of small dimension but it appears not to be interesting for large sized matrices. In fact, his method uses a discretization of the unit sphere in $\mathbb{R}^3$ and for each grid point a maximum eigenvalue of a certain associated Hermitian matrix has to be computed. Hochstenbach's Algorithm [6] applies only for Hermitian positive definite matrices B (or any positive definite linear combination of A and B). We have also focused on spectral inclusion regions for matrix pencils based on fields of values.

It would be of interest to obtain accurate and fast algorithms to plot $W(A, B)$ whenever neither A nor B are Hermitian.

Acknowledgements The authors wish to thank the Referees for most valuable comments. This work was partially supported by the Centre for Mathematics of the University of Coimbra – UID/MAT/00324/2013, funded by the Portuguese Government through FCT/MEC and co-funded by the European Regional Development Fund through the Partnership Agreement PT2020.

References

1. Bebiano, N., da Providência, J., Nata, A., da Providência, J.P.: Computing the numerical range of Krein space operators. Open Math. **13**, 2391–5455 (2014)
2. Bebiano, N., da Providência, J., Ismaeili, F.: The Characteristic Polynomial of Linear Pencils of Small Size and the Numerical Range, in this volume
3. Chien, M.-T., Nakazato, H.: The numerical range of linear pencils of 2-by-2 matrices. Linear Algebra Appl. **341**, 69–100 (2002)
4. Davis, C.: The Toeplitz-Hausdorff theorem explained. Canad. Math. Bull. **14**, 245–246 (1971)
5. Gustafson, K.E., Rao, D.K.M.: Numerical Range, the Field of Values of Linear Operators and Matrices. Springer, New York (1997)

6. Hochstenbach, M.E.: Fields of values and inclusion regions for matrix pencils. Electron. Trans. Numer. Anal. **38**, 98–112 (2011)
7. Kippenhahn, R.: Über den wertevorrat einer matrix. Math. Nachr. **6**, 193–228 (1951)
8. Li, C.-K., Rodman, L.: Numerical range of matrix polynomials. SIAM J. Matrix Anal. Appl. **15**, 1256–1265 (1994)
9. Li, C.-K., Rodman, L.: Shapes and computer generation of numerical ranges of Krein space operators. Electron. J. Linear Algebra **3**, 31–47 (1998)
10. Loghin, D., van Guzen, M., Jonkers, E.: Bounds on the eigenvalue range and on the field of values of non-Hermitian and indefinite finite element matrices. J. Comput. Appl. Mat. **189**(1–2), 304–323 (2006)
11. Marcus, M., Pesce, C.: Computer generated numerical ranges and some resulting theorems. Linear Multilinear Algebra **20**, 121–157 (1987)
12. Psarrakos, P.J.: Numerical range of linear pencils. Linear Algebra Appl. **317**, 127–141 (2000)
13. The Matrix Market, a repository for test matrices. http://math.nist.gov/MatrixMarket

The Characteristic Polynomial of Linear Pencils of Small Size and the Numerical Range

Natália Bebiano, João da Providência and Fatemeh Esmaeili

Abstract The numerical range of a linear pencil (A, B) of matrices of size n, of which either A or B is Hermitian, may be characterized in terms of a certain algebraic curve of class n, called the boundary generating curve. This curve is explicitly given by the characteristic polynomial of the pencil. For $n = 2$ and $n = 3$, each possible type of boundary generating curve can be completely described. For $n = 3$, the curve type is given by Newton's classification of cubic curves. Illustrative examples of the different possibilities are given.

Keywords Linear pencil · Numerical range · Characteristic polynomial

1 Introduction

Let $A, B \in M_n$, the algebra of $n \times n$ complex matrices. The linear pencil (A, B) is the set of matrices $A - \lambda B$, where λ is a real or complex parameter. A pencil is said to be *regular* if the polynomial $\det(A - \lambda B)$ does not identically vanish, otherwise it is *singular*. If the matrix B is nonsingular, the *spectrum* of the regular pencil denoted by $\sigma(A, B)$ consists of all the zeros λ of the polynomial $\det(A - \lambda B)$. The spectral theory of pencils is an important issue in pure mathematics as well as in applications (e.g., see [3, 8, 12, 13] and their references). An informative containment region for the spectrum of (A, B) is the numerical range [4, 8].

N. Bebiano (✉) · F. Esmaeili
Department of Mathematics, CMUC, University of Coimbra, 3001-454 Coimbra, Portugal
e-mail: bebiano@mat.uc.pt

F. Esmaeili
e-mail: esmaeili.3143@gmail.com

J. da Providência
Department of Physics, CFisUC, University of Coimbra, 3001-516 Coimbra, Portugal
e-mail: providencia@teor.fis.uc.pt

N. Bebiano (ed.), *Applied and Computational Matrix Analysis*,
Springer Proceedings in Mathematics & Statistics 192,
DOI 10.1007/978-3-319-49984-0_13

The *numerical range* (also called the *field of values*) of a linear pencil is defined and denoted as

$$W(A, B) = \{\lambda \in \mathbb{C} : x^*(A - \lambda B)x = 0, \quad \text{for some } 0 \neq x \in \mathbb{C}^n\} \tag{1}$$

(cf. [10, 13]). If B is singular, then the pencil (A, B) may have an infinite eigenvalue λ, nevertheless (1) does not contain the point at infinity. So, from the above definition, $W(A, B)$ is not necessarily a spectral inclusion region for the generalized eigenvalue problem $Ax = \lambda Bx$. Indeed, we consider a slightly modified definition: if A, B have a common null space, then

$$W(A, B) = \mathbb{C} \cup \{\infty\};$$

otherwise

$$W(A, B) = \left\{ \frac{x^*Ax}{x^*Bx} : 0 \neq x \in \mathbb{C}^n \right\}. \tag{2}$$

where $1/0$ is understood as the point at infinity. When B is the identity matrix, (2) reduces to the (classical) field of values of the $n \times n$ matrix A,

$$W(A) = \{x^*Ax : \ x \in \mathbb{C}^n, \ \|x\| = 1\},$$

where $\|x\| = \langle x, x\rangle^{1/2} = (x^*x)^{1/2}$ is the usual Euclidean norm in $\mathbb{C}^n$. This concept has been extensively investigated; see, for instance, [7] and references therein.

Throughout, we shall be concerned with regular pencils (A, B) of which either A or B is Hermitian. Let us assume that B is Hermitian. The *characteristic polynomial* of the pencil (A, B) is defined as

$$f(u, v, w) = \det(uH + vK + wB),$$

where $A = H + iK$, and

$$H = (A + A^*)/2, \quad K = (A - A^*)/(2i)$$

are Hermitian matrices.

The main goal of this article is to investigate connections between the characteristic polynomial $f(u, v, w)$ and the shape of $W(A, B)$. The paper is organized as follows. In Sect. 2 we recall some properties of algebraic curves used subsequently. In Sect. 3 we characterize the field of values of 2×2 linear pencils, distinguishing the cases of B being definite, indefinite and singular. These results allow simple direct proofs of the convexity of $W(A, B)$ for B Hermitian definite or semidefinite as well as the pseudo-convexity of $W(A, B)$ for B indefinite. In Sect. 4, each possible boundary generating curve is described for 3×3 matrices of which one of them is Hermitian. In Sect. 5 illustrative examples are given.

2 The Polynomial $f(u, v, w)$ and $W(A, B)$

As we shall see in the sequel, the characteristic polynomial of (A, B) gives rise to the *boundary generating curve* of the numerical range $W(A, B)$. To investigate this relation and for the sake of completeness, we present some prerequisites concerning plane algebraic curves.

An ordered pair of complex numbers (x, y) is a (complex) point in *nonhomogeneous point coordinates*. If x and y are real numbers, (x, y) is called a *real* point. A point in *homogeneous point coordinates* is a triple of complex numbers (x, y, z), not all zero. If r is any non zero complex number, then (x, y, z) and (rx, ry, rz) represent the same point. We identify the point (x, y, z) in homogeneous coordinates with the point $(x/z, y/z)$ in nonhomogeneous coordinates. On the other hand, the point (x, y) becomes $(x, y, 1)$ in homogeneous coordinates. Any point with $z = 0$ is a *point at infinity*.

If B is Hermitian positive definite (HPD), we clearly have

$$W(A, B) = W(B^{-1/2}AB^{-1/2}),$$

and so the numerical range of the pencil reduces to the classical numerical range. Toeplitz and Hausdorff have proven that the classical field of values is a convex set [7]. So, assuming that B is positive definite, then $W(A, B)$ is convex.

A *supporting line* of a convex set $S \subset \mathbb{C}$ is a line that intersects S at least in one point and that defines two half-planes, such that one of them does not contain any point of S. It can be shown, using similar reasoning to the one in [9, Theorem 10] that

Theorem 1 *Let B be positive definite and let A be arbitrary. If $ux + vy + w = 0$ is the equation of a supporting line of $W(A, B)$, then*

$$f(u, v, w) = \det(uH + vK + wB) = 0. \tag{3}$$

It can be easily proved that a similar result to the above one holds for B indefinite or semi-definite. Since $f(u, v, w)$ is a homogeneous polynomial of degree n, (3) may be viewed as the line equation of an algebraic curve in the complex projective plane $P\mathbb{C}^2$. The set of lines $(u : v : w)$ (with equation $ux + vy + wz = 0$) such that $f(u, v, w) = 0$, may be regarded as a set of lines in the plane whose envelope is a certain curve. Considering the *dual curve*, i.e., the curve in line coordinates,

$$\Gamma^* = \{(u : v : w) \in P\mathbb{C}^2 : f(u, v, w) = 0\},$$

by dualization, we may easily determine:

$$\Gamma = \{(x : y : z) \in P\mathbb{C}^2 : xu + yv + zw = 0 \text{ is a tangent of } \Gamma^*\}.$$

The real affine view of Γ, say

$$C(A, B) = \{(x, y) \in \mathbb{R}^2 \colon (x : y : 1) \in \Gamma\},$$

is called the *associated curve* or *boundary generating curve* of $W(A, B)$.

For $(A, B) \in M_n$, with B positive definite, it is a simple consequence of an extension of a result of Kippenhahn (see [9]) that the curve $C(A, B)$ generates $W(A, B)$ as its convex hull.

We recall that an usual procedure to find the point equation of the boundary generating curve $C(A, B)$ is to eliminate one of the indeterminates, say u, from (3) and $ux + vy + w = 0$, dehomogenize the result by setting $w = 1$, and to eliminate the remaining parameter v from the equations

$$F(v, x, y) = f(-(1 + vy)x^{-1}, v, 1) = 0 \quad \text{and} \quad \frac{\partial F(v, x, y)}{\partial v} = 0.$$

The curve $f(u, v, w) = 0$ has *class n* (because the defining polynomial has degree n), that is, through a general point in the plane there are n lines (may be complex) tangent to the curve.

A point P, not equal to the *circular points at infinity* $(1 : i : 0)$ and $(1 : -i : 0)$, is called a *focus* of a curve C if the line l_1 through P and $(1 : i : 0)$ and the line l_2 through P and $(1 : -i : 0)$ are tangent to C at points other than the circular points at infinity. The coefficients of the polynomial $f(u, v, w)$ are real, as it can be easily seen. A curve of class n with real coefficients has n real foci, according to proper multiplicities, and $n^2 - n$ foci which are not real [14].

As a consequence of a result, independently obtained by Murnaghan [11] and Kippenhahn [9], the real foci of the algebraic curve defined by $\det(uH + vK + wB) = 0$, where B is positive definite, are the eigenvalues of the matrix $B^{-1}A$, with $A = H + iK$. The corresponding result for B indefinite is as follows [3].

Theorem 2 *Let $A, B \in M_n$, with B indefinite nonsingular. The n real foci of the algebraic curve defined by the equation $f(u, v, w) = \det(uH + vK + wB) = 0$ are the eigenvalues of the pencil (A, B), where $A = H + iK$ with H and K Hermitian.*

For details on plane algebraic curves, we refer the interested reader to [5].

3 Linear Pencils Generated by 2 × 2 Matrices

For matrices A and B of dimension two, the boundary generating curve $C(A, B)$ is a curve of class two, more concretely, a conic. The three theorems that characterize the boundary of $W(A, B)$, for B Hermitian, in terms of the invariants of the pencil (A, B) are stated below. The case 2 by 2 is specially important, since the numerical range of an $n \times n$ pencil may be characterized by compression to the bidimensional setting [4, 12].

Theorem 3 (Elliptical Range Theorem) *Let A, B be 2×2 matrices with B positive definite. Then $W(A, B)$ is a (possibly degenerate) closed elliptical disc, whose foci are the eigenvalues of $B^{-1}A$, λ_1 and λ_2. and the lengths of the major and minor axis are, respectively,*

$$M = \sqrt{\mathrm{Tr}\left(A^*B^{-1}AB^{-1}\right) - 2\mathrm{Re}\left(\overline{\lambda_1}\lambda_2\right)},$$

and

$$N = \sqrt{\mathrm{Tr}\left(A^*B^{-1}AB^{-1}\right) - |\lambda_1|^2 - |\lambda_2|^2}.$$

In the case of degeneracy, $W(A, B)$ may reduce to a line segment whose endpoints are λ_1 and λ_2, or to a singleton if and only if $\lambda_1 = \lambda_2$.

Theorem 4 (Hyperbolical Range Theorem) *Let A, B be 2×2 matrices with B indefinite. Then $W(A, B)$ is bounded by a hyperbola with foci at λ_1 and λ_2, the eigenvalues of $B^{-1}A$, and transverse and non-transverse axis of length*

$$M = \sqrt{\mathrm{Tr}(B^{-1}A^*B^{-1}A) - 2\mathrm{Re}(\lambda_1\bar{\lambda_2})}$$

and

$$N = \sqrt{|\lambda_1|^2 + |\lambda_2|^2 - \mathrm{Tr}(B^{-1}A^*B^{-1}A)}.$$

If $\mathrm{Tr}(B^{-1}A^*B^{-1}A) - 2\mathrm{Re}(\lambda_1\bar{\lambda_2}) < 0$, the hyperbola degenerates and $W(A, B)$ is the whole complex plane. In the case of degeneracy of the hyperbola, $W(A, B)$ may reduce to two half-lines of the line defined by λ_1 and λ_2, and with these endpoints.

Now, we consider $W(A, B)$ for $A, B \in M_2$, with B positive (negative) semidefinite. Observing that

$$W(e^{i\phi}(A + \zeta B), kB) = \frac{1}{k}e^{i\phi}(W(A, B) + \zeta), \quad k, \phi \in \mathbb{R}, \ \zeta \in \mathbb{C},$$

and using the invariance of $W(A, B)$ under unitary similarities, we may take

$$B = \mathrm{diag}(1, 0), \quad A = \begin{bmatrix} ae^{i\gamma} & ce^{i\gamma} \\ d & b \end{bmatrix}, \quad c, d \geq 0, \ b > 0, \ a = \frac{cd}{b}. \tag{4}$$

Notice that $W(A, B) = \mathbb{C}$ if $b = 0$.

Theorem 5 (Parabolical Range Theorem) *Let A, B be of the form (4). Then $W(A, B)$ is bounded by the (possibly degenerate) parabola with focus $\lambda_0 = 0$ and equation*

$$\frac{y^2}{4p^2} - \frac{x}{p} = 1,$$

where

$$p = \frac{a^2b^2 + c^4 - 2abc^2\cos\gamma}{4bc^2}.$$

In the case of degeneracy of the parabola, $W(A, B)$ reduces to one half-line with $\lambda_0 = 0$ as endpoint.

We remark that for $A = (a_{ij}) \in M_2$, with $a_{22} \neq 0$ and $B = \text{diag}(1, 0)$, the slope of the axis of the parabolic boundary, relatively to the positive semi real axis, is equal to $\theta_0 = \text{Arg}(a_{22})$, and the focus of the parabola is the (finite) eigenvalue of the pencil (A, B). The vertex of the parabola is the point $u_0^* A u_0 / u_0^* B u_0$, where u_0 is an eigenvector of the Hermitian pencil

$$\left(\frac{1}{2}(A\mathrm{e}^{-i\theta_0} + A^* \mathrm{e}^{i\theta_0}), B\right)$$

associated with its single (finite) eigenvalue.

4 Characterization of $W(A, B)$ for $A, B \in M_3$ with B Hermitian

4.1 $C(A, B)$ for B Positive Definite and A Arbitrary

Under the present assumptions, $W(A, B)$ is convex, bounded and closed, since it reduces to $W(B^{-1/2} A B^{-1/2})$, and so inherits the properties of the classical numerical range. Following the arguments in [9, Theorem 10], we can prove the following

Theorem 6 *The convex hull of $C(A, B)$ is $W(A, B)$.*

Kippenhahn classified the associated curve in this context, considering the factorizability of the polynomial $f(u, v, w)$. Adopting this procedure, we easily conclude that the following possibilities may occur.

1st Case: The polynomial $f(u, v, w)$ factorizes into three linear factors. Each one of these factors corresponds to an eigenvalue of $B^{-1}A$ and $C(A, B)$ reduces to these eigenvalues. This property is still valid for $A, B \in M_n$ with B Hermitian positive definite.

2nd Case: Suppose that $B = \text{diag}(b_1, b_2, b_3)$ and that $A \in M_3$ is a B-decomposable matrix, i.e., there exists a nonsingular matrix $V \in M_3$, such that

$$V^* B V = B, \quad V^* A V = \begin{bmatrix} cb_1 & 0 \\ 0 & A_1 \end{bmatrix}, \tag{5}$$

where $c \in \mathbb{C}$ and $A_1 \in M_2$. Thus, $f(u, v, w)$ factorizes into a linear and an irreducible quadratic factor, and so $C(A, B)$ consists of the point c and of the boundary of the elliptical disc $W(A_1, \text{diag}(b_2, b_3))$.

3rd Case: The matrix A is B-indecomposable, but the polynomial $f(u, v, w)$ factorizes into a linear and a quadratic factor. The linear factor corresponds to an eigenvalue of $B^{-1}A$. The quadratic factor corresponds to an ellipse. In fact, the conic can not be neither a parabola, because one of its real foci is a point at infinity and this contradicts Theorem 2, nor an hyperbola because this curve is unbounded. Therefore, $C(A, B)$ consists of an ellipse and a point.

4th Case: Finally, suppose that the polynomial $f(u, v, w)$ is irreducible. The number of real cusps of an (irreducible) class three curve is 1 or 3, and the order of the boundary generating curve is 4 or 6. By Newton's classification of cubic curves [1] and dual considerations, there are the following possibilities for the associated curve:

C1. $C(A, B)$ is a sextic, consisting of an oval and a closed tricuspid curve lying in its interior;
C2. $C(A, B)$ is a quartic, with one cusp and an ordinary double tangent at two of its points.

There are examples showing that all these types of curves appear as boundary generating curves of $W(A, B)$.

4.2 *$C(A, B)$ for B Indefinite and A Arbitrary*

A set $S \subset \mathbb{C}$ is said to be *pseudo-convex* if, for any $x, y \in S$, either the line segment $px + (1 - p)y$, $0 \leq p \leq 1$, is contained in S, or the halflines $px + (1 - p)y$, with $p \geq 1$, and $px + (1 - p)y$, with $p \leq 0$, are there contained.

Theorem 7 *Let $A, B \in M_n$ with B indefinite. Then $W(A, B)$ is pseudo-convex.*

Proof Let us consider $\lambda_1 \neq \lambda_2 \in W(A, B)$. Then, there exist $0 \neq v_1, \ 0 \neq v_2 \in W(A, B)$ such that $v_i^* A v_i = \lambda_i v_i^* B v_i, \ i = 1, 2$. Let $\tilde{v}_1, \ \tilde{v}_2$ be orthonormal vectors belonging to the subspace $\mathcal{H}_2$ spanned by $v_1, \ v_2$. Assume first that $\mathcal{H}_2$ is non degenerate. Let $A_{\tilde{v}_1,\tilde{v}_2}$ and $B_{\tilde{v}_1,\tilde{v}_2}$ be the compressions of A and B, respectively, to $\mathcal{H}_2$. Obviously, $W(A_{\tilde{v}_1,\tilde{v}_2}, B_{\tilde{v}_1,\tilde{v}_2})$ is either an elliptical, parabolical or hyperbolical domain, depending on $B_{\tilde{v}_1,\tilde{v}_2}$ being definite, semidefinite or indefinite. If $W(A_{\tilde{v}_1,\tilde{v}_2}, B_{\tilde{v}_1,\tilde{v}_2})$ is an elliptical or parabolical disc, it is convex. In this case, we have that

$$\{\lambda_1 + x(\lambda_2 - \lambda_1) : 0 \leq x \leq 1\} \subseteq W(A_{\tilde{v}_1,\tilde{v}_2}, B_{\tilde{v}_1,\tilde{v}_2}) \subseteq W(A, B).$$

If $W(A_{\tilde{v}_1,\tilde{v}_2}, B_{\tilde{v}_1,\tilde{v}_2})$ is hyperbolical, it is pseudo-convex. In this case, either

$$\{\lambda_1 + x(\lambda_2 - \lambda_1) : 0 \leq x \leq 1\} \subseteq W(A_{\tilde{v}_1,\tilde{v}_2}, B_{\tilde{v}_1,\tilde{v}_2}) \subseteq W(A, B).$$

or

$$\{\lambda_1 + x(\lambda_2 - \lambda_1) : x \leq 0 \ \text{ or } \ x \geq 1\} \subseteq W(A_{\tilde{v}_1,\tilde{v}_2}, B_{\tilde{v}_1,\tilde{v}_2}) \subseteq W(A, B),$$

This completes the proof when $\mathcal{H}_2$ is non degenerate.

If $\mathcal{H}_2$ is degenerate, replace v_2 by $v_2 + \varepsilon v_3$, where v_3 is such that the space spanned by v_1, v_3, is non degenerate. For ε sufficiently small, the point generated by $v_2 + \varepsilon v_3$ is in the neighborhood of λ_2 and the result follows. □

For B indefinite nonsingular, consider $\mathbb{C}^n$ endowed with the *B-inner product* $\langle Bx, y\rangle = y^*Bx$, and corresponding *B-norm* $\|x\|_B^2 = \langle Bx, x\rangle$ [6]. For arbitrary $A \in M_3$, $W(A, B)$ has been characterized in [3], following Kippenhahn's approach in the classical case.

Let us consider

$$W(A, B) = \left\{\frac{\langle Au, u\rangle}{\langle Bu, u\rangle} : u \in \mathbb{C}^n,\ \langle Bu, u\rangle \neq 0\right\}.$$

For convenience, we also consider the sets

$$W_+(A, B) = \left\{\frac{\langle Au, u\rangle}{\langle Bu, u\rangle} : u \in \mathbb{C}^n,\ \langle Bu, u\rangle > 0\right\},$$

and

$$W_-(A, B) = \left\{\frac{\langle Au, u\rangle}{\langle Bu, u\rangle} : u \in \mathbb{C}^n,\ \langle Bu, u\rangle < 0\right\}.$$

Obviously,

$$W(A, B) = W_+(A, B) \cup W_-(A, B).$$

In our analysis, when A and B are both Hermitian, we shall consider the eigenvalues of *positive* and *negative* type, that is, the eigenvalues with associated eigenvectors with positive and negative B-norm, respectively. We shall denote by $\sigma_+(A, B)$ ($\sigma_-(A, B)$) the set of eigenvalues of positive (negative) type.

Let X^+ (X^-) be a set of points in $W_+(A, B)$ ($W_-(A, B)$) and let Ξ^+ (Ξ^-) be the convex hull of X^+ (X^-). Consider the lines defined by points z_+, z_- with $z_+ \in \Xi^+$ and $z_- \in \Xi^-$. The union of all half-lines with z_+ as endpoint not containing z_- and the half-lines with z_- as endpoint not containing z_+, is the so called *pseudo-convex hull* of $X^+$and X^-.

The curve $C(A, B)$ has branches of a well defined sign type, either *positive* or *negative*, say $C_+(A, B)$ and $C_-(A, B)$. The sign is determined by considering for the corresponding root w of (3), an associated eigenvector ξ, such that

$$(uH + vK + wB)\xi = 0.$$

The type of each branch of $C(A, B)$ is characterized by the sign of the B-norm $\langle B\xi, \xi\rangle$.

For pencils of the class $\mathcal{N}\mathcal{D}$ (see [4, Section 5]) the following holds. The proof follows analogous steps to those in [9, Theorem 10].

Theorem 8 *Let $A, B \in M_n$ with B Hermitian indefinite non singular. If the pencil (A, B) is in $\mathcal{N}\mathcal{D}$, then the pseudo-convex hull of $C_+(A, B)$ and $C_-(A, B)$ is $W(A, B)$.*

We classify the associated curve $C(A, B)$, considering the factorizability of the polynomial $f(u, v, w)$. Without loss of generality, we may assume that $B = \text{diag}(b_1, b_2, -b_3)$, $b_1, b_2, b_3 > 0$. The following possibilities may occur.

1st Case: The polynomial $f(u, v, w)$ factorizes into three linear factors. Each one of these factors corresponds to an eigenvalue of $B^{-1}A$ and $C(A, B)$ reduces to the eigenvalues. This result still holds for matrices A, B of arbitrary size, under the above conditions.

2nd Case: Suppose that $A \in M_3$ is B-decomposable, i.e., there exists a nonsingular matrix V, such that $V^*BV = B = \text{diag}(b_1, b_2, -b_3)$ and

$$V^*AV = \begin{bmatrix} cb_1 & 0 \\ 0 & A_1 \end{bmatrix}, \tag{6}$$

or

$$V^*AV = \begin{bmatrix} A_1 & 0 \\ 0 & -cb_3 \end{bmatrix}, \tag{7}$$

where $c \in \mathbb{C}$ and $A_1 \in M_2$.

If A is of the form (6), then $C(A_1, \text{diag}(b_2, -b_3))$ is an hyperbola with one branch in $W_+(A, B)$ and the other one in $W_-(A, B)$. We may write

$$C(A_1, \text{diag}(b_2, -b_3)) = C_+(A_1, \text{diag}(b_2, -b_3)) \cup C_-(A_1, \text{diag}(b_2, -b_3)),$$

where $C_\pm(A_1, \text{diag}(b_2, -b_3)) \subset W_\pm(A, B)$. Clearly, $c \in W_+(A, B)$. Let $X_+ = \text{conv}(c, C_+(A_1, \text{diag}(b_2, -b_3)))$. The *pseudo-convex hull* of X_+ and $C_-(A_1, \text{diag}(b_2, -b_3))$ coincides with $W(A, B)$.

Suppose, now, that A is of the form (7). Notice that $c \in W_-(A, B)$ and $C(A_1, \text{diag}(b_1, b_2)) \subset W_+(A, B)$. Then $W(A, B)$ is the pseudo-convex hull of c and an ellipse (possibly degenerate): $C(A_1, \text{diag}(b_1, b_2))$.

3rd Case: The matrix A is B-indecomposable, but the polynomial $f(u, v, w)$ factorizes into a linear and an irreducible quadratic factor. The quadratic factor corresponds to an hyperbola or to an ellipse. The conic can not be a parabola, because one of its real foci is a point at infinity and this contradicts Theorem 2.

Therefore, $C(A, B)$ consists of: 1) one point, produced by vectors with a negative B-norm, and an ellipse produced by vectors with a positive B-norm, 2) one point, produced by vectors with a positive B-norm, and an hyperbola, with one branch produced by vectors with a negative B-norm and the other branch produced by vectors with a positive B-norm.

In case 1), $W(A, B) = \mathbb{C}$. In case 2), $W(A, B) = \mathbb{C}$, whenever the point lies inside the hyperbolic disc of negative type, otherwise $W(A, B)$ is a hyperbolical disc.

4th Case: Finally, suppose that the polynomial $f(u, v, w)$ is irreducible. The number of real cusps of an (irreducible) class three curve is 1 or 3, and the order of the boundary generating curve is 4 or 6. By Newton's classification of cubic curves and dual considerations, there are the following possibilities for the associated curve:

C1. $C(A, B)$ is a sextic, with three real cusps and at least one oval component;
C2. $C(A, B)$ is a quartic, with three real cusps and a real double tangent (at two complex points of the curve);
C3. $C(A, B)$ is a quartic with one real cusp and a real double tangent (at two real points of the curve);
C4. $C(A, B)$ is a cubic with a real cusp and a real flex;
C5. $C(A, B)$ is a sextic, with three real cusps and not containing neither oval components nor double tangents.

There are examples showing that all the above curves may occur as boundary generating curves [3]. The characterization of $W(A, B)$ requires the determination of the signs of each branch of $C(A, B)$, in order to obtain the pseudo-convex hull of the boundary generating curve.

4.3 $C(A, B)$ for B Positive Semidefinite and A Arbitrary

Theorem 9 *Let $A, B \in M_n$ with B positive semidefinite. Then $W(A, B)$ is convex.*

Proof Let us consider $\lambda_1 \neq \lambda_2 \in W(A, B)$. Then, there exist $0 \neq v_1,\ 0 \neq v_2 \in W(A, B)$ such that $v_i^* A v_i = \lambda_i v_i^* B v_i,\ i = 1, 2$. Let $\tilde{v}_1,\ \tilde{v}_2$ be orthonormal vectors belonging to the subspace $\mathscr{H}_2$ spanned by $v_1,\ v_2$. Let $A_{\tilde{v}_1,\tilde{v}_2}$ and $B_{\tilde{v}_1,\tilde{v}_2}$ be the compressions of A and B, respectively, to $\mathscr{H}_2$. Obviously, $W(A_{\tilde{v}_1,\tilde{v}_2}, B_{\tilde{v}_1,\tilde{v}_2})$ is either a parabolical or elliptical disc, so it is convex. Thus, $[\lambda_1, \lambda_2] \in W(A_{\tilde{v}_1,\tilde{v}_2}, B_{\tilde{v}_1,\tilde{v}_2}) \subseteq W(A, B)$, which completes the proof. □

We next characterize $W(A, B)$, for B positive semi-definite and an arbitrary $A \in M_3$, using again Kippenhahn's approach. We classify the associated curve $C(A, B)$, considering the factorizability of the polynomial $f(u, v, w)$.

Assume that $A \in M_3$ and B is positive semidefinite. The following possibilities for $C(A, B)$ may occur.

1st Case: Suppose that $B = \text{diag}(b_1, b_2, 0),\ b_1, b_2 > 0$, and $A \in M_3$ is a B-decomposable matrix, i.e., there exists a nonsingular matrix V such that $V^*BV = B$ and V^*AV is as in (6). Then, $W(A, B)$ is the convex hull of c and $C(A_1, \text{diag}(b_2, 0))$.

2nd Case: Suppose that $B = \text{diag}(b_1, b_2, 0),\ b_1, b_2 > 0$, and A is a 3×3 B-decomposable matrix, i.e., there exists a non-singular matrix V, such that $V^*BV = B$ and

$$V^*AV = \begin{bmatrix} A_1 & 0 \\ 0 & c \end{bmatrix}, \tag{8}$$

where $c \in \mathbb{C}$ and A_1 is a 2×2 matrix. Thus, $W(A, B)$ is the convex hull of a certain point at infinity and $C(A_1, \text{diag}(b_1, b_2))$ (cf. Example 4).

3rd Case: Suppose that $B = \text{diag}(b_1, b_2, 0)$, $b_1, b_2 > 0$, and the matrix A is B-indecomposable, but the polynomial $f(u, v, w)$ factorizes into a linear and an irreducible quadratic factor. The linear factor corresponds to an eigenvalue of the pencil, and the quadratic factor corresponds to a parabola. Therefore, $C(A, B)$ consists of one real point and a parabola (cf. Example 3), being $W(A, B)$ its convex hull.

4th Case: Suppose that $B = \text{diag}(b_1, b_2, 0)$, $b_1, b_2 > 0$, and the polynomial $f(u, v, w)$ is irreducible. By Newton's classification of cubic curves and dual considerations, there are the following possibilities for the associated curve:

C1. $C(A, B)$ is a sextic, with three real cusps and at least one oval component (cf. Example 1);
C2. $C(A, B)$ is a quartic, with one cusp and an ordinary double tangent at two of its real points (cf. Example 2).

5th Case: Suppose that $B = \text{diag}(b_1, 0, 0)$, $b_1 > 0$. There exists a non-singular matrix V, such that $V^*BV = B$ and

$$V^*AV = \begin{bmatrix} a_{11} & a_{12} & a_{13} \\ 0 & a_{22} & a_{23} \\ 0 & 0 & a_{33} \end{bmatrix}.$$

If the existence of vectors $\xi \neq 0$ such that $\xi^*A\xi = \xi^*B\xi = 0$ is excluded, it follows that

$$0 \notin W\left(\begin{bmatrix} a_{22} & a_{23} \\ 0 & a_{33} \end{bmatrix}\right)$$

and then $W(A, B)$ is a proper subset of the complex plane bounded by a certain algebraic curve, which is a quartic, if the characteristic polynomial is irreducible (cf. Example 5), and a conic if the characteristic polynomial is factorizable (cf. Example 7). However, if

$$0 \in W\left(\begin{bmatrix} a_{22} & a_{23} \\ 0 & a_{33} \end{bmatrix}\right),$$

then $W(A, B)$ is the whole complex plane (cf. Example 6)

4.4 $C(A, B)$ *for B Indefinite Singular and A Arbitrary*

Let A be arbitrary, $B = \text{diag}(b_1, -b_2, 0)$, with $b_1, b_2 > 0$. We say that $\theta \in [0, 2\pi[$ is an *admissible* direction if the Hermitian pencil $(H(\mathrm{e}^{-i\theta}A), B)$ has real eigenvalues with associated non-isotropic eigenvectors, and for $\sigma_+(H(\mathrm{e}^{-i\theta}A), B) = \{\alpha_\theta\}$, $\sigma_-(H(\mathrm{e}^{-i\theta}A), B) = \{\beta_\theta\}$, we have $(\alpha_\theta - \beta_\theta)\, u^*Au > 0$, where $u = (0, 0, 1)^T$. The condition $(\alpha_\theta - \beta_\theta)\ u^*Au > 0$, ensures that $W(H(e^{-i\theta}A), B) \neq \mathbb{R}$. If admissible directions do not exist, $W(A, B) = \mathbb{C}$ (see Theorem 2.1 of [4]).

Proposition 1 *Let* (A, B) *be a* 3×3 *self-adjoint pencil with* $B = \text{diag}(b_1, -b_2, 0)$, $b_1, b_2 > 0$, *such that* $W(A, B) \neq \mathbb{C}$. *Let* $u = (0, 0, 1)^T$, $\sigma_+(A, B) = \{\alpha\}$, $\sigma_-(A, B) = \{\beta\}$.
(i) If $(\alpha - \beta)\ u^*Au > 0$, *then* $W(A, B) =]-\infty, \min(\alpha, \beta)] \cup [\max(\alpha, \beta), +\infty[$.
(ii) If $(\alpha - \beta)\ u^*Au < 0$, *then* $W(A, B) = \mathbb{R}$.

For $A \in M_3$ and B indefinite singular, the different possibilities that may occur for $C(A, B)$ can be identified according with the procedures in the previous sections (cf. Example 8).

5 Examples

The figures presented in this section have been produced with *Mathematica 5.1*, also used to determine the point equation of $C(A, B)$. The associated curve is represented in the figures. The boundaries of $W(A, B)$ are represented by thick lines.

Example 1 Let

$$A = \begin{bmatrix} 1 & 1 & 4/5 \\ 0 & 1 & 4/5 \\ 0 & 0 & 1 \end{bmatrix}, \quad B = \text{diag}(1, 1, 0).$$

The characteristic polynomial of the pencil is

$$f(u, v, w) = \frac{1}{100}(71u^3 - 29uv^2 + 192u^2w - 8v^2w + 100uw^2).$$

The Cartesian equation of the boundary generating curve of $W(A, B)$ is

$$\begin{aligned} &-1731619 + 6115752x - 6709556x^2 + 3123808x^3 - 655104x^4 + 51200x^5 \\ &-1891452y^2 + 7557408xy^2 - 17370208x^2y^2 + 9142400x^3y^2 - 160000x^4y^2 \\ &-15865104y^4 + 51091200xy^4 - 21320000x^2y^4 - 21160000y^6 = 0. \end{aligned}$$

The boundary of $W(A, B)$ is represented in Fig. 1 by the outer curve.

Example 2 Let

$$A_4 = \begin{bmatrix} 1 & 1/2 & 1 \\ 0 & 1 & 1 \\ 0 & 0 & 1 \end{bmatrix}, \quad B = \text{diag}(1, 1, 0)$$

The characteristic polynomial of the pencil (A_4, B) is

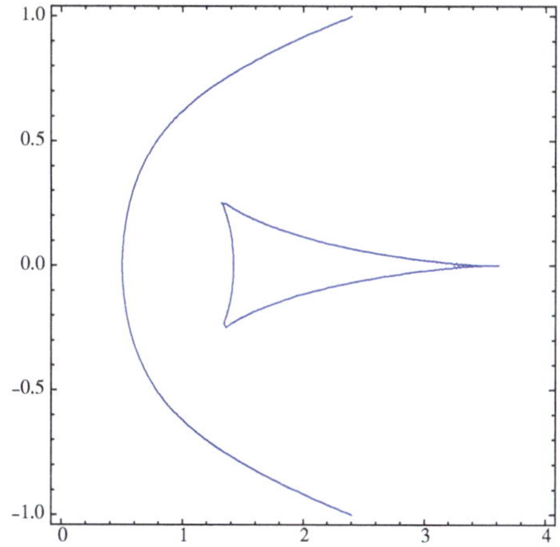

Fig. 1 Boundary generating curve of $W(A, B)$ (Example 1)

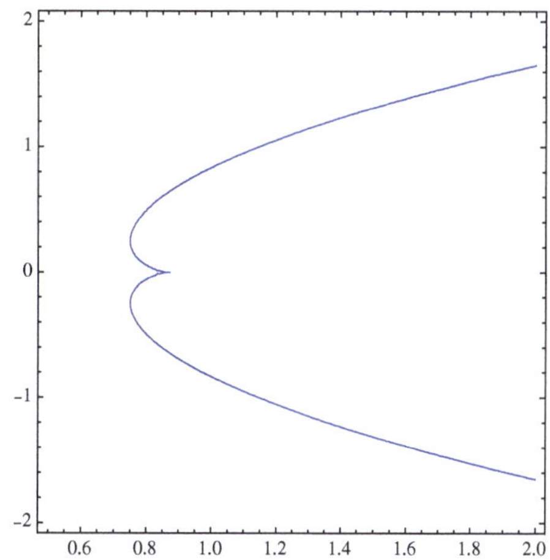

Fig. 2 Boundary generating curve of $W(A_4, B)$ (Example 2)

$$f(u, v, w) = \frac{1}{16}(9u^3 - 7uv^2 + 24u^2w - 8v^2w + 16uw^2).$$

The Cartesian equation of the boundary generating curve of $W(A_4, B)$ is

$$-343 + 1176x - 1344x^2 + 512x^3 - 592y^2 + 1024xy^2 - 256x^2y^2 - 256y^4 = 0.$$

$W(A_4, B)$ is the convex hull of $C(A_4, B)$, represented in Fig. 2, and has a flat portion on the boundary parallel to the imaginary axis.

Example 3 Let

$$A_1 = \begin{bmatrix} 1 & 1 & 1 \\ 0 & 1 & 1 \\ 0 & 0 & 1 \end{bmatrix}, \quad B = \operatorname{diag}(1, 1, 0).$$

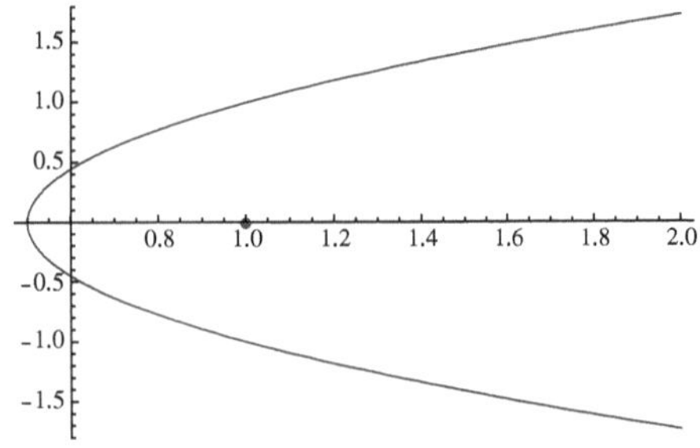

Fig. 3 Boundary of $W(A_1, B)$ (Example 3)

The characteristic polynomial of the pencil (A_1, B) is

$$f(u, v, w) = \frac{1}{2}(u + w)(u^2 - v^2 + 2uw).$$

The Cartesian equation of the boundary generating curve of $W(A_1, B)$ is

$$(y^2 - 2x + 1)((x - 1)^2 + y^2) = 0.$$

(cf. Fig. 3).

Example 4 Let

$$A = \begin{bmatrix} 1 & 1 & 0 \\ 0 & 1 & 0 \\ 0 & 0 & 1 \end{bmatrix}, \quad B = \mathrm{diag}(1, 1, 0)$$

The characteristic polynomial of (A, B) is

$$f(u, v, w) = \frac{1}{4}u(3u^2 - v^2 + 8uw + 4w^2).$$

The Cartesian equation of $C(A, B)$ is $(x - 1)^2 + y^2 = \frac{1}{4}$ and is represented in Fig. 4. There are two flat portions, extending to infinity, on the boundary of $W(A, B)$.

Example 5 Let $B = \mathrm{diag}(1, 0, 0)$ and $A = A_1$ in Example 3. The characteristic polynomial of the pencil is $f(u, v, w) = (2u^3 - 2uv^2 + 3u^2w - v^2w)/4$. The Cartesian equation of the boundary of $W(A, B)$ is

$$16 - 48x + 48x^2 - 20x^3 + 3x^4 + 36y^2 - 36xy^2 - 18x^2y^2 + 27y^4 = 0$$

and is represented in Fig. 5.

Example 6 Let $B = \mathrm{diag}(1, 0, 0)$ and

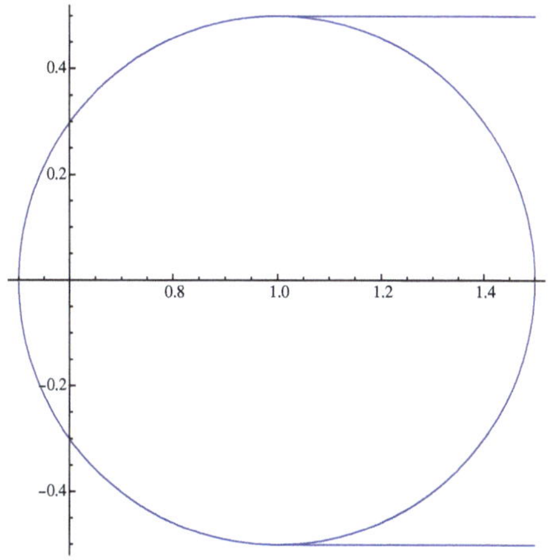

Fig. 4 Boundary of $W(A, B)$ (Example 4)

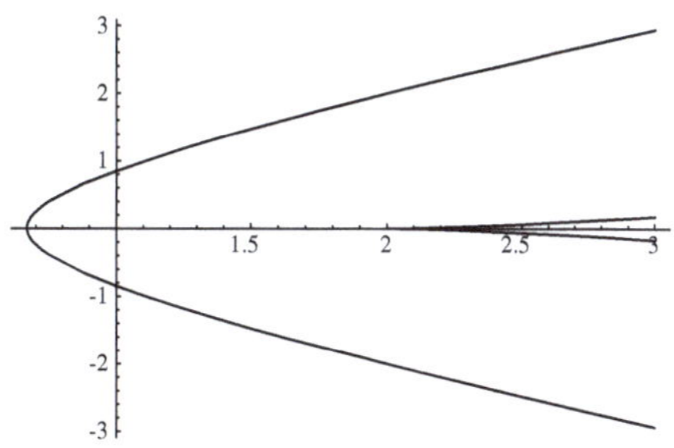

Fig. 5 Boundary of $W(A, B)$ (Example 5)

$$A = \begin{pmatrix} 1 & 1 & 1 \\ 0 & 1 & 1 \\ 0 & 0 & -1 \end{pmatrix}.$$

The characteristic polynomial of (A, B) is $f(u, v, w) = 1/4(-4u^3 - 5u^2w - v^2w)$. The Cartesian equation of $C(A, B)$ is the deltoid

$$-4x^3 + 5x^4 + 108y^2 - 180xy^2 + 50x^2y^2 + 125y^4 = 0.$$

Since

$$0 \in W\left(\begin{pmatrix} 1 & 1 \\ 0 & -1 \end{pmatrix}\right),$$

it follows that $W(A, B) = C$.

Example 7 Let $B = \text{diag}(1, 0, 0)$, and

$$A = \begin{pmatrix} 1 & 1 & 1 \\ 0 & 1 & 0 \\ 0 & 0 & 1 \end{pmatrix}.$$

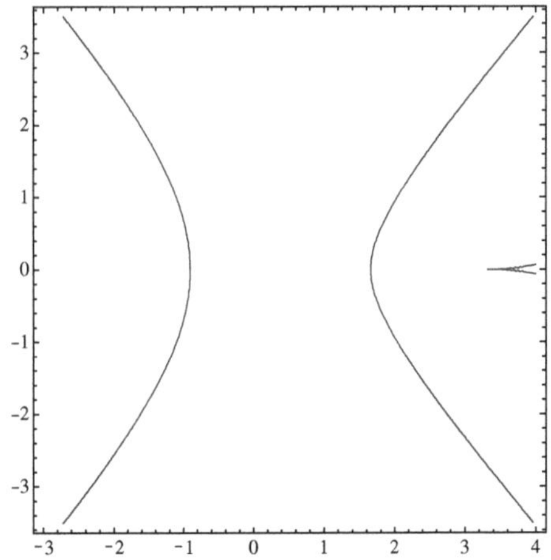

Fig. 6 $C(A, B)$ for Example 8

The characteristic polynomial of (A, B) is given by $f(u, v, w) = 1/2u(u^2 - v^2 + 2uw)$. The boundary of $W(A, B)$ is parabolic and its Cartesian equation is

$$y^2 - 2x + 1 = 0.$$

Example 8 Let $B = \text{diag}(1, -1, 0)$, and

$$A = \begin{pmatrix} 2 & 2 & 1 \\ 0 & 2 & 2 \\ 0 & 0 & 1 \end{pmatrix}.$$

The characteristic polynomial of (A, B) is given by

$$f(u, v, w) = \frac{1}{2}3u^3 - \frac{1}{2}5uv^2 - \frac{1}{4}3u^2w - \frac{1}{4}3v^2w - uw^2.$$

The boundary generating curve $C(A, B)$ is represented in Fig. 6, it has Cartesian equation

$$6000 - 2400x - 5080x^2 + 4248x^3 - 1161x^4 + 108x^5 + 2808y^2 + 1752xy^2 + 1678x^2y^2$$

$$-2184x^3y^2 + 36x^4y^2 + 2007y^4 + 2316xy^4 - 568x^2y^4 + 420y^6 = 0$$

and is constituted of 2 branches, $C_+(A, B)$ for $x \leq (3 - \sqrt{105})/8$ and $C_-(A, B)$ for $x \geq (3 + \sqrt{105})/8$. The pseudo-convex hull of $C_+(A, B)$ and $C_-(A, B)$ is $W(A, B)$.

6 Final Remarks

We presented the classification of the boundary generating curves of $W(A, B)$ for 2×2 and 3×3 matrices A, B, following Kipenhann's approach for the classical numerical range of a matrix. We have considered linear pencils generated by a pair (A, B) of which at least one of the matrices is Hermitian. It would be challenging to drop this constraint. The systematic investigation of the existence of flat portions on the boundary, as well as its implications on the matrix structure, are open problems deserving the attention of researchers. The interplay between the algebraic properties of the polynomial $f(u, v, w)$ and the geometric properties of $W(A, B)$ must be stressed and deserves further investigation.

Acknowledgements This work was partially supported by the Centre for Mathematics of the University of Coimbra – UID/MAT/00324/2013, funded by the Portuguese Government through FCT/MEC and co-funded by the European Regional Development Fund through the Partnership Agreement PT2020.

References

1. Ball, W.W.R.: On Newton's classification of cubic curves. Proc. Lond. Math. Soc. **22**, 104–143 (1890)
2. Bebiano, N., Lemos, R., da ProvidĹencia, J., Soares, G.: On the geometry of numerical ranges in spaces with an indefinite inner product. Linear Algebr. Appl. **399**, 17–34 (2005)
3. Bebiano, N., da Providencia, J., Teixeira, R.: Indefinite numerical range of 3 × 3 matrices. Czechoslov. Math. J. **59**, 221–239 (2009)
4. Bebiano, N., da Providencia, J., Nata, A., Providencia, J.P.: Fields of values of linear pencils and spectral inclusion regions, in this volume
5. Brieskorn, E., Knörrer, H.: Plane Algebraic Curves. Birkhäuser Verlag, Basel (1986)
6. Gohberg, I., Lancaster, P., Rodman, L.: Indefinite Linear Algebra and Applications. Birkhäuser Verlag, Basel (2005)
7. Gustafson, K.E., Rao, D.K.M.: Numerical Range, The Field of Values of Linear Operators and Matrices. Springer, New York (1997)
8. Hochstenbach, M.E.: Fields of values and inclusion regions for matrix pencils. Electron. Trans. Numer. Anal. **38**, 98–112 (2011)
9. Kippenhahn, R.: Über der wertevorrat einer matrix. Math. Nachr. **6**, 193–228 (1951)
10. Li, C.-K., Rodman, L.: Numerical range of matrix polynomials. SIAM J. Matrix Anal. Appl. **15**, 1256–1265 (1994)
11. Murnaghan, F.D.: On the field of values of a square matrix. Proc. Nat. Acad. Sci. USA **18**, 246–248 (1932)
12. Nakazato, H., Psarrakos, P.: On the shape of numerical range of matrix polynomials. Linear Algebr. Appl. **338**, 105–123 (2001)
13. Psarrakos, P.: Numerical range of linear pencils. Linear Algebr. Appl. **317**, 127–141 (2000)
14. Shapiro, H.: A conjecture of Kippenhahn about the characteristic polynomial of a pencil generated by two Hermitian matrices II. Linear Algebr. Appl. **45**, 97–108 (1982)

Integer Powers of Certain Complex Pentadiagonal Toeplitz Matrices

Hatice Kübra Duru and Durmuş Bozkurt

Abstract In this paper, we obtain a general expression for the entries of the rth ($r \in \mathbb{Z}$) power of a certain $n \times n$ pentadiagonal Toeplitz matrix. Additionally, we present the complex factorizations of Fibonacci polynomials.

Keywords Pentadiagonal Toeplitz matrices · Fibonacci polynomials · Fibonacci numbers · Pell numbers

1 Introduction

Band matrices are used in many areas and are also included in the solution of many systems. In particular they appear in numerical analysis, differential equations, difference equations, in the solution of boundary value problems, in the numerical solution of ordinary and partial differential equations, delay differential equations, interpolation problems, and in many applied fields. Lately, the calculations of integer powers and of the eigenvalues of band matrices have been well studied in the literature. In [10–12] Rimas obtained the positive integer powers of certain symmetric pentadiagonal matrices and symmetric anti-pentadiagonal matrices in terms of the Chebyshev polynomials. The characteristic polynomial and eigenvectors for pentadiagonal matrices are derived in [4]. Arslan et al. [1] investigated the general expression of the powers of even order symmetric pentadiagonal matrices. Öteleş and Akbulak [9] presented the general expression for the entries of the powers of certain $n \times n$ complex tridiagonal matrices, in terms of the Chebyshev polynomials of the first kind and two

H. Kübra Duru (✉) · D. Bozkurt
Science Faculty Department of Mathematic, Selcuk University, Konya, Turkey
email: hkduru@selcuk.edu.tr

D. Bozkurt
email: dbozkurt@selcuk.edu.tr

N. Bebiano (ed.), *Applied and Computational Matrix Analysis*,
Springer Proceedings in Mathematics & Statistics 192,
DOI 10.1007/978-3-319-49984-0_14

complex factorizations for Fibonacci and Pell numbers. The result for all positive integer powers of a Toeplitz matrix is restated in [13]. Duru and Bozkurt [3] obtained the general expression of the powers of some tridiagonal matrices. The powers of odd order circulant matrices are calculated in [6].

In this paper, we present a general expression for the entries of the rth power of a certain $n \times n$ complex pentadiagonal Toeplitz matrix.

2 Eigenvalues and eigenvectors of A_n

Theorem 1 *Let A_n be the $n \times n$ ($n = 2t, 2 \leq t \in \mathbb{N}$) pentadiagonal Toeplitz matrix*

$$A_n := \begin{bmatrix} a & 0 & b & & & & \\ 0 & a & 0 & b & & & \\ c & 0 & a & 0 & \ddots & & \\ & c & \ddots & \ddots & \ddots & b & \\ & & \ddots & 0 & a & 0 & b \\ & & & c & 0 & a & 0 \\ & & & & c & 0 & a \end{bmatrix}, \tag{1}$$

where $a \in \mathbb{C}$ and $b, c \in \mathbb{C} \setminus \{0\}$. Then the eigenvalues and eigenvectors of the matrix A_n are

$$\lambda_k = a - 2\sqrt{bc}\cos\left(\frac{2k\pi}{n+2}\right), \quad k = 1, 2, \ldots, \frac{n}{2} \tag{2}$$

and

$$\begin{bmatrix} U_0(\alpha_k) \\ 0 \\ \mu^{1/2} U_1(\alpha_k) \\ 0 \\ \mu U_2(\alpha_k) \\ 0 \\ \vdots \\ \mu^{(n-2)/4} U_{\frac{n-2}{2}}(\alpha_k) \\ 0 \end{bmatrix}, \quad j = 1, 3, 5, \ldots, n-3, n-1; \; k = \frac{j+1}{2},$$

$$\begin{bmatrix} 0 \\ U_0(\alpha_k) \\ 0 \\ \mu^{1/2}U_1(\alpha_k) \\ 0 \\ \mu U_2(\alpha_k) \\ 0 \\ \vdots \\ \mu^{(n-4)/4}U_{\frac{n-4}{2}}(\alpha_k) \\ 0 \\ \mu^{(n-2)/4}U_{\frac{n-2}{2}}(\alpha_k) \end{bmatrix}; \; j = 2, 4, 6, \ldots, n-2, n; k = \frac{j}{2},$$

where $\mu = \frac{c}{b}$, $\alpha_k = \frac{\lambda_k - a}{2\sqrt{bc}}$ *and* $U_n(.)$ *is the nth degree Chebyshev polynomial of the second kind.*

Proof Let

$$\det(Q_n) := \begin{vmatrix} x-a & b & & & & \\ c & x-a & b & & & \\ & c & x-a & b & & \\ & & \ddots & \ddots & \ddots & \\ & & & c & x-a & b \\ & & & & c & x-a \end{vmatrix}.$$

For the initial conditions $\det(Q_0) = 1$ and $\det(Q_1) = x - a$, we have

$$\det(Q_n) = (x-a)\det(Q_{n-1}) - bc\det(Q_{n-2}), \quad n \geq 2. \tag{3}$$

The solution of the difference equation in (3) is

$$\det(Q_n(x)) = (bc)^{\frac{n}{2}} U_n\left(\frac{x-a}{2\sqrt{bc}}\right), \tag{4}$$

where $U_n(.)$ is the nth degree Chebyshev polynomial of the second kind [8]:

$$U_n(x) = \frac{\sin((n+1)\theta)}{\sin(\theta)},$$

with $x = \cos(\theta)$. All the roots of $U_n(x)$ are in the interval $[-1, 1]$. Let

$$|\lambda I_n - A_n| = \Delta_{A_n}(\lambda)$$

and due to (3), we have

$$\begin{aligned}
\Delta_{A_4}(\lambda) &= (\lambda^2 - 2\lambda a + a^2 - bc)^2 = (Q_2)^2 \\
\Delta_{A_6}(\lambda) &= (a-\lambda)^2(\lambda^2 - 2\lambda a + a^2 - 2bc)^2 = (Q_3)^2 \\
\Delta_{A_8}(\lambda) &= ((\lambda - a)^4 - 3\lambda^2 bc - 3a^2 bc + 6\lambda abc + b^2c^2)^2 = (Q_4)^2 \\
&\ \vdots \\
\Delta_{A_n}(\lambda) &= \left(Q_{\frac{n}{2}}(\lambda)\right)^2 .
\end{aligned}$$

Then we have

$$\Delta_{A_n}(\lambda) = (bc)^{\frac{n}{2}} \left(U_{\frac{n}{2}}\left(\frac{\lambda - a}{2\sqrt{bc}}\right)\right)^2 . \tag{5}$$

The eigenvalues of A_n are obtained as

$$\lambda_k = a - 2\sqrt{bc}\cos\left(\frac{2k\pi}{n+2}\right), \quad k = 1, 2, \ldots, \frac{n}{2},$$

from (5).

The multiplicity of all the eigenvalues λ_k ($k = 1, 2, \ldots, \frac{n}{2}$) of the matrix A_n are 2. Since $rank(\lambda_k I_n - A_n) = n - 2$, to each eigenvalue λ_k correspond two Jordan cells $J_k(\lambda_k)$ in the matrix J. That is,

$$J_n = \text{diag}(\lambda_1, \lambda_1, \lambda_2, \lambda_2, \ldots, \lambda_{\frac{n}{2}}, \lambda_{\frac{n}{2}}). \tag{6}$$

Considering the relations $K^{-1}A_nK = J_n$ [5], we obtain the matrices K and K^{-1} and derive the expression of the matrix A_n^r for $r \in \mathbb{N}$. Let us denote the j-th column of K by K_j ($j = 1, \ldots, n$). Then

$$A_nK = (K_1\lambda_1\ K_2\lambda_1\ K_3\lambda_2\ K_4\lambda_2 \ldots K_{n-1}\lambda_{\frac{n}{2}}\ K_n\lambda_{\frac{n}{2}}). \tag{7}$$

From Eq. (7), we have the system of linear equations as follows:

$$\begin{aligned}
A_nK_1 &= K_1\lambda_1 \\
A_nK_2 &= K_2\lambda_1 \\
A_nK_3 &= K_3\lambda_2 \\
A_nK_4 &= K_4\lambda_2 \\
&\ \vdots \\
A_nK_{n-3} &= K_{n-3}\lambda_{\frac{n-2}{2}} \\
A_nK_{n-2} &= K_{n-2}\lambda_{\frac{n-2}{2}} \\
A_nK_{n-1} &= K_{n-1}\lambda_{\frac{n}{2}} \\
A_nK_n &= K_n\lambda_{\frac{n}{2}} .
\end{aligned} \tag{8}$$

Solving the system of linear equations in (8), we obtain

$$K_j = \begin{bmatrix} U_0(\alpha_k) \\ 0 \\ \mu^{1/2}U_1(\alpha_k) \\ 0 \\ \mu U_2(\alpha_k) \\ 0 \\ \vdots \\ \mu^{(n-2)/4}U_{\frac{n-2}{2}}(\alpha_k) \\ 0 \end{bmatrix}, \; j = 1, 3, 5, \ldots, n-3, n-1; k = \frac{j+1}{2}, \tag{9}$$

and

$$K_j = \begin{bmatrix} 0 \\ U_0(\alpha_k) \\ 0 \\ \mu^{1/2}U_1(\alpha_k) \\ 0 \\ \mu U_2(\alpha_k) \\ 0 \\ \vdots \\ \mu^{(n-4)/4}U_{\frac{n-4}{2}}(\alpha_k) \\ 0 \\ \mu^{(n-2)/4}U_{\frac{n-2}{2}}(\alpha_k) \end{bmatrix}; \; j = 2, 4, 6, \ldots, n-2, n; k = \frac{j}{2}, \tag{10}$$

where $\mu = \frac{c}{b}$, $\alpha_k = \frac{\lambda_k - a}{2\sqrt{bc}}$ and $U_n(.)$ is the nth degree Chebyshev polynomial of the second kind. □

Theorem 2 *Let A_n be the $n \times n$ ($n = 2t+1$, $t \in \mathbb{N}$) pentadiagonal Toeplitz matrix in (1). Then the eigenvalues and eigenvectors of the matrix A_n are*

$$\beta_m = \begin{cases} a - 2\sqrt{bc}\cos\left(\frac{(m+1)\pi}{n+3}\right), & m = 1, 3, 5, \ldots, n \\ a - 2\sqrt{bc}\cos\left(\frac{m\pi}{n+1}\right), & m = 2, 4, 6, \ldots, n-1 \end{cases} \tag{11}$$

and

$$\begin{bmatrix} U_0(\delta_j) \\ 0 \\ \mu^{1/2}U_1(\delta_j) \\ 0 \\ \mu U_2(\delta_j) \\ \vdots \\ 0 \\ \mu^{(n-1)/4}U_{\frac{n-1}{2}}(\delta_j) \end{bmatrix}; \; j = 1, 3, 5, \ldots, n-2, n;$$

$$\begin{bmatrix} 0 \\ U_0\left(\delta_j\right) \\ 0 \\ \mu^{1/2}U_1\left(\delta_j\right) \\ 0 \\ \mu U_2\left(\delta_j\right) \\ 0 \\ \vdots \\ \mu^{(n-3)/4}U_{\frac{n-3}{2}}\left(\delta_j\right) \\ 0 \end{bmatrix} ; \quad j = 2, 4, 6, \ldots, n-3, n-1,$$

where $\mu = \frac{c}{b}$, $\delta_j = \frac{\beta_j - a}{2\sqrt{bc}}$ *and* $U_n(.)$ *is the nth degree Chebyshev polynomial of the second kind.*

Proof Let

$$|\beta I_n - A_n| = \Delta_{A_n}(\beta)$$

and owing to (3), we obtain

$$\begin{aligned} \Delta_{A_3}(\beta) &= a(a^2 - bc) = Q_1(\beta)\, Q_2(\beta) \\ \Delta_{A_5}(\beta) &= a(a^2 - 2bc)(a^2 - bc) = Q_2(\beta)\, Q_3(\beta) \\ \Delta_{A_7}(\beta) &= a(a^2 - 2bc)(a^4 - 3a^2bc - b^2c^2) = Q_3(\beta)\, Q_4(\beta) \\ &\vdots \\ \Delta_{A_n}(\beta) &= Q_{\frac{n-1}{2}}(\beta)\, Q_{\frac{n+1}{2}}(\beta)\,. \end{aligned}$$

From Eq. (4), we have

$$\Delta_{A_n}(\beta) = (bc)^{\frac{n}{2}}\, U_{\frac{n-1}{2}}\left(\frac{\beta - a}{2\sqrt{bc}}\right) U_{\frac{n+1}{2}}\left(\frac{\beta - a}{2\sqrt{bc}}\right).$$

The eigenvalues of A_n $(n = 2t+1, t \in \mathbb{N})$ are

$$\beta_m = \begin{cases} a - 2\sqrt{bc}\cos\left(\frac{(m+1)\pi}{n+3}\right), & m = 1, 3, 5, \ldots, n \\ a - 2\sqrt{bc}\cos\left(\frac{m\pi}{n+1}\right), & m = 2, 4, 6, \ldots, n-1. \end{cases}$$

All the eigenvalues β_m $(m = 1, 2, \ldots, n)$ of the matrix A_n are simple. Since $rank(\beta_m I_n - A_n) = n - 1$, to each eigenvalue β_m correspond Jordan cells $J_m^{\dagger}(\beta_m)$ in the matrix $J^{\dagger}$. That is,

$$J^{\dagger} = \operatorname{diag}(\beta_1, \beta_2, \beta_3, \ldots, \beta_n)\,. \tag{12}$$

Using the well known equality $S^{-1}A_nS = J^{\dagger}$, we obtain the matrices S and S^{-1}. Let us denote the j-th column of S by S_j $(j = 1, \ldots, n)$. Then

$$A_n S = (S_1\beta_1\ S_2\beta_2\ S_3\beta_3\ S_4\beta_4 \ldots S_{n-1}\beta_{n-1}\ S_n\beta_n). \tag{13}$$

We have the system of linear equations

$$\begin{array}{rcl} A_n S_1 &=& S_1\beta_1 \\ A_n S_2 &=& S_2\beta_2 \\ A_n S_3 &=& S_3\beta_3 \\ A_n S_4 &=& S_4\beta_4 \\ &\vdots& \\ A_n S_{n-3} &=& S_{n-3}\beta_{n-3} \\ A_n S_{n-2} &=& S_{n-2}\beta_{n-2} \\ A_n S_{n-1} &=& S_{n-1}\beta_{n-1} \\ A_n S_n &=& S_n\beta_n. \end{array} \tag{14}$$

Solving the system of linear equations in (14), we have

$$S_j = \begin{bmatrix} U_0\left(\delta_j\right) \\ 0 \\ \mu^{1/2}U_1\left(\delta_j\right) \\ 0 \\ \mu U_2\left(\delta_j\right) \\ \vdots \\ 0 \\ \mu^{(n-1)/4}U_{\frac{n-1}{2}}\left(\delta_j\right) \end{bmatrix};\ j = 1, 3, 5, \ldots, n-2, n; \tag{15}$$

and

$$S_j = \begin{bmatrix} 0 \\ U_0\left(\delta_j\right) \\ 0 \\ \mu^{1/2}U_1\left(\delta_j\right) \\ 0 \\ \mu U_2\left(\delta_j\right) \\ 0 \\ \vdots \\ \mu^{(n-3)/4}U_{\frac{n-3}{2}}\left(\delta_j\right) \\ 0 \end{bmatrix};\ \ j = 2, 4, 6, \ldots, n-3, n-1, \tag{16}$$

where $\mu = \frac{c}{b}$, $\delta_j = \frac{\beta_j - a}{2\sqrt{bc}}$ and $U_n(.)$ is the nth degree Chebyshev polynomial of the second kind. □

3 The Integer Powers of the Matrix A_n

Firstly, we suppose n a positive even integer ($n = 2t$, $2 \leq t \in \mathbb{N}$).
Considering (9) and (10), we write the matrix K

$$K = \left[\begin{array}{ccc} U_0(\alpha_1) & 0 & U_0(\alpha_2) \\ 0 & U_0(\alpha_1) & 0 \\ \mu^{1/2}U_1(\alpha_1) & 0 & \mu^{1/2}U_1(\alpha_2) \\ 0 & \mu^{1/2}U_1(\alpha_1) & 0 \\ \mu U_2(\alpha_1) & 0 & \mu U_2(\alpha_2) \\ 0 & \mu U_2(\alpha_1) & 0 \\ \vdots & \vdots & \vdots \\ \mu^{(n-4)/4}U_{\frac{n-4}{2}}(\alpha_1) & 0 & \mu^{(n-4)/4}U_{\frac{n-4}{2}}(\alpha_2) \\ 0 & \mu^{(n-4)/4}U_{\frac{n-4}{2}}(\alpha_1) & 0 \\ \mu^{(n-2)/4}U_{\frac{n-2}{2}}(\alpha_1) & 0 & \mu^{(n-2)/4}U_{\frac{n-2}{2}}(\alpha_2) \\ 0 & \mu^{(n-2)/4}U_{\frac{n-2}{2}}(\alpha_1) & 0 \end{array} \right.$$

$$\left. \begin{array}{cccc} 0 & \dots & U_0\left(\alpha_{\frac{n}{2}}\right) & 0 \\ U_0(\alpha_2) & \dots & 0 & U_0\left(\alpha_{\frac{n}{2}}\right) \\ 0 & \dots & \mu^{1/2}U_1\left(\alpha_{\frac{n}{2}}\right) & 0 \\ \mu^{1/2}U_1(\alpha_2) & \dots & 0 & \mu^{1/2}U_1\left(\alpha_{\frac{n}{2}}\right) \\ 0 & \dots & \mu U_2\left(\alpha_{\frac{n}{2}}\right) & 0 \\ \vdots & \ddots & \vdots & \vdots \\ 0 & \dots & \mu^{(n-4)/4}U_{\frac{n-4}{2}}\left(\alpha_{\frac{n}{2}}\right) & 0 \\ \mu^{(n-4)/4}U_{\frac{n-4}{2}}(\alpha_2) & \dots & 0 & \mu^{(n-4)/4}U_{\frac{n-4}{2}}\left(\alpha_{\frac{n}{2}}\right) \\ 0 & \dots & \mu^{(n-2)/4}U_{\frac{n-2}{2}}\left(\alpha_{\frac{n}{2}}\right) & 0 \\ \mu^{(n-2)/4}U_{\frac{n-2}{2}}(\alpha_2) & \dots & 0 & \mu^{(n-2)/4}U_{\frac{n-2}{2}}\left(\alpha_{\frac{n}{2}}\right) \end{array} \right]. \quad (17)$$

Now, let us find the inverse matrix K^{-1} of the matrix K. If we denote the i-th row of the inverse matrix K^{-1} by K_i^{-1}, then we have

$$K_i^{-1} := \left[\begin{array}{c} q_k U_0(\alpha_k) \\ 0 \\ q_k \mu^{-1/2} U_1(\alpha_k) \\ 0 \\ q_k \mu^{-1} U_2(\alpha_k) \\ 0 \\ q_k \mu^{-3/2} U_3(\alpha_k) \\ \vdots \\ 0 \\ q_k \mu^{-(n-2)/4} U_{\frac{n-2}{2}}(\alpha_k) \\ 0 \end{array} \right]^T ; i = 1, 3, 5, \dots, n-1;\ k = \frac{i+1}{2} \quad (18)$$

and

$$K_i^{-1} := \begin{bmatrix} 0 \\ q_k U_0(\alpha_k) \\ 0 \\ q_k \mu^{-1/2} U_1(\alpha_k) \\ 0 \\ q_k \mu^{-1} U_2(\alpha_k) \\ 0 \\ \vdots \\ q_k \mu^{-(n-4)/4} U_{\frac{n-4}{2}}(\alpha_k) \\ 0 \\ q_k \mu^{-(n-2)/4} U_{\frac{n-2}{2}}(\alpha_k) \end{bmatrix}^T ; i = 2, 4, 6, \ldots, n;\ k = \frac{i}{2}, \tag{19}$$

where $\mu = \frac{c}{b}$, $q_k = \frac{4-4\alpha_k^2}{n+2}$ and $\alpha_k = \frac{\lambda_k - a}{2\sqrt{bc}}$ for $k = 1, 2, \ldots, \frac{n}{2}$. Thus, we obtain

$$K^{-1} = \left[\begin{array}{cccccc} q_1 U_0(\alpha_1) & 0 & q_1 \mu^{-1/2} U_1(\alpha_1) & \cdots & q\mu^{-(n-2)/4} U_{\frac{n-2}{2}}(\alpha_1) & 0 \\ 0 & q_1 U_0(\alpha_1) & 0 & \cdots & 0 & q_1 \mu^{-(n-2)/4} U_{\frac{n-2}{2}}(\alpha_1) \\ q_2 U_0(\alpha_2) & 0 & q_2 \mu^{-1/2} U_1(\alpha_2) & \cdots & q\mu^{-(n-2)/4} U_{\frac{n-2}{2}}(\alpha_2) & 0 \\ 0 & q_2 U(\alpha_2) & 0 & \cdots & 0 & q_2 \mu^{-(n-2)/4} U_{\frac{n-2}{2}}(\alpha_2) \\ q_3 U_0(\alpha_3) & 0 & q_3 \mu^{-1/2} U_1(\alpha_3) & \cdots & q_3 \mu^{-(n-2)/4} U_{\frac{n-2}{2}}(\alpha_3) & 0 \\ 0 & q_3 U_0(\alpha_3) & 0 & \cdots & 0 & q_3 \mu^{-(n-2)/4} U_{\frac{n-2}{2}}(\alpha_3) \\ \vdots & \vdots & \vdots & \ddots & \vdots & \vdots \\ q_{\frac{n-2}{2}} U_0(\alpha_{\frac{n-2}{2}}) & 0 & q_{\frac{n-2}{2}} \mu^{-1/2} U_1(\alpha_{\frac{n-2}{2}}) & \cdots & q_{\frac{n-2}{2}} \mu^{-(n-2)/4} U_{\frac{n-2}{2}}(\alpha_{\frac{n-2}{2}}) & 0 \\ 0 & q_{\frac{n-2}{2}} U_0(\alpha_{\frac{n-2}{2}}) & 0 & \cdots & 0 & q_{\frac{n-2}{2}} \mu^{-(n-2)/4} U_{\frac{n-2}{2}}(\alpha_{\frac{n-2}{2}}) \\ q_{\frac{n}{2}} U_0(\alpha_{\frac{n}{2}}) & 0 & q_{\frac{n}{2}} \mu^{-1/2} U_1(\alpha_{\frac{n}{2}}) & \cdots & q_{\frac{n}{2}} \mu^{-(n-2)/4} U_{\frac{n-2}{2}}(\alpha_{\frac{n}{2}}) & 0 \\ 0 & q_{\frac{n}{2}} U_0(\alpha_{\frac{n}{2}}) & 0 & \cdots & 0 & q_{\frac{n}{2}} \mu^{-(n-2)/4} U_{\frac{n-2}{2}}(\alpha_{\frac{n}{2}}) \end{array}\right]. \tag{20}$$

By combining (6), (17) and (20) and using the equality $A_n^r = KJ^rK^{-1}$ $(r \in \mathbb{N})$ [5], we compute the rth powers of the matrix A_n

$$A_n^r = KJ^rK^{-1} = W(r) = \left(w_{ij}(r)\right). \tag{21}$$

So for $i, j = \overline{1, n}$

$$w_{ij}(r) = \begin{cases} 0, & if\ (-1)^{i+j} = -1, \\ \sum_{k=1}^{\frac{n}{2}} \lambda_k^r q_k \mu^{\frac{i-j}{4}} U_{\frac{i\varepsilon_{ij}-\varphi_{ij}}{2}}(\alpha_k)\, U_{\frac{j\varepsilon_{ij}-\varphi_{ij}}{2}}(\alpha_k), & if\ (-1)^{i+j} = 1, \end{cases} \tag{22}$$

$$\varepsilon_{ij} = \begin{cases} 1, & if\ (-1)^i = (-1)^j = -1, \\ -1, & if\ (-1)^i = (-1)^j = 1, \end{cases} \tag{23}$$

$$\varphi_{ij} = \begin{cases} 1, & if\ (-1)^i = (-1)^j = -1, \\ -n, & if\ (-1)^i = (-1)^j = 1, \end{cases} \tag{24}$$

where $\mu = \frac{c}{b}$, $q_k = \frac{4-4\alpha_k^2}{n+2}$, $\alpha_k = \frac{\lambda_k - a}{2\sqrt{bc}}$, and λ_k $\left(k = 1, 2, 3, \ldots, \frac{n}{2}\right)$ are the eigenvalues of the matrix A_n $(n = 2t, 2 \le t \in \mathbb{N})$.

Example 1 Setting $n = 6$ in Theorem 1, we have

$$\begin{aligned} J &= \mathrm{diag}(\lambda_1, \lambda_1, \lambda_2, \lambda_2, \lambda_3, \lambda_3) \\ &= \mathrm{diag}(a - 1.414\sqrt{bc}, a - 1.414\sqrt{bc}, a, a, a + 1.414\sqrt{bc}, a + 1.414\sqrt{bc}) \end{aligned}$$

and

$$\begin{aligned} A_6^r &= KJ^rK^{-1} = W(r) \\ &= \left(w_{ij}(r)\right) = \begin{bmatrix} w_{11}(r) & w_{12}(r) & w_{13}(r) & w_{14}(r) & w_{15}(r) & w_{16}(r) \\ w_{21}(r) & w_{22}(r) & w_{23}(r) & w_{24}(r) & w_{25}(r) & w_{26}(r) \\ w_{31}(r) & w_{32}(r) & w_{33}(r) & w_{34}(r) & w_{35}(r) & w_{36}(r) \\ w_{41}(r) & w_{42}(r) & w_{33}(r) & w_{44}(r) & w_{45}(r) & w_{46}(r) \\ w_{51}(r) & w_{52}(r) & w_{53}(r) & w_{54}(r) & w_{55}(r) & w_{56}(r) \\ w_{61}(r) & w_{62}(r) & w_{63}(r) & w_{64}(r) & w_{65}(r) & w_{66}(r) \end{bmatrix}, \end{aligned}$$

$$w_{ij}(r) = 0\ for\ (-1)^{i+j} = -1,$$

$$\begin{aligned}
w_{11}(r) &= w_{22}(r) = w_{55}(r) = w_{66}(r) \\
&= 0.25\left(a - 1.414\sqrt{bc}\right)^r + 0.5a^r + 0.25\left(a + 1.414\sqrt{bc}\right)^r; \\
w_{13}(r) &= w_{24}(r) = w_{35}(r) = w_{46}(r) \\
&= -0.354\left(a - 1.414\sqrt{bc}\right)^r \mu^{-1/2} + 0.354\left(a + 1.414\sqrt{bc}\right)^r \mu^{-1/2}; \\
w_{15}(r) &= w_{26}(r) \\
&= 0.25\left(a - 1.414\sqrt{bc}\right)^r \mu^{-1} - 0.5a^r\mu^{-1} + 0.25\left(a + 1.414\sqrt{bc}\right)^r \mu^{-1}; \\
w_{31}(r) &= w_{42}(r) = w_{53}(r) = w_{64}(r) \\
&= -0.354\left(a - 1.414\sqrt{bc}\right)^r \mu^{1/2} + 0.354\left(a + 1.414\sqrt{bc}\right)^r \mu^{1/2}; \\
w_{33}(r) &= w_{44}(r) = 0.5\left(a - 1.414\sqrt{bc}\right)^r + 0.5\left(a + 1.414\sqrt{bc}\right)^r; \\
w_{51}(r) &= w_{62}(r) \\
&= 0.25\left(a - 1.414\sqrt{bc}\right)^r \mu - 0.5a^r\mu + 0.25\left(a + 1.414\sqrt{bc}\right)^r \mu.
\end{aligned}$$

Example 2 Setting $r = 4,\ n = 8,\ a = i + 1,\ b = 2$ and $c = 1$ in Theorem 1, we get

$$\begin{aligned}
J &= \operatorname{diag}(\lambda_1, \lambda_1, \lambda_2, \lambda_2, \lambda_3, \lambda_3, \lambda_4, \lambda_4) \\
&= \operatorname{diag}(-1.288 + i, -1.288 + i, 0.126 + i, 0.126 + i, \\
&\quad 1.874 + i, 1.874 + i, 3.288 + i, 3.288 + i)
\end{aligned}$$

and

$$\begin{aligned}
A_8^4 &= \left(w_{ij}(4)\right) \\
&= \begin{bmatrix}
4+24i & 0 & 16+48i & 0 & 24+48i & 0 & 32+32i & 0 \\
0 & 4+24i & 0 & 16+48i & 0 & 24+48i & 0 & 32+32i \\
8+24i & 0 & 16+48i & 0 & 32+64i & 0 & 24+48i & 0 \\
0 & 8+24i & 0 & 16+48i & 0 & 32+64i & 0 & 24+48i \\
6+12i & 0 & 16+32i & 0 & 16+48i & 0 & 16+48i & 0 \\
0 & 6+12i & 0 & 16+32i & 0 & 16+48i & 0 & 16+48i \\
4+4i & 0 & 6+12i & 0 & 8+24i & 0 & 4+24i & 0 \\
0 & 4+4i & 0 & 6+12i & 0 & 8+24i & 0 & 4+24i
\end{bmatrix}.
\end{aligned}$$

Secondly, we suppose n a positive odd integer ($n = 2t + 1,\ t \in \mathbb{N}$). From (15) and (16) we write S as:

$$S = \begin{bmatrix} U_0(\delta_1) & 0 & U_0(\delta_3) & U_0(\delta_4) & \cdots & 0 & U_0(\delta_n) \\ 0 & U_0(\delta_2) & 0 & 0 & \cdots & U_0(\delta_{n-1}) & 0 \\ \mu^{1/2}U_1(\delta_1) & 0 & \mu^{1/2}U_1(\delta_3) & \mu^{1/2}U_1(\delta_4) & \cdots & 0 & \mu^{1/2}U_1(\delta_n) \\ 0 & \mu^{1/2}U_1(\delta_2) & 0 & 0 & \cdots & \mu^{1/2}U_1(\delta_{n-1}) & 0 \\ \mu U_2(\delta_1) & 0 & \mu U_2(\delta_3) & \mu U_2(\delta_4) & \cdots & 0 & \mu U_2(\delta_n) \\ 0 & \mu U_2(\delta_2) & 0 & 0 & \cdots & \mu U_2(\delta_{n-1}) & 0 \\ \vdots & \vdots & \vdots & \vdots & \ddots & \vdots & \vdots \\ 0 & \mu^{(n-5)/4}U_{\frac{n-5}{2}}(\delta_2) & 0 & 0 & \cdots & \mu^{(n-5)/4}U_{\frac{n-5}{2}}(\delta_{n-1}) & 0 \\ \mu^{(n-3)/4}U_{\frac{n-3}{2}}(\delta_1) & 0 & \mu^{(n-3)/4}U_{\frac{n-3}{2}}(\delta_3) & \mu^{(n-3)/4}U_{\frac{n-3}{2}}(\delta_4) & \cdots & 0 & \mu^{(n-3)/4}U_{\frac{n-3}{2}}(\delta_n) \\ 0 & \mu^{(n-3)/4}U_{\frac{n-3}{2}}(\delta_2) & 0 & 0 & \cdots & \mu^{(n-3)/4}U_{\frac{n-3}{2}}(\delta_{n-1}) & 0 \\ \mu^{(n-1)/4}U_{\frac{n-1}{2}}(\delta_1) & 0 & \mu^{(n-1)/4}U_{\frac{n-1}{2}}(\delta_3) & \mu^{(n-1)/4}U_{\frac{n-1}{2}}(\delta_4) & \cdots & 0 & \mu^{(n-1)/4}U_{\frac{n-1}{2}}(\delta_n) \end{bmatrix}. \tag{25}$$

Now let us find the inverse matrix S^{-1} of the matrix S. If we denote the ith row of the inverse matrix S^{-1} by S_i^{-1}, then we obtain

$$S_i^{-1} := \begin{bmatrix} y_i U_0(\delta_i) \\ 0 \\ y_i \mu^{-1/2} U_1(\delta_i) \\ 0 \\ y_i \mu^{-1} U_2(\delta_i) \\ 0 \\ y_i \mu^{-3/2} U_3(\delta_i) \\ \vdots \\ y_i \mu^{-(n-3)/4} U_{\frac{n-3}{2}}(\delta_i) \\ 0 \\ y_i \mu^{-(n-1)/4} U_{\frac{n-1}{2}}(\delta_i) \end{bmatrix}^T ; i = 1, 3, 5, \ldots, n \tag{26}$$

and

$$S_i^{-1} := \begin{bmatrix} 0 \\ y_i U_0(\delta_i) \\ 0 \\ y_i \mu^{-1/2} U_1(\delta_i) \\ 0 \\ y_i \mu^{-1} U_2(\delta_i) \\ 0 \\ \vdots \\ 0 \\ y_i \mu^{-(n-3)/4} U_{\frac{n-3}{2}}(\delta_i) \\ 0 \end{bmatrix}^T ; i = 2, 4, 6, \ldots, n-1, \qquad (27)$$

where $\mu = \frac{c}{b}$, $\delta_m = \frac{\beta m - a}{2\sqrt{bc}}$ and

$$y_i = \begin{cases} \frac{4-4\delta_i^2}{n+3}, & if \;\; i = 1, 3, 5, \ldots, n \\ \frac{4-4\delta_i^2}{n+1}, & if \;\; i = 2, 4, 6, \ldots, n-1, \end{cases}$$

for $m = 1, 2, \ldots, n$. Then we have

$$S^{-1} = \left[\begin{matrix} y_1 U_0(\delta_1) & 0 & y_1 \mu^{-1/2} U_1(\delta_1) \\ 0 & y_2 U_0(\delta_2) & 0 \\ y_3 U_0(\delta_3) & 0 & y_3 \mu^{-1/2} U_1(\delta_3) \\ 0 & y_4 U_0(\delta_4) & 0 \\ y_5 U_0(\delta_5) & 0 & y_5 \mu^{-1/2} U_1(\delta_5) \\ 0 & y_6 U_0(\delta_6) & 0 \\ \vdots & \vdots & \vdots \\ 0 & y_{n-3} U_0(\delta_{n-3}) & 0 \\ y_{n-2} U_0(\delta_{n-2}) & 0 & y_{n-2} \mu^{-1/2} U_1(\delta_{n-2}) \\ 0 & y_{n-1} U_0(\delta_{n-1}) & 0 \\ y_n U_0(\delta_n) & 0 & y_n \mu^{-1/2} U_1(\delta_n) \end{matrix} \right.$$

$$\left. \begin{matrix} \cdots & 0 & y_1 \mu^{-(n-1)/4} U_{\frac{n-1}{2}}(\delta_1) \\ \cdots & y_2 \mu^{-(n-3)/4} U_{\frac{n-3}{2}}(\delta_2) & 0 \\ \cdots & 0 & y_3 \mu^{-(n-1)/4} U_{\frac{n-1}{2}}(\delta_3) \\ \cdots & y_4 \mu^{-(n-3)/4} U_{\frac{n-3}{2}}(\delta_4) & 0 \\ \cdots & 0 & y \mu^{-(n-1)/4} U_{\frac{n-1}{2}}(\delta_5) \\ \cdots & y_6 \mu^{-(n-3)/4} U_{\frac{n-3}{2}}(\delta_6) & 0 \\ \ddots & \vdots & \vdots \\ \cdots & y_{n-3} \mu^{-(n-3)/4} U_{\frac{n-3}{2}}(\delta_{n-3}) & 0 \\ \cdots & 0 & y_{n-2} \mu^{-(n-1)/4} U_{\frac{n-1}{2}}(\delta_{n-2}) \\ \cdots & y_{n-1} \mu^{-(n-3)/4} U_{\frac{n-3}{2}}(\delta_{n-1}) & 0 \\ \cdots & 0 & y_n \mu^{-(n-1)/4} U_{\frac{n-1}{2}}(\delta_n) \end{matrix} \right] . \qquad (28)$$

By combining (12), (25) and (28) and using the equality $A_n^r = S\left(J^{\dagger}\right)^r S^{-1}$ $(r \in \mathbb{N})$ [5], we compute the rth powers of the matrix A_n

$$A_n^r = S\left(J^{\dagger}\right)^r S^{-1} = Z(r) = \left(z_{ij}(r)\right). \tag{29}$$

Therefore

$$z_{ij}(r) = \begin{cases} 0, & \text{if } (-1)^{i+j} = -1, \\ \sum_{\omega=1}^{v_{ij}} \beta_{2\omega-\psi_{ij}}^r y_{2\omega-\psi_{ij}} \mu^{\frac{i-j}{4}} U_{\frac{i+\psi_{ij}-2}{2}}\left(\delta_{2\omega-\psi_{ij}}\right) U_{\frac{j+\psi_{ij}-2}{2}}\left(\delta_{2\omega-\psi_{ij}}\right), & \text{if } (-1)^{i+j} = 1, \end{cases} \tag{30}$$

$$\psi_{ij} = \begin{cases} 1, & if \ (-1)^i = (-1)^j = -1, \\ 0, & if \ (-1)^i = (-1)^j = 1, \end{cases} \tag{31}$$

$$v_{ij} = \begin{cases} \frac{n+1}{2}, & if \ (-1)^i = (-1)^j = -1, \\ \frac{n-1}{2}, & if \ (-1)^i = (-1)^j = 1, \end{cases} \tag{32}$$

$$y_i = \begin{cases} \frac{4-4\delta_i^2}{n+3}, & if \ i = 1, 3, 5, \ldots, n \\ \frac{4-4\delta_i^2}{n+1}, & if \ i = 2, 4, 6, \ldots, n-1, \end{cases}$$

$\mu = \frac{c}{b}$, $\delta_m = \frac{\beta_m - a}{2\sqrt{bc}}$ and β_m $\left(m = \overline{1,n}\right)$ are the eigenvalues of the matrix A_n ($n = 2t+1, t \in \mathbb{N}$).

Example 3 Taking $n = 5$ in Theorem 2, we obtain

$$J^{\dagger} = \text{diag}(\beta_1, \beta_2, \beta_3, \beta_4, \beta_5) = \text{diag}\left(a - \sqrt{2bc}, a - \sqrt{bc}, a, a + \sqrt{bc}, a + \sqrt{2bc}\right)$$

and

$$A_5^r = S\left(J^{\dagger}\right)^r S^{-1} = \left(z_{ij}(r)\right) = \begin{bmatrix} z_{11}(r) & z_{12}(r) & z_{13}(r) & z_{14}(r) & z_{15}(r) \\ z_{21}(r) & z_{22}(r) & z_{23}(r) & z_{24}(r) & z_{25}(r) \\ z_{31}(r) & z_{32}(r) & z_{33}(r) & z_{34}(r) & z_{35}(r) \\ z_{41}(r) & z_{42}(r) & z_{43}(r) & z_{44}(r) & z_{45}(r) \\ z_{51}(r) & z_{52}(r) & z_{53}(r) & z_{54}(r) & z_{55}(r) \end{bmatrix},$$

$$z_{ij}(r) = 0 \ for \ (-1)^{i+j} = -1,$$

and

$$z_{11}(r) = z_{55}(r) = \frac{\left(a - \sqrt{2bc}\right)^r + 2a^r + \left(a + \sqrt{2bc}\right)^r}{4};$$

$$z_{13}(r) = z_{35}(r) = \frac{\left(a + \sqrt{2bc}\right)^r - \left(a - \sqrt{2bc}\right)^r}{2\sqrt{2}\mu^{1/2}};$$

$$z_{15}(r) = \frac{\left(a - \sqrt{2bc}\right)^r - 2a^r + \left(a + \sqrt{2bc}\right)^r}{4\mu};$$

$$z_{22}(r) = z_{33}(r) = z_{44}(r) = \frac{\left(a+\sqrt{2bc}\right)^r + \left(a-\sqrt{2bc}\right)^r}{2};$$

$$z_{24}(r) = \frac{\left(a+\sqrt{2bc}\right)^r - \left(a-\sqrt{2bc}\right)^r}{2\mu^{1/2}};$$

$$z_{31}(r) = z_{53}(r) = \frac{\left(\left(a+\sqrt{2bc}\right)^r - \left(a-\sqrt{2bc}\right)^r\right)\mu^{1/2}}{2\sqrt{2}};$$

$$z_{42}(r) = \frac{\left(\left(a+\sqrt{2bc}\right)^r - \left(a-\sqrt{2bc}\right)^r\right)\mu^{1/2}}{2};$$

$$z_{51}(r) = \frac{\left(\left(a-\sqrt{2bc}\right)^r - 2a^r + \left(a+\sqrt{2bc}\right)^r\right)\mu}{4}.$$

Example 4 Taking $r=5,\ n=9,\ a=i+2;\ b=-i$ and $c=i$ in Theorem 2, we get

$$\begin{aligned} J^{\dagger} &= \operatorname{diag}(\beta_1,\beta_2,\beta_3,\beta_4,\beta_5,\beta_6,\beta_7,\beta_8,\beta_9) \\ &= \operatorname{diag}(0.268+i, 0.382+i, 1+i, 1.382+i, 2+i, \\ &\qquad 2.618+i, 3+i, 3.618+i, 3.732+i) \end{aligned}$$

and

$$A_9^5 = S\left(J^{\dagger}\right)^5 S^{-1} = \left(z_{ij}(5)\right)$$

$$= \begin{bmatrix} 2+161i & 0 & 200-30i & 0 & -50-125i & 0 & -40+34i & 0 & 10+5i \\ 0 & 2+161i & 0 & 200-30i & 0 & -50-125i & 0 & -40+33i & 0 \\ -200+30i & 0 & 52+286i & 0 & 240-64i & 0 & -60-130i & 0 & -40+34i \\ 0 & -200+30i & 0 & 52+286i & 0 & 240-63i & 0 & -50-125i & 0 \\ -50-125i & 0 & -240+64i & 0 & 62+291i & 0 & 240-64i & 0 & -50-125i \\ 0 & -50-125i & 0 & -240+63i & 0 & 52+286i & 0 & 200-30i & 0 \\ 40-34i & 0 & -60-130i & 0 & -240+64i & 0 & 52+286i & 0 & 200-30i \\ 0 & 40-33i & 0 & -50-125i & 0 & -200+30i & 0 & 2+161i & 0 \\ 10+5i & 0 & 40-34i & 0 & -50-125i & 0 & -200+30i & 0 & 2+161i \end{bmatrix}.$$

Corollary 1 *Let the matrix A_n be the $n \times n$ pentadiagonal matrix in (1) ($a, b, c \in \mathbb{C} \setminus \{0\}$), let for Theorem 1*

$$a \neq 2\sqrt{bc}cos\left(\frac{2k\pi}{n+2}\right)$$

($n = 2t, 2t \in \mathbb{N}$), and for Theorem 2

$$a \neq 2\sqrt{bc}cos\left(\frac{(m+1)\pi}{n+3}\right), a \neq 2\sqrt{bc}cos\left(\frac{m\pi}{n+1}\right)$$

($n = 2t+1, t \in \mathbb{N}$). Then, there exists the inverse of the matrix A_n, and there are negative integer powers of the matrix A_n.

Example 5 Setting $r = -4,\ n = 4,\ a = 8,\ b = 7$ and $c = 9$ in Theorem 1, we get

$$\begin{aligned} J &= \text{diag}(\lambda_1, \lambda_1, \lambda_2, \lambda_2) \\ &= \text{diag}(0.063, 0.063, 15.937, 15.937) \end{aligned}$$

and

$$A_4^{-4} = \left(w_{ij}(-4)\right) = \begin{bmatrix} 32257 & 0 & -28448 & 0 \\ 0 & 32257 & 0 & -28448 \\ -36576 & 0 & 32257 & 0 \\ 0 & -36576 & 0 & 32257 \end{bmatrix}.$$

Example 6 Taking $r = -3,\ n = 7,\ a = 1;\ b = 1$ and $c = 2$ in Theorem 2, we get

$$\begin{aligned} J^{\dagger} &= \text{diag}\left(\beta_1, \beta_2, \beta_3, \beta_4, \beta_5, \beta_6, \beta_7\right) \\ &= \text{diag}\left(-1.288, -1, 0.126, 1, 1.874, 3, 3.288\right) \end{aligned}$$

and

$$A_7^{-3} = \left(z_{ij}(-3)\right) = \begin{bmatrix} 181 & 0 & -79 & 0 & -56 & 0 & 64 \\ 0 & \frac{7}{27} & 0 & \frac{7}{27} & 0 & -\frac{10}{27} & 0 \\ -158 & 0 & 69 & 0 & 49 & 0 & -56 \\ 0 & \frac{14}{27} & 0 & -\frac{13}{27} & 0 & \frac{7}{27} & 0 \\ -224 & 0 & 98 & 0 & 69 & 0 & -79 \\ 0 & -\frac{40}{27} & 0 & \frac{14}{27} & 0 & \frac{7}{27} & 0 \\ 512 & 0 & -224 & 0 & -158 & 0 & 181 \end{bmatrix}.$$

Remark 1 Note that our results in this paper are more general forms of the results obtained in [1, 2]. One can easily see this, taking $a := 0,\ b := 1$ and $c := 1$ in Theorem 1 and Theorem 2, respectively.

4 Complex Factorization

The well-known Fibonacci polynomials $F(x) = \{F_n(x)\}_{n=1}^{\infty}$ are defined in [7] by the recurrence relation

$$F_n(x) = xF_{n-1}(x) + F_{n-2}(x),$$

where $F_0(x) = 0$, $F_1(x) = 1$, $F_2(x) = x$ and $n \geq 3$. Notice that $F_1(2) = 1$, $F_2(2) = 2$ and $F_n(2) = 2F_{n-1}(2) + F_{n-2}(2)$, where $n \geq 3$. So $P_n = F_n(2)$ defines the well-known Pell numbers [7].

Corollary 2 *Let the matrix A_n be the $n \times n$ $(n = 2t,\ 2 \leq t \in \mathbb{N})$ pentadiagonal matrix as in (1) with $a := x,\ b := \mathbf{i}$ and $c := \mathbf{i}$ where $\mathbf{i} = \sqrt{-1}$. Then*

$$\det(A_n) = \left(F_{\frac{n}{2}+1}(x)\right)^2, \tag{33}$$

where $F_n(x)$ is the nth Fibonacci polynomial.

Proof The determinant of A_n $(n = 2t, 2 \leq t \in \mathbb{N})$ can be written as

$$\det(A_n) = \left[\left|\text{tridiag}_{\frac{n}{2}}(c, a, b)\right|\right]^2. \tag{34}$$

In [2], authors acquired that

$$F_{n+1} = \left|\text{tridiag}_n(i, 1, i)\right|. \tag{35}$$

If we choose $a := x,\ b := \mathbf{i}$ and $c := \mathbf{i}$ in (34), and substituting (35) into (34), we obtain

$$\begin{aligned}\det(A_n) &= \left[\left|\text{tridiag}_{\frac{n}{2}}(i, x, i)\right|\right]^2 \\ &= \left[F_{\frac{n}{2}+1}(x)\right]^2.\end{aligned}$$

□

Corollary 3 *Let the matrix A_n be the $n \times n$ $(n = 2t,\ 2 \leq t \in \mathbb{N})$ pentadiagonal matrix in (1). Then*

$$\det(A_n) = \begin{cases} \left(F_{\frac{n}{2}+1}\right)^2 & if\ a = 1,\ b = i\ and\ c = i \\ \left(P_{\frac{n}{2}+1}\right)^2 & if\ a = 2,\ b = i\ and\ c = i, \end{cases} \tag{36}$$

where F_n and P_n denote the nth Fibonacci and the nth Pell numbers, respectively.

Theorem 3 *Let the matrix A_n be as in (1) with $a := x,\ b := \mathbf{i}$ and $c := \mathbf{i}$. Then the complex factorization of the generalized Fibonacci polynomial is of the following form:*

$$F_{\frac{n}{2}+1}(x) = \prod_{k=1}^{n}\left(x + 2\cos\left(\frac{2k\pi}{n+2}\right)\right). \tag{37}$$

Proof The eigenvalues of the matrix A_n in (2) are

$$\lambda_k = x + 2\cos\left(\frac{2k\pi}{n+2}\right), \quad k = 1, 2, \ldots, \frac{n}{2},$$

so the determinant of the matrix A_n is

$$\det(A_n) = \prod_{k=1}^{n}\left(x + 2\cos\left(\frac{2k\pi}{n+2}\right)\right)^2.$$

From Eq. (37) and Corollary 2, the complex factorization of the generalized Fibonacci polynomial is provided. □

Corollary 4 *Let the matrix A_n, $n \times n$ $(n = 2t+1,\ t \in \mathbb{N})$ be a pentadiagonal matrix as in (1) with $a := x,\ b := \mathbf{i}$ and $c := \mathbf{i}$ where $\mathbf{i} = \sqrt{-1}$. Then*

$$\det(A_n) = F_{\frac{n+1}{2}}(x)F_{\frac{n+1}{2}+1}(x), \tag{38}$$

where $F_n(x)$ is nth Fibonacci polynomial.

Proof The determinant of A_n $(n = 2t+1, t \in \mathbb{N})$ can be written as

$$\det(A_n) = \left|\text{tridiag}_{\frac{n-1}{2}}(c, a, b)\right| \left|\text{tridiag}_{\frac{n+1}{2}}(c, a, b)\right|. \tag{39}$$

If we choose $a := x,\ b := \mathbf{i}$ and $c := \mathbf{i}$ in (39), and substituting (35) into (39), we have

$$\begin{aligned}\det(A_n) &= \left|\text{tridiag}_{\frac{n-1}{2}}(i, x, i)\right| \left|\text{tridiag}_{\frac{n+1}{2}}(i, x, i)\right| \\ &= F_{\frac{n+1}{2}}(x)F_{\frac{n+1}{2}+1}(x).\end{aligned}$$

□

Corollary 5 *Let the matrix A_n, $n \times n$ $(n = 2t+1,\ t \in \mathbb{N})$ be a pentadiagonal matrix as in (1). Then*

$$\det(A_n) = \begin{cases} F_{\frac{n+1}{2}}F_{\frac{n+1}{2}+1} & if\ a = 1,\ b = i\ and\ c = i \\ P_{\frac{n+1}{2}}P_{\frac{n+1}{2}+1} & if\ a = 2,\ b = i\ and\ c = i, \end{cases} \tag{40}$$

where F_n and P_n denote the nth Fibonacci and the nth Pell numbers.

Acknowledgements The authors are partially supported by TUBITAK and the Office of Selçuk University Research Project (BAP).

References

1. Arslan, S., Köken, F., Bozkurt, D.: Positive integer powers and inverse for one type of even ordersymmetric pentadiagonal matrices. Appl. Math. Comput. **219**, 5241–5248 (2013)
2. Cahill, N.D., D'Errico, J.R., Spence, J.P.: Complex factorizations of the Fibonacci and Lucas numbers. Fibonacci Quart. **41**(1), 13–19 (2003)
3. Duru, H.K., Bozkurt, D.: Powers of complex tridiagonal matrices. Alabama J. Math. **38** (2014)
4. Hadj, A.D.A., Elouafi, M.: On the characteristic polynomial, eigenvectors and determinant of a pentadiagonal matrix. Appl. Math. Comput. **198**, 634–642 (2008)
5. Horn, R.A., Johnson, C.R.: Matrix Analysis, pp. 40–41. Cambridge University Press, Cambridge (2013)
6. Köken, F., Bozkurt, D.: Positive integer powers for one type of odd order circulant matrices. Appl. Math. Comput. **217**, 4377–4381 (2011)
7. Koshy, T.: Fibonacci and Lucas Numbers with Applications, pp. 443–445. Wiley, NY (2001)
8. Mason, J.C., Handscomb, D.C.: Chebyshev Polynomials, pp. 3–5. CRC Press, Washington (2003)
9. Öteleş, A., Akbulak, M.: Positive integer powers of certain complex tridiagonal matrices. Math. Sci. Lett. **2**(1), 63–72 (2013)
10. Rimas, J.: On computing of arbitrary positive integer powers for one type of symmetric pentadiagonal matrices of odd order. Appl. Math. Comput. **204**, 120–129 (2008)
11. Rimas, J.: On computing of arbitrary positive integer powers for one type of symmetric pentadiagonal matrices of even order. Appl. Math. Comput. **203**, 582–591 (2008)
12. Rimas, J.: On computing of arbitrary positive integer powers for one type of even order symmetric anti-pentadiagonal matrices. Appl. Math. Comput. **211**, 54–74 (2009)
13. Wu, H.: On positive integer powers of Toeplitz matrices. J. Math. Res. **5**(4), 52–57 (2013)

Chains and Antichains in the Bruhat Order for Classes of (0, 1)-Matrices

Ricardo Mamede

Abstract Let $\mathscr{A}(R, S)$ denote the set of all matrices of zeros and ones with row sum vector R and column sum vector S. This set can be ordered by a generalization of the usual Bruhat order for permutations. Contrary to the classical Bruhat order on permutations, where permutations can be seen as permutation matrices, the Bruhat order on the class $\mathscr{A}(R, S)$ is not, in general, graded, and an interesting problem is the determination of bounds for the maximal length of chains and antichains in this poset. In this survey we aim to provide a self-contained account of the recent developments involving the determination of maximum lengths of chains and antichains in the Bruhat order on some classes of matrices in $\mathscr{A}(R, S)$.

Keywords (0, 1)-Matrices · Majorization · Bruhat order · Row and column sum vector

1 Introduction

Matrices whose entries are just zeros and ones occur naturally in many different contexts, both in mathematics, in connection with graphs and, more generally, families of subsets of a finite set, and in other areas including educational tests, ecological studies, and social networks. A special class amongst these are the zero-one matrices with a prescribed row sum vector R and a prescribed column sum vector S, denoted by $\mathscr{A}(R, S)$. This class of zero-one matrices was object of intensive study during the 1950s and 1960s by H.J. Ryser, D.R. Fulkerson, R.M. Haber, and D. Gale (see [5–7, 16, 17, 24, 32] and the references therein), and has since then attracted the attention of many combinatorists.

One of the fundamental results involving the class $\mathscr{A}(R, S)$ is the beautiful characterization, in terms of majorization, of the existence of a matrix in this class, obtained independently by D. Gale [17], using the theory of network flows, and by

R. Mamede (✉)
Department of Mathematics, CMUC, University of Coimbra, 3001-454 Coimbra, Portugal
e-mail: mamede@mat.uc.pt

N. Bebiano (ed.), *Applied and Computational Matrix Analysis*,
Springer Proceedings in Mathematics & Statistics 192,
DOI 10.1007/978-3-319-49984-0_15

H.J. Ryser [31], using induction and a direct combinatorial reasoning. An interesting case in which the nonemptiness is guaranteed emerges when $R = S = (k, \ldots, k)$ is the constant vector having each of its n components equal to k. In this case we simply write $\mathscr{A}(n, k)$ for $\mathscr{A}(R, S)$. In particular, $\mathscr{A}(n, 1)$ is the class of all permutation matrices of order n, which can be identified with the symmetric group $\mathscr{S}_n$.

This identification inspired Brualdi and Hwang [8] to define a Bruhat partial order $\preccurlyeq$ on a nonempty class $\mathscr{A}(R, S)$, which generalizes the classical Bruhat order on the symmetric group. Nevertheless, the characterization of this order seems much harder than the classical order: for instance, a characterization of the cover relations for the Bruhat order in $\mathscr{A}(R, S)$ is not known. Since, in general, this is not a graded poset, an interesting problem is the determination of bounds for the maximal length of chains and antichains for the Bruhat order in $\mathscr{A}(R, S)$. The aim of the present article is to contribute to the clarification of this problem by providing a self-contained account of the recent developments involving chains and antichains in the Bruhat order of the matrix classes $\mathscr{A}(2k, k)$ and $\mathscr{A}(n, 2)$.

2 The Class $\mathscr{A}(R, S)$

Let $\mathbb{N}$ denote the set of non-negative integers. A weak composition with sum $\tau \in \mathbb{N}$ is a finite sequence $R = (r_1, \ldots, r_m)$ of non-negative integers with $\sum_i r_i = \tau$. A partition is a weakly decreasing weak composition. It is convenient to not distinguish between two partitions which only differ by a string of zeros at the end. We identify a partition $P = (p_1, \ldots, p_m)$ with its Ferrers diagram, obtained by placing p_i left justified ones in the ith row, for $1 \leq i \leq m$. For example, if $P = (3, 3, 2, 2, 1)$, its Ferrers diagram is

$$\begin{matrix} 1 & 1 & 1 \\ 1 & 1 & 1 \\ 1 & 1 & \\ 1 & 1 & \\ 1 & & \end{matrix} .$$

The conjugate partition $P^* = (p_1^*, \ldots, p_n^*)$ of P is the partition corresponding to the transpose of the Ferrers diagram of P. In other words, each entry p_i^* of P^* satisfy

$$p_i^* = |\{k : p_k \geq i\}|.$$

For instance, the conjugate of $P = (3, 3, 2, 2, 1)$ is the partition $P^* = (5, 4, 2)$ and its Ferrers diagram is

$$\begin{matrix} 1 & 1 & 1 & 1 & 1 \\ 1 & 1 & 1 & 1 & \\ 1 & 1 & & & \end{matrix} .$$

The dominance or majorization order on partitions $R = (r_1, \ldots, r_m)$ and $S = (s_1, \ldots, s_n)$ with the same sum τ is defined by setting $R \leq S$ if

$$r_1 + \cdots + r_i \leq s_1 + \cdots + s_i,$$

for $i = 1, \ldots, \min\{m, n\}$. The set of all partitions with sum τ ordered by majorization is a lattice with maximum element (n) and minimum element $(1^n) = (1, 1, \ldots, 1)$, and is self dual under the map which sends each partition to its conjugate. Graphically, $R \leq S$ if and only if the diagram of R is obtained by "lowering" at least one 1 in the diagram of S. Clearly $R \leq S$ if and only if $S^* \leq R^*$. Moreover, S covers R, written as $R \lhd S$, if and only if S is obtained from R by lifting exactly one 1 in the diagram of R to the next available position such that the transfer must be from some r_k to r_j with $j < k$ and either $k = j + 1$ or $r_k = r_j$ [9]. In this case we say that S is obtained from R by a transfer from r_k to r_j.

Lemma 1 *Let R and S be partitions with sum τ. Then $R \leq S$ if and only if S can be obtained from R by a finite sequence of transfers.*

Let m and n be two positive integers and let $R = (r_1, \ldots, r_m)$ and $S = (s_1, \ldots, s_n)$ be compositions with the same sum

$$r_1 + r_2 + \cdots + r_m = s_1 + s_2 + \cdots + s_n.$$

The set of all $m \times n$ matrices over $\{0, 1\}$ with ith row sum equal to r_i, for $1 \leqslant i \leqslant m$, and jth column sum equal to s_j, for $1 \leqslant j \leqslant n$, is commonly denoted by $\mathscr{A}(R, S)$. For the characterization of $\mathscr{A}(R, S)$ we may assume that all entries r_i and s_j are positive, since otherwise each matrix in $\mathscr{A}(R, S)$ has a row of 0's or a column of 0's. Moreover, without loss of generality we may also assume that R and S are partitions, since otherwise for permutation matrices P and Q of orders m and n, respectively, we have

$$\mathscr{A}(RP, SQ) = \{PAQ : A \in \mathscr{A}(R, S)\}.$$

When $S = R^*$ it is easy to check that the set $\mathscr{A}(R, R^*)$ has only one element, namely the matrix $A(R, n)$ of size $m \times n$ obtained by completing with zeros the Ferrers diagram of R, placed in the upper left corner. For instance, if $R = (3, 3, 2, 2, 1)$, then

$$A(R, 3) = \begin{pmatrix} 1 & 1 & 1 \\ 1 & 1 & 1 \\ 1 & 1 & 0 \\ 1 & 1 & 0 \\ 1 & 0 & 0 \end{pmatrix} \in \mathscr{A}(R, R^*).$$

The general characterization of the set $\mathscr{A}(R, S)$ was obtained independently by D. Gale [17], using the theory of network flows, and by H.J. Ryser [31], using induction and a direct combinatorial reasoning. Since then various proofs were obtained

[5]. The proof we present here is due to M. Krause [25] and uses properties of dominance order (see also [3, 4, 13, 26]).

Theorem 1 (Gale–Ryser theorem) *Let R and S be partitions with the same sum τ. Then, $\mathscr{A}(R, S)$ is nonempty if and only if $S \leq R^*$.*

Proof Assume there is a (0, 1)-matrix with row sum R and columns sum S. Since the ones in $A(R, n)$ are left-justified, any matrix $A = [a_{ij}] \in \mathscr{A}(R, S)$ has at most as many ones in the first k columns as $A(R, n)$ has, for all $k \leq n$, that is,

$$\sum_{j=1}^{k} s_j = \sum_{j=1}^{k}\sum_{i=1}^{m} a_{ij} \leq \sum_{j=1}^{k}\sum_{i=1}^{m} A(R, n)_{ij} = \sum_{j=1}^{k} r_j^*.$$

It follows that $S \leq R^*$.

Reciprocally, assume that $S \leq R^*$. Then, by Lemma 1, R^* can be obtained from S by a finite number of transfers

$$S = R^t \lhd R^{t-1} \lhd \cdots \lhd R^1 = R^*,$$

where R^k is obtained from R^{k-1} by a transfer, for $i = 2, \ldots, t$. We proceed by induction over $t \geq 1$. When $t = 1$ we have $S = R^*$ and in this case $\mathscr{A}(R, R^*)$ is nonempty and has only the matrix $A(R, n)$. The conclusion follows by induction after we have proven the following claim:

Claim: If $\mathscr{A}(R, P)$ is nonempty, and $P' \lhd P$ is obtained from P by a transfer, then also $\mathscr{A}(R, P')$ is nonempty.

Proof of Claim: Let $A = [a_{ij}] \in \mathscr{A}(R, P)$ and assume that P' is obtained from $P = (p_1, \ldots, p_n)$ by a transfer from p_i to p_j, for some $i < j$. Then $p_i > p_j$ and there is a row k in A where $a_{ki} = 1$ and $a_{kj} = 0$. Consider $A' = [a'_{pq}]$ where $a'_{ki} = 0$ and $a'_{kj} = 1$, while all other entries of A' agree with those from A. Clearly, $A' \in \mathscr{A}(R, P')$.
This proves the claim, and therefore the theorem. □

For example, by the Gale–Ryser theorem there exists a matrix A in the set $\mathscr{A}(R, S)$, with $R = (3, 3, 2, 2, 1)$ and $S = (4, 4, 2, 1)$, since $R^* = (5, 4, 2)$ majorizes S. Starting with the matrix $A(R, 4)$ and the sequence of cover relations

$$S = (4, 4, 2, 1) \lhd (5, 3, 2, 1) \lhd (5, 4, 1, 1) \lhd (5, 4, 2) = R^*,$$

the procedure obtained from the proof of the theorem above leads to a solution for A as follows:

$$A(R, 4) = \begin{pmatrix} 1\,1\,1\,0 \\ 1\,1\,1\,0 \\ 1\,1\,0\,0 \\ 1\,1\,0\,0 \\ 1\,0\,0\,0 \end{pmatrix} \to \begin{pmatrix} 1\,1\,1\,0 \\ 1\,1\,0\,1 \\ 1\,1\,0\,0 \\ 1\,1\,0\,0 \\ 1\,0\,0\,0 \end{pmatrix} \to \begin{pmatrix} 1\,1\,1\,0 \\ 1\,0\,1\,1 \\ 1\,1\,0\,0 \\ 1\,1\,0\,0 \\ 1\,0\,0\,0 \end{pmatrix} \to$$

$$\rightarrow \begin{pmatrix} 1\,1\,1\,0 \\ 0\,1\,1\,1 \\ 1\,1\,0\,0 \\ 1\,1\,0\,0 \\ 1\,0\,0\,0 \end{pmatrix} = A \in \mathscr{A}(R,S).$$

An important class of matrices in which nonemptiness is assured by the Gale–Ryser theorem occurs when $m = n$, $k \in \mathbb{N}$ such that $0 \leq k \leq n$, and

$$R = S = (k, \ldots, k)$$

is the constant vector having each component equal to k: in this case we write $\mathscr{A}(n,k)$ instead of $\mathscr{A}(R,S)$.

While the proof of the nomemptiness of the set $\mathscr{A}(R,S)$ is constructive, as Ryser predicted "the exact number of them is undoubtedly an extremely intricate function of the row and column sums". In 1988 Wang [33] presented such a formula which involves $2^n - 2n$ variables. Several improvements have since then been achieved with a substancial reduction of the number of variables. Nevertheless, computing a closed manageable formula for such sequence is still an open problem which looks quite hard (cf., e.g., [1, 10, 21–23, 27, 30, 34] and the references therein for some partial results). For the case $\mathscr{A}(n,k)$, an asymptotic formula was obtained by O'Neil [29] (see also [15]):

$$\#\mathscr{A}(n,k) \sim \frac{(kn)!}{(k!)^{2n}} e^{-(k-1)^2/2}. \tag{1}$$

3 The Bruhat Order on $\mathscr{A}(R,S)$

Amongst the various ways to define a partial order on the symmetric group $\mathscr{S}_n$, the Bruhat order is the most prominent of all as it can be generalized to any Coxeter group. Identifying permutations with permutation matrices, Brualdi and Hwang [8] generalized further this partial order to the class of matrices $\mathscr{A}(R,S)$. In this section we describe this process and analyse some characteristics of the Bruhat order on $\mathscr{A}(R,S)$.

An inversion of a permutation $p = p_1 p_2 \cdots p_n \in \mathscr{S}_n$ is a pair (p_i, p_j) such that $i < j$ but $p_i > p_j$. The Bruhat order on the symmetric group $\mathscr{S}_n$ can then be defined by declaring that permutation p is less than or equal to permutation q, denoted $p \preceq q$, if and only if either $p = q$, or p can be obtained from q by a series of operations, each of which interchanges the two entries of an inversion. An operation of this type reduces the number of inversions in a permutation. The identity permutation $12 \cdots n$ is the unique minimal element in the Bruhat order on $\mathscr{S}_n$, and the unique maximal element is the permutation $n(n-1)\cdots 1$.

The symmetric group can be identified in a natural way with the class of permutation matrices $\mathscr{A}(n,1)$ of order n. Using this identification, an inversion on a permutation $p \in \mathscr{S}_n$ corresponds to a pair of ones in the corresponding permutation matrix P, one of which is located to the top-right of the other. More precisely, if $P = [p_{ij}] \in \mathscr{A}(n, 1)$, an inversion in P consists of any two entries $p_{ij} = p_{k\ell} = 1$ such that $(i - k)(j - k) < 0$. We denote the total number of inversions in P by $\nu(P)$ [19].

For permutation matrices P and Q of order n, corresponding to permutations p and q, we say that P is less than or equal to Q in the Bruhat order, and write $P \preceq Q$, whenever $p \preceq q$.

An alternative, but equivalent, way to define the Bruhat order on the symmetric group is to use the Gale order [2] on subsets of a fixed size of $[n] := \{1, \dots, n\}$. Given two nonempty subsets $X = \{a_1, \dots, a_k\}$ and $Y = \{b_1, \dots, b_k\}$ of $[n]$, written in increasing order, we say that X is less than, or equal to Y in the Gale order, denoted $X \leq_G Y$, if and only if $a_1 \leq b_1, a_2 \leq b_2, \dots, a_k \leq b_k$. For $p = p_1 p_2 \cdots p_n$, let $p[k] = \{p_1, p_2, \dots, p_k\}$. The following lemma is a straightforward consequence of the definitions [8].

Lemma 2 *Let p and q be permutations in $\mathscr{S}_n$. Then,*

$$p \preceq q \text{ if and only if } p[k] \leq_G q[k], \quad (1 \leq k \leq n).$$

For a $m \times n$ matrix $A = [a_{ij}]$, let $\Sigma_A = (\sigma_{ij}(A))$ denote the $m \times n$ matrix whose (i, j)-entry equals

$$\sigma_{ij}(A) = \sum_{k=1}^{i} \sum_{\ell=1}^{j} a_{k,\ell} \qquad (1 \leq i \leq m, 1 \leq j \leq n).$$

That is, $\sigma_{i,j}(A)$ is the sum of the entries in the leading i by j submatrix of A. Using the Gale order, it is easy to check that for permutation matrices P and Q of order n, one has $P \preceq Q$ if and only if $\Sigma_P \geq \Sigma_Q$, where this latter order is the entrywise order (see [5] for a proof).

Lemma 3 *If P and Q are permutation matrices of order n, then $P \preccurlyeq Q$ if and only if $\Sigma_P \geq \Sigma_Q$.*

This result, which is equivalent to Lemma 2, can be used to extend to the class $\mathscr{A}(R, S)$ the Bruhat order on permutation matrices. If A_1 and A_2 are matrices in the class $\mathscr{A}(R, S)$, then we say that A_1 is less than, or equal to A_2 in the Bruhat order, denoted $A_1 \preceq A_2$, if and only if $\Sigma_{A_1} \geq \Sigma_{A_2}$ in the entrywise order, i.e. $\sigma_{ij}(A_1) \geq \sigma_{ij}(A_2)$ for all $1 \leq i \leq m$ and $1 \leq j \leq n$.

It is well known that the Bruhat order on permutation matrices is graded, that is all maximal chains have the same length, with rank function given by the number of inversions. But in general, the Bruhat order on $\mathscr{A}(R, S)$ is not graded. For instance, consider the following matrices in $\mathscr{A}(4, 2)$:

$$A = \begin{pmatrix} 1 & 1 & 0 & 0 \\ 0 & 0 & 1 & 1 \\ 0 & 1 & 0 & 1 \\ 1 & 0 & 1 & 0 \end{pmatrix} \text{ and } C = \begin{pmatrix} 1 & 1 & 0 & 0 \\ 0 & 1 & 0 & 1 \\ 1 & 0 & 1 & 0 \\ 0 & 0 & 1 & 1, \end{pmatrix}$$

and also

$$X_1 = \begin{pmatrix} 1 & 1 & 0 & 0 \\ 0 & 1 & 0 & 1 \\ 0 & 0 & 1 & 1 \\ 1 & 0 & 1 & 0 \end{pmatrix}, \quad Y_1 = \begin{pmatrix} 1 & 1 & 0 & 0 \\ 0 & 0 & 1 & 1 \\ 1 & 0 & 0 & 1 \\ 0 & 1 & 1 & 0 \end{pmatrix},$$

$$Y_2 = \begin{pmatrix} 1 & 1 & 0 & 0 \\ 0 & 0 & 1 & 1 \\ 1 & 0 & 1 & 0 \\ 0 & 1 & 0 & 1 \end{pmatrix} \text{ and } Y_3 = \begin{pmatrix} 1 & 1 & 0 & 0 \\ 0 & 0 & 1 & 1 \\ 1 & 1 & 0 & 0 \\ 0 & 0 & 1 & 1, \end{pmatrix}.$$

In [7], it shown that A covers X_1, X_1 covers C, A covers Y_1, Y_1 covers Y_2, Y_2 covers Y_3, and Y_3 covers C. That is, there are maximal chains from A to C of lengths 2 and 4 in the Bruhat order, proving that the class $\mathscr{A}(4, 2)$, under the Bruhat order, is not graded.

Also, as pointed before, in the Bruhat order on the class $\mathscr{A}(n, 1)$ there is a unique minimal matrix, the identity I_n, and a unique maximal matrix, the permutation matrix D_n with 1's in the positions $(1, n), (2, n-1), \ldots, (n, 1)$. In general, however, there can be many minimal and maximal matrices in a nonempty class $\mathscr{A}(R, S)$. Brualdi and Hwang showed in [8] that the following algorithm constructs a minimal matrix in the Bruhat order on the class $\mathscr{A}(n, k)$. Note that if $A \preccurlyeq B$ in $\mathscr{A}(n, k)$, and A' and B' are obtained from A and B respectively by reversing the order of their columns, then $B' \preccurlyeq A'$. Therefore, maximal matrices in the Bruhat order of $\mathscr{A}(n, k)$ are obtained by reversing the order of the columns in minimal matrices.

As usual, we let J_{mn} denote the m by n matrix of all 1's, abbreviated to J_n when $m = n$.

Algorithm to Construct a Minimal Matrix in the Bruhat Order on $\mathscr{A}(n, k)$.

1. Let $n = qk + r$ where $0 \leq r < k$.
2. If $r = 0$, then $A = J_k \oplus \cdots \oplus J_k$, ($q$ J_k's) is a minimal matrix, where $\oplus$ denotes the direct sum of matrices.
3. Else, $r \neq 0$.

   a. If $q \geq 2$, let

   $$A = X \oplus J_k \oplus \cdots \oplus J_k, \quad (q-1 \ J_k\text{'s}, X \text{ has order } k+r),$$

   and let $n \leftarrow k + r$.

   b. Else, $q = 1$, and let

$$A = \left(\begin{array}{c|c} J_{r,k} & O_r \\ \hline X & J_{k,r} \end{array}\right), \quad (X \text{ has order } k),$$

and let $n \leftarrow k$ and $k \leftarrow k - r$.

c. Proceed recursively with the current values of n and k to determine X.

For example, with $n = 9$ and $k = 2$ the algorithm above construct the following minimal matrix in $\mathscr{A}(9, 2)$:

$$A = \left(\begin{array}{c|c} J_{1,2} & O_1 \\ \hline J_1 \oplus J_1 & J_{2,1} \end{array}\right) \oplus J_2 \oplus J_2 \oplus J_2.$$

We start by writing $9 = 4 \cdot 2 + 1$ to get, by step $3(a)$, $A = X \oplus J_2 \oplus J_2 \oplus J_2$, where the matrix X has order 3, and we set $n = 3$ and $k = 2$. Next, since $3 = 1 \cdot 2 + 1$, by step $3(b)$, we get

$$X = \left(\begin{array}{c|c} J_{1,2} & O_1 \\ \hline Y & J_{2,1} \end{array}\right),$$

where Y is of order 2, and we set $n = 2$ and $k = 1$. Finally, since $2 = 2 \cdot 1$, by step 2 we get $Y = J_1 \oplus J_1$.

If we let F_n denote the matrix of order n with 0's in positions $(1, n)$, $(2, n-2), \ldots, (n, 1)$ and 1's elsewhere, then the minimal matrix for $\mathscr{A}(9, 2)$ obtained in the example above is a direct sum of matrices equal to J_2 and $F_3 = \begin{pmatrix} 1 & 1 & 0 \\ 1 & 0 & 1 \\ 0 & 1 & 1 \end{pmatrix}$. As Brualdi and Hwang proved in [8], this is part of a general property of the classes $\mathscr{A}(n, 2)$.

Theorem 2 [8] *Let n be an integer greater than or equal to 2. Then a matrix in $\mathscr{A}(n, 2)$ is a minimal matrix in the Bruhat order if and only if it is the direct sum of matrices equal to J_2 and F_3.*

Hence, when n is odd, we can construct a minimal matrix P_n in $\mathscr{A}(n, 2)$ as the direct sum of $n/2$ copies of J_2, and the corresponding maximal matrix Q_n:

$$P_n = \begin{pmatrix} J_2 & 0 & \cdots & 0 \\ 0 & J_2 & \cdots & 0 \\ \vdots & \vdots & \ddots & \vdots \\ 0 & 0 & \cdots & J_2 \end{pmatrix} \text{ and } Q_n = \begin{pmatrix} 0 & \cdots & 0 & J_2 \\ 0 & \cdots & J_2 & 0 \\ \vdots & \ddots & \vdots & \vdots \\ J_2 & \cdots & 0 & 0 \end{pmatrix}, \tag{2}$$

and when n is odd, we construct a minimal matrix P_n as the direct sum of $(n-3)/2$ copies of J_2 and one copy of F_3, and the corresponding maximal matrix Q_n:

$$P_n = \begin{pmatrix} J_2 & 0 & \cdots & 0 & 0 \\ 0 & J_2 & \cdots & 0 & 0 \\ \vdots & \vdots & \ddots & \vdots & \vdots \\ 0 & 0 & \cdots & J_2 & 0 \\ 0 & 0 & \cdots & 0 & F_3 \end{pmatrix} \text{ and } Q_n = \begin{pmatrix} 0 & 0 & \cdots & 0 & J_2 \\ 0 & 0 & \cdots & J_2 & 0 \\ \vdots & \vdots & \ddots & \vdots & \vdots \\ 0 & J_2 & \cdots & 0 & 0 \\ F_3' & 0 & \cdots & 0 & 0 \end{pmatrix}. \quad (3)$$

In [7] it was proved that $\mathscr{A}(n, k)$ contains a unique minimal element in the Bruhat order if and only if $k \in \{0, 1, n-1, n\}$ or $n = 2k$. Notice also that when $n = 2k$, step 2 of the algorithm above produces the minimal matrix $J_k \oplus J_k$, so this is the only minimal matrix in $\mathscr{A}(2k, k)$.

Theorem 3 [7] *Let n and k be integers with $0 \le k \le n$. Then the class $\mathscr{A}(n, k)$ has a unique minimal element in the Bruhat order if and only if $k \in \{0, 1, n-1, n\}$ or $n = 2k$. The unique minimal and maximal matrices in $\mathscr{A}(2k, k)$ are, respectively*

$$P_k = \begin{pmatrix} J_k & O_k \\ O_k & J_k \end{pmatrix} \text{ and } Q_k = \begin{pmatrix} O_k & J_k \\ J_k & O_k \end{pmatrix}.$$

Since $\mathscr{A}(n, k) \simeq \mathscr{A}(n, n-k)$ (the map $A \longmapsto J_n - A$ does the job), $\#\mathscr{A}(n, 0) = 1$ and $\mathscr{A}(n, 1) \simeq S_n$, the most interesting case in which there is uniqueness of minimal and maximal matrices is $\mathscr{A}(2k, k)$.

4 Chains in $\mathscr{A}(2k, k)$ and $\mathscr{A}(n, 2)$

In this section we address the problem of finding the maximum length of a chain in the Bruhat order on the classes $\mathscr{A}(2k, k)$ and $\mathscr{A}(n, 2)$, giving algorithms to construct such chains, following [11, 19]. We start by proving that the maximum length of a chain in the Bruhat order on the class $\mathscr{A}(2k, k)$ is k^4, giving an algorithm that construct such a sequence.

Theorem 4 [11] *For any positive integer k, the maximal length of a chain in the Bruhat order in $\mathscr{A}(2k, k)$ equals k^4.*

Proof For any $A, B \in \mathscr{A}(R, S)$ such that $A \preceq B$, as an immediate consequence of the definition of Bruhat order, an upper bound for the length of any admissible chain between A and B is clearly given by

$$\varphi(A, B) := \sum_{i=1}^{m} \sum_{j=1}^{n} [\sigma_{ij}(A) - \sigma_{ij}(B)].$$

Since by Theorem 3 the poset $(\mathscr{A}(2k, k), \preceq)$ admits a unique minimum P_k and a unique maximum Q_k, any chain between two pairwise comparable elements can be extended to a chain between P_k and Q_k.

After some lengthy but rather straightforward computations we get

$$\sigma_{ij}(P_k) = \begin{cases} ij & \text{if } i, j \leq k \\ ik & \text{if } i \leq k \leq j \\ jk & \text{if } i \geq k \geq j \\ ij - k(i+j-2k) & \text{if } i, j \geq k \end{cases},$$

$$\sigma_{ij}(Q_k) = \begin{cases} 0 & \text{if } i, j \leq k \\ i(j-k) & \text{if } i \leq k \leq j \\ j(i-k) & \text{if } i \geq k \geq j \\ k(i+j-2k) & \text{if } i, j \geq k \end{cases},$$

and $\varphi(P_k, Q_k) = k^4$.

Hence it suffices to present an instance of a chain between P_k and Q_k having exactly such length. We do that in an algorithmic way, presenting a procedure to generate an order preserving path in the Hasse diagram of $\mathscr{A}(2k, k)$.

Procedure [Switch(t, r)] $1 \leq t, r \leq 2k - 1$.

Input: $A = (a_{ij}) \in \mathscr{A}(2k, k)$ such that the submatrix

$$\begin{pmatrix} a_{t,r} & a_{t,r+1} \\ a_{t+1,r} & a_{t+1,r+1} \end{pmatrix} = \begin{pmatrix} 1 & 0 \\ 0 & 1 \end{pmatrix}.$$

Output: $B = (b_{ij}) \in \mathscr{A}(2k, k)$ such that $b_{ij} = a_{ij}$ if $1 \leq i, j \leq 2k$ and $\{i, j\} \notin \{\{t, r\}, \{t, r+1\}, \{t+1, r\}, \{t+1, r+1\}\}$, and

$$\begin{pmatrix} b_{t,r} & b_{t,r+1} \\ b_{t+1,r} & b_{t+1,r+1} \end{pmatrix} = \begin{pmatrix} 0 & 1 \\ 1 & 0 \end{pmatrix}.$$

It is easy to see that executing procedure Switch(t, r) the output covers the input in the Bruhat order for any choice of parameters. Our chain will be made by repeated applications of the procedure Switch(t, r).

Procedure [Switch-rows(t)] $1 \leq t \leq 2k - 1$.

Input: $A = (a_{ij}) \in \mathscr{A}(2k, k)$ such that rows $t, t+1$ equal

$$\begin{pmatrix} 1, \ldots, 1, 0, \ldots, 0 \\ 0, \ldots, 0, 1, \ldots, 1 \end{pmatrix}.$$

For $\alpha = k$ down to 1 do
  Begin
    For $\beta = \alpha$ to $\alpha + k - 1$ do Switch(t, β).
  End.

Output: $B = (b_{ij}) \in \mathscr{A}(2k, k)$ such that $b_{i,j} = a_{i,j}$ for any $1 \leq i, j \leq 2k$ such that $i \neq t, t+1$, and rows $t, t+1$ equal

$$\begin{pmatrix} 0, \ldots, 0, 1, \ldots, 1 \\ 1, \ldots, 1, 0, \ldots, 0 \end{pmatrix}.$$

Algorithm [Chain(k)] $k \in \mathbb{N} \setminus \{0\}$.
Input: P_k.
For $\alpha = k$ down to 1 do
Begin
For $\beta = \alpha$ to $\alpha + k - 1$ do Switch-rows(β).
End.
Output: Q_k.

We can see that, for any choice of parameters, the procedure "Switch" is invoked k^2 times by procedure "Switch-rows", and that algorithm "Chain" recalls procedure "Switch-rows" k^2 times as well, so there are k^4 application of procedure "Switch". Since, as already remarked, each time that procedure "Switch" is recalled we are moving up (by one cover relation) in the Hasse diagram of the poset $(\mathscr{A}(2k, k), \preceq)$, all the constructed elements are pairwise distinct members of the desired chain, and the result follows. □

For the sake of clarity, we present in detail our construction of the chain for the case $k = 2$.

Example 1 A chain of maximal length in the class $\mathscr{A}(4, 2)$. Dots represent zeros.

$$P_2 = \begin{pmatrix} 1 & 1 & \cdot & \cdot \\ 1 & 1 & \cdot & \cdot \\ \cdot & \cdot & 1 & 1 \\ \cdot & \cdot & 1 & 1 \end{pmatrix} \mapsto \begin{pmatrix} 1 & 1 & \cdot & \cdot \\ 1 & \cdot & 1 & \cdot \\ \cdot & 1 & \cdot & 1 \\ \cdot & \cdot & 1 & 1 \end{pmatrix} \mapsto \begin{pmatrix} 1 & 1 & \cdot & \cdot \\ 1 & \cdot & \cdot & 1 \\ \cdot & 1 & 1 & \cdot \\ \cdot & \cdot & 1 & 1 \end{pmatrix} \mapsto \begin{pmatrix} 1 & 1 & \cdot & \cdot \\ \cdot & 1 & \cdot & 1 \\ 1 & \cdot & 1 & \cdot \\ \cdot & \cdot & 1 & 1 \end{pmatrix}$$

$$\mapsto \begin{pmatrix} 1 & 1 & \cdot & \cdot \\ \cdot & \cdot & 1 & 1 \\ 1 & 1 & \cdot & \cdot \\ \cdot & \cdot & 1 & 1 \end{pmatrix} \mapsto \begin{pmatrix} 1 & 1 & \cdot & \cdot \\ \cdot & \cdot & 1 & 1 \\ 1 & \cdot & 1 & \cdot \\ \cdot & 1 & \cdot & 1 \end{pmatrix} \mapsto \begin{pmatrix} 1 & 1 & \cdot & \cdot \\ \cdot & \cdot & 1 & 1 \\ 1 & \cdot & \cdot & 1 \\ \cdot & 1 & 1 & \cdot \end{pmatrix} \mapsto \begin{pmatrix} 1 & 1 & \cdot & \cdot \\ \cdot & \cdot & 1 & 1 \\ \cdot & 1 & \cdot & 1 \\ 1 & \cdot & 1 & \cdot \end{pmatrix}$$

$$\mapsto \begin{pmatrix} 1 & 1 & \cdot & \cdot \\ \cdot & \cdot & 1 & 1 \\ \cdot & \cdot & 1 & 1 \\ 1 & 1 & \cdot & \cdot \end{pmatrix} \mapsto \begin{pmatrix} 1 & \cdot & 1 & \cdot \\ \cdot & 1 & \cdot & 1 \\ \cdot & \cdot & 1 & 1 \\ 1 & 1 & \cdot & \cdot \end{pmatrix} \mapsto \begin{pmatrix} 1 & \cdot & \cdot & 1 \\ \cdot & 1 & 1 & \cdot \\ \cdot & \cdot & 1 & 1 \\ 1 & 1 & \cdot & \cdot \end{pmatrix} \mapsto \begin{pmatrix} \cdot & 1 & \cdot & 1 \\ 1 & \cdot & 1 & \cdot \\ \cdot & \cdot & 1 & 1 \\ 1 & 1 & \cdot & \cdot \end{pmatrix}$$

$$\mapsto \begin{pmatrix} \cdot & \cdot & 1 & 1 \\ 1 & 1 & \cdot & \cdot \\ \cdot & \cdot & 1 & 1 \\ 1 & 1 & \cdot & \cdot \end{pmatrix} \mapsto \begin{pmatrix} \cdot & \cdot & 1 & 1 \\ 1 & \cdot & 1 & \cdot \\ \cdot & 1 & \cdot & 1 \\ 1 & 1 & \cdot & \cdot \end{pmatrix} \mapsto \begin{pmatrix} \cdot & \cdot & 1 & 1 \\ 1 & \cdot & \cdot & 1 \\ \cdot & 1 & 1 & \cdot \\ 1 & 1 & \cdot & \cdot \end{pmatrix} \mapsto \begin{pmatrix} \cdot & \cdot & 1 & 1 \\ \cdot & 1 & \cdot & 1 \\ 1 & \cdot & 1 & \cdot \\ 1 & 1 & \cdot & \cdot \end{pmatrix}$$

$$\mapsto \begin{pmatrix} \cdot & \cdot & 1 & 1 \\ \cdot & \cdot & 1 & 1 \\ 1 & 1 & \cdot & \cdot \\ 1 & 1 & \cdot & \cdot \end{pmatrix} = Q_2.$$

We turn now our attention to the problem of finding the maximum length of a chain in the Bruhat order on the class $\mathscr{A}(n, 2)$, following closely M. Ghebleh [19]. The key factor for the construction of such a maximal chain is the number of inversions of a (0, 1)-matrix, which was shown by Ghebleh to be monotonic with respect to the

Bruhat order on the class $\mathscr{A}(n, 2)$ using a variation of the Bruhat order on $\mathscr{A}(R, S)$, called the secondary Bruhat order, which coincides with the Bruhat order on the class $\mathscr{A}(n, 2)$.

Lemma 4 *Let $A, C \in \mathscr{A}(n, 2)$. If $A \precneqq C$, then $\nu(A) < \nu(C)$.*

Note that from this result we can conclude that if $\nu(A) = \nu(C)$ for two distinct $A, C \in \mathscr{A}(n, 2)$ then A and C are incomparable in the Bruhat order. Therefore, the set $\nu^{-1}(t)$ of all matrices $A \in \mathscr{A}(n, 2)$ with $\nu(A) = t$ is an antichain in the Bruhat order of $\mathscr{A}(n, 2)$, for any integers $n \geq 2$ and $t \geq 0$.

In the next results, we present Ghebleh's construction of chains of lengths $2n(n-2)$ if $n \geq 4$ is even, or $2n(n-2) - 1$ if $n \geq 5$ is odd, respectively, in the Bruhat order of $\mathscr{A}(n, 2)$, starting at the minimal matrices P_n and ending at the maximal matrices Q_n given in (2) and (3), respectively.

Proposition 1 *If $n \geq 4$ is even, then there is a chain of length $2n(n-2)$ from P_n to Q_n in the Bruhat order of $\mathscr{A}(n, 2)$.*

Proof The chain is constructed recursively by induction on n. For $n = 4$ the chain was given by Theorem 4 and presented in Example 1. So, let $n \geq 6$ be even, and note that by (2), $P_n = P_{n-2} \oplus J_2$. By the induction hypothesis, there is a chain of length $2(n-2)(n-4)$ from P_{n-2} to Q_{n-2}. Taking the direct sum of the matrices in such chain with J_2, we obtain a chain of the same length from P_n to $A_1 = Q_{n-2} \oplus J_2$. This chain can be extended to one from P_n to Q_n as follows. Let E_1 be the submatrix of A_1 induced by rows $1, 2, n-1, n$ and columns $n-3, n-2, n-1, n$. Then, $E_1 = P_4$. We extend the current chain by keeping all entries outside E_1 constant, and applying the chain of case $n = 4$ in the positions corresponding to E_1. This extends the current chain by 16. Let A_2 denote the end of this chain. We procced by applying the same procedure to the submatrix $E_2 = P_4$ of A_2 induced by rows $3, 4, n-1, n$, and columns $n-5, n-4, n-3, n-2$. The process is repeated for a total of $n/2 - 1$ times, after which the resulting chain ends at $A_{n/2} = Q_n$. The length of this chain is

$$2(n-2)(n-4) + 16(n/2 - 1) = 2n(n-2),$$

as required. □

Example 2 A chain of maximal length in the class $\mathscr{A}(5, 2)$. Dots represent zeros.

$$P_5 = \begin{pmatrix} 1 & 1 & \cdot & \cdot & \cdot \\ 1 & 1 & \cdot & \cdot & \cdot \\ \cdot & \cdot & 1 & 1 & \cdot \\ \cdot & \cdot & 1 & \cdot & 1 \\ \cdot & \cdot & \cdot & 1 & 1 \end{pmatrix} \mapsto \begin{pmatrix} 1 & 1 & \cdot & \cdot & \cdot \\ 1 & 1 & \cdot & \cdot & \cdot \\ \cdot & \cdot & 1 & 1 & \cdot \\ \cdot & \cdot & \cdot & 1 & 1 \\ \cdot & \cdot & 1 & \cdot & 1 \end{pmatrix} \mapsto \begin{pmatrix} 1 & 1 & \cdot & \cdot & \cdot \\ 1 & \cdot & 1 & \cdot & \cdot \\ \cdot & 1 & \cdot & 1 & \cdot \\ \cdot & \cdot & \cdot & 1 & 1 \\ \cdot & \cdot & 1 & \cdot & 1 \end{pmatrix}$$

$$\mapsto \begin{pmatrix} 1 & 1 & \cdot & \cdot & \cdot \\ 1 & \cdot & \cdot & 1 & \cdot \\ \cdot & 1 & 1 & \cdot & \cdot \\ \cdot & \cdot & \cdot & 1 & 1 \\ \cdot & \cdot & 1 & \cdot & 1 \end{pmatrix} \mapsto \begin{pmatrix} 1 & 1 & \cdot & \cdot & \cdot \\ 1 & \cdot & \cdot & 1 & \cdot \\ \cdot & 1 & \cdot & 1 & \cdot \\ \cdot & \cdot & 1 & \cdot & 1 \\ \cdot & \cdot & 1 & \cdot & 1 \end{pmatrix} \mapsto \begin{pmatrix} 1 & 1 & \cdot & \cdot & \cdot \\ 1 & \cdot & \cdot & 1 & \cdot \\ \cdot & 1 & \cdot & \cdot & 1 \\ \cdot & \cdot & 1 & 1 & \cdot \\ \cdot & \cdot & 1 & \cdot & 1 \end{pmatrix}$$

$$\mapsto \begin{pmatrix} 1&1&\cdot&\cdot&\cdot\\ 1&\cdot&\cdot&1&\cdot\\ \cdot&1&\cdot&\cdot&1\\ \cdot&\cdot&1&\cdot&1\\ \cdot&\cdot&1&1&\cdot \end{pmatrix} =: Z \mapsto \begin{pmatrix} 1&1&\cdot&\cdot&\cdot\\ 1&\cdot&\cdot&\cdot&1\\ \cdot&1&\cdot&1&\cdot\\ \cdot&\cdot&1&\cdot&1\\ \cdot&\cdot&1&1&\cdot \end{pmatrix} \mapsto \begin{pmatrix} 1&1&\cdot&\cdot&\cdot\\ 1&\cdot&\cdot&\cdot&1\\ \cdot&\cdot&1&1&\cdot\\ \cdot&1&\cdot&\cdot&1\\ \cdot&\cdot&1&1&\cdot \end{pmatrix}$$

$$\mapsto \begin{pmatrix} 1&1&\cdot&\cdot&\cdot\\ 1&\cdot&\cdot&\cdot&1\\ \cdot&\cdot&1&1&\cdot\\ \cdot&\cdot&1&\cdot&1\\ \cdot&1&\cdot&1&\cdot \end{pmatrix} \mapsto \begin{pmatrix} 1&1&\cdot&\cdot&\cdot\\ 1&\cdot&\cdot&\cdot&1\\ \cdot&\cdot&1&1&\cdot\\ \cdot&\cdot&\cdot&1&1\\ \cdot&1&1&\cdot&\cdot \end{pmatrix} \mapsto \begin{pmatrix} 1&\cdot&1&\cdot&\cdot\\ 1&\cdot&\cdot&\cdot&1\\ \cdot&1&\cdot&1&\cdot\\ \cdot&\cdot&\cdot&1&1\\ \cdot&1&1&\cdot&\cdot \end{pmatrix}$$

$$\mapsto \begin{pmatrix} 1&\cdot&\cdot&1&\cdot\\ 1&\cdot&\cdot&\cdot&1\\ \cdot&1&1&\cdot&\cdot\\ \cdot&\cdot&\cdot&1&1\\ \cdot&1&1&\cdot&\cdot \end{pmatrix} \mapsto \begin{pmatrix} 1&\cdot&\cdot&1&\cdot\\ 1&\cdot&\cdot&\cdot&1\\ \cdot&1&\cdot&1&\cdot\\ \cdot&\cdot&1&\cdot&1\\ \cdot&1&1&\cdot&\cdot \end{pmatrix} \mapsto \begin{pmatrix} 1&\cdot&\cdot&1&\cdot\\ 1&\cdot&\cdot&\cdot&1\\ \cdot&1&\cdot&\cdot&1\\ \cdot&\cdot&1&1&\cdot\\ \cdot&1&1&\cdot&\cdot \end{pmatrix}$$

$$\mapsto \begin{pmatrix} 1&\cdot&\cdot&\cdot&1\\ 1&\cdot&\cdot&1&\cdot\\ \cdot&1&\cdot&\cdot&1\\ \cdot&\cdot&1&1&\cdot\\ \cdot&1&1&\cdot&\cdot \end{pmatrix} \mapsto \begin{pmatrix} 1&\cdot&\cdot&\cdot&1\\ 1&\cdot&\cdot&1&\cdot\\ \cdot&\cdot&1&\cdot&1\\ \cdot&1&\cdot&1&\cdot\\ \cdot&1&1&\cdot&\cdot \end{pmatrix} \mapsto \begin{pmatrix} 1&\cdot&\cdot&\cdot&1\\ \cdot&1&\cdot&1&\cdot\\ \cdot&\cdot&1&\cdot&1\\ 1&\cdot&\cdot&1&\cdot\\ \cdot&1&1&\cdot&\cdot \end{pmatrix}$$

$$\mapsto \begin{pmatrix} 1&\cdot&\cdot&\cdot&1\\ \cdot&1&\cdot&1&\cdot\\ \cdot&\cdot&1&\cdot&1\\ \cdot&1&\cdot&1&\cdot\\ 1&\cdot&1&\cdot&\cdot \end{pmatrix} \mapsto \begin{pmatrix} 1&\cdot&\cdot&\cdot&1\\ \cdot&1&\cdot&1&\cdot\\ \cdot&\cdot&1&\cdot&1\\ \cdot&\cdot&1&1&\cdot\\ 1&1&\cdot&\cdot&\cdot \end{pmatrix} \mapsto \begin{pmatrix} 1&\cdot&\cdot&\cdot&1\\ \cdot&\cdot&1&1&\cdot\\ \cdot&1&\cdot&\cdot&1\\ \cdot&\cdot&1&1&\cdot\\ 1&1&\cdot&\cdot&\cdot \end{pmatrix}$$

$$\mapsto \begin{pmatrix} 1&\cdot&\cdot&\cdot&1\\ \cdot&\cdot&1&1&\cdot\\ \cdot&\cdot&1&\cdot&1\\ \cdot&1&\cdot&1&\cdot\\ 1&1&\cdot&\cdot&\cdot \end{pmatrix} \mapsto \begin{pmatrix} 1&\cdot&\cdot&\cdot&1\\ \cdot&\cdot&1&1&\cdot\\ \cdot&\cdot&\cdot&1&1\\ \cdot&1&1&\cdot&\cdot\\ 1&1&\cdot&\cdot&\cdot \end{pmatrix} \mapsto \begin{pmatrix} \cdot&\cdot&1&\cdot&1\\ 1&\cdot&\cdot&1&\cdot\\ \cdot&\cdot&\cdot&1&1\\ \cdot&1&1&\cdot&\cdot\\ 1&1&\cdot&\cdot&\cdot \end{pmatrix}$$

$$\mapsto \begin{pmatrix} \cdot&\cdot&\cdot&1&1\\ 1&\cdot&1&\cdot&\cdot\\ \cdot&\cdot&\cdot&1&1\\ \cdot&1&1&\cdot&\cdot\\ 1&1&\cdot&\cdot&\cdot \end{pmatrix} \mapsto \begin{pmatrix} \cdot&\cdot&\cdot&1&1\\ 1&\cdot&\cdot&1&\cdot\\ \cdot&\cdot&1&\cdot&1\\ \cdot&1&1&\cdot&\cdot\\ 1&1&\cdot&\cdot&\cdot \end{pmatrix} \mapsto \begin{pmatrix} \cdot&\cdot&\cdot&1&1\\ 1&\cdot&\cdot&\cdot&1\\ \cdot&\cdot&1&1&\cdot\\ \cdot&1&1&\cdot&\cdot\\ 1&1&\cdot&\cdot&\cdot \end{pmatrix}$$

$$\mapsto \begin{pmatrix} \cdot&\cdot&\cdot&1&1\\ \cdot&\cdot&1&\cdot&1\\ 1&\cdot&\cdot&1&\cdot\\ \cdot&1&1&\cdot&\cdot\\ 1&1&\cdot&\cdot&\cdot \end{pmatrix} \mapsto \begin{pmatrix} \cdot&\cdot&\cdot&1&1\\ \cdot&\cdot&\cdot&1&1\\ 1&\cdot&1&\cdot&\cdot\\ \cdot&1&1&\cdot&\cdot\\ 1&1&\cdot&\cdot&\cdot \end{pmatrix} \mapsto \begin{pmatrix} \cdot&\cdot&\cdot&1&1\\ \cdot&\cdot&\cdot&1&1\\ \cdot&1&1&\cdot&\cdot\\ 1&\cdot&1&\cdot&\cdot\\ 1&1&\cdot&\cdot&\cdot \end{pmatrix} = Q_5$$

Proposition 2 *If $n \geq 5$ is odd, then there is a chain of length $2n(n-2)-1$ from P_n to Q_n in the Bruhat order of $\mathscr{A}(n,2)$.*

Proof As for the even case, the chain is constructed recursively by induction on n. For $n = 5$ the chain is shown in Example 2. Let $n = 2k+5$ with $k \geq 1$. Then $P_n = P_{2k} \oplus P_5$. By applying the chain for the case $n = 5$ of Example 2 to the submatrix of P_n formed by its last five rows and its last five columns, we obtain a chain of length 29 from P_n to $A = P_{2k} + Q_5$. Let E be the 5×5 submatrix of A induced by the rows $2k-1, 2k, n-1, n-1, n$ and columns $2k-1, \ldots, 2k+3$. Then $E =$

$J_2 \oplus F_3'$. Note that $\Sigma_E \geq \Sigma_Z$, where Z is the seventh matrix of the chain in $\mathscr{A}(5, 2)$ constructed in Example 2. Thus, $E \preceq Z$ and we may apply the subchain of Example 2 that start with Z and ends with Q_5 to extend the current chain by 24. We repeat this procedure k times to obtain a chain ending at

$$C = \begin{pmatrix} 0 & P_{n-3} \\ F_3' & 0 \end{pmatrix}.$$

Using now Proposition 1, we may extend this chain to end at

$$Q_n = \begin{pmatrix} 0 & Q_{n-3} \\ F_3' & 0 \end{pmatrix}.$$

This chain has length $29 + 24k + 2(n-3)(n-5) = 2n(n-2) - 1$. □

The next result shows that the chains constructed in Propositions 1 and 2 are indeed the longest possible chains in the Bruhat order of $\mathscr{A}(n, 2)$.

Theorem 5 [19] *Let $n \geq 4$ and let $\delta(n)$ denote the maximum length of a chain in the Bruhat order of $\mathscr{A}(n, 2)$. Then,*

$$\delta(n) = \begin{cases} 2n(n-2), & \text{if } n \equiv 0 (mod\, 2) \\ 2n(n-2) - 1, & \text{if } n \equiv 1 (mod\, 2) \end{cases}.$$

Proof Let $A_0 \preceq A_1 \preceq \cdots \preceq A_k$ be a chain in the Bruhat order of $\mathscr{A}(n, 2)$. By Lemma 4 we have $\nu(A_0) < \nu(A_1) < \cdots < \nu(A_k)$, from which it follows that $k \leq \nu(A_k) - \nu(A_0)$. Since a chain of maximum length $\delta(n)$ begins with a minimal element and ends with a maximal element, we obtain

$$\delta(n) \leq \max\{\nu(Q)\} - \min\{\nu(P)\},$$

where the maximum is over all maximal matrices Q and the minimum is over all minimal matrices P in the Bruhat order of $\mathscr{A}(n, 2)$.

On the other hand, since by Theorem 2 any minimal matrix in $\mathscr{A}(n, 2)$ is a direct sum of matrices equal to J_2 and F_3, and $\nu(J_2) = 1$ and $\nu(F_3) = 2$, a minimal matrix with the smallest number of inversions cannot have more than one direct sum component F_3. Therefore, $\nu(P_n)$ is the smallest possible value of $\nu(P)$, for any $P \in \mathscr{A}(n, 2)$, and similarly, $\nu(Q_n)$ is the largest possible value of $\nu(Q)$, for any $Q \in \mathscr{A}(n, 2)$. Therefore, we obtain

$$\delta(n) \leq \nu(Q_n) - \nu(P_n).$$

Note that there are no inversions in P_n involving entries from different J_2 and F_3 direct sum components. Thus, $\nu(P_n) = \lceil n/2 \rceil$. In the maximal matrix Q_n, however, every pair of ones in different J_2 and F_3' sum components gives an inversion, while there are $\nu(J_2) = 1$ and $\nu(F_3') = 7$ inversions within each block. A simple calculation gives $\nu(Q_n) = \lfloor (4n^2 - 7n)/2 \rfloor$. Hence,

$$\delta(n) \leq \lfloor (4n^2 - 7n)/2 \rfloor - \lceil n/2 \rceil = \begin{cases} 2n(n-2), & \text{if } n \equiv 0 (mod\, 2) \\ 2n(n-2) - 1, & \text{if } n \equiv 1 (mod\, 2) \end{cases}.$$

The constructions obtained in the proofs of Propositions 1 and 2 proves the lower bound. □

5 Antichains in $\mathscr{A}(2k, k)$ and $\mathscr{A}(n, 2)$

Dilworth's theorem [14] states that the maximum number of elements in any antichain in a partially ordered set equals the minimum number of chains into which the set may be partitioned. Mirsky's dual of this theorem [28] states that the maximum number of elements in any chain in a partially ordered set equals the minimum number of antichains into which the set may be partitioned. Denoting by $h(n, k)$ (respectively $w(n, k)$) the maximum number of elements in a antichain (respectively chain) in the Bruhat order of $\mathscr{A}(n, k)$, Dilworth's and Mirsky's theorems imply

$$h(n, k)w(n, k) \geq \#\mathscr{A}(n, k). \tag{4}$$

By Theorems 4 and 5 we have

$$h(2k, k) = k^4 + 1 \text{ and } h(n, 2) = 2n(n-2) + \varepsilon,$$

for $k \geq 1$, $n \geq 4$, where ε is 0 if n is odd, and 1 if n is even. These values, together with Eqs. (1) and (4), indicates that the maximum number of elements in any antichain on the Bruhat order of the sets $\mathscr{A}(2k, k)$ and $\mathscr{A}(n, 2)$ have exponentially large sizes. In what follows we derive lower bound for the number $w(n, k)$ of the Bruhat order on the classes $\mathscr{A}(2k, k)$ and $\mathscr{A}(n, 2)$.

In the first result we explain the construction made in [12] of antichains of size $(\frac{k}{2}^4 + 1)^2$ on the class $\mathscr{A}(2k, k)$, a result which was improved in [20], as we will see. Nevertheless, the constructive nature of these antichains make it worth to include it in this survey.

Theorem 6 *For any integer $k \geqslant 2$, let $w(2k, k)$ be the largest size of an antichain in the Bruhat order in $\mathscr{A}(2k, k)$. Then*

$$w(2k, k) \geq \left(\frac{k}{2}^4 + 1\right)^2.$$

Proof We start by proving the bound for $w(2k, k)$ when $k \equiv 0 (mod 2)$. Recalling that by Theorem 3, $\mathscr{A}(2k, k)$ admits a minimum P_k and a maximum Q_k, let us consider the matrix

$$A = \left(\begin{array}{cc|cc} J_{\frac{k}{2}} & J_{\frac{k}{2}} & O_{\frac{k}{2}} & O_{\frac{k}{2}} \\ O_{\frac{k}{2}} & O_{\frac{k}{2}} & J_{\frac{k}{2}} & J_{\frac{k}{2}} \\ \hline O_{\frac{k}{2}} & O_{\frac{k}{2}} & J_{\frac{k}{2}} & J_{\frac{k}{2}} \\ J_{\frac{k}{2}} & J_{\frac{k}{2}} & O_{\frac{k}{2}} & O_{\frac{k}{2}} \end{array}\right) = \left(\begin{array}{c|c|c} J_{\frac{k}{2}} & \multirow{2}{*}{$P_{\frac{k}{2}}$} & O_{\frac{k}{2}} \\ O_{\frac{k}{2}} & & J_{\frac{k}{2}} \\ \hline O_{\frac{k}{2}} & \multirow{2}{*}{$Q_{\frac{k}{2}}$} & J_{\frac{k}{2}} \\ J_{\frac{k}{2}} & & O_{\frac{k}{2}} \end{array}\right)$$

$$= \left(\begin{array}{c|c|c} J^*_{\frac{k}{2}} & \multirow{2}{*}{$P^\bullet_{\frac{k}{2}}$} & O^*_{\frac{k}{2}} \\ O^*_{\frac{k}{2}} & & J^*_{\frac{k}{2}} \\ \hline O^\dagger_{\frac{k}{2}} & \multirow{2}{*}{$Q^\odot_{\frac{k}{2}}$} & J^\dagger_{\frac{k}{2}} \\ J^\dagger_{\frac{k}{2}} & & O^\dagger_{\frac{k}{2}} \end{array}\right) \tag{5}$$

which satisfies is actually the matrix generated at step $\frac{k^4}{2}$ by the algorithm in Theorem 4. We use symbols $^\bullet$, $^\odot$, *, and † just to mark and indicate the corresponding submatrices of A. Note that $^\bullet \simeq {}^* \simeq P_{\frac{k}{2}}$ and $^\odot \simeq {}^\dagger \simeq Q_{\frac{k}{2}}$.

The *Chain* algorithm of Theorem 4 generates a chain of maximal length n^4 between P_n and Q_n, for any integer $n \geqslant 2$, and it is straightforward to see that it can be reverted, viz. we can consider the *Rev–Chain* algorithm which generates the same chain backwards from Q_n and P_n.

Applying simultaneously Chain and Rev–Chain algorithms to $^\bullet$ and $^\odot$, and denoting this operation as *central-antichain* algorithm, we get $\left(\frac{k}{2}\right)^4 + 1$ incomparable elements. In fact, let

$$A^{ca}_\ell = \left(\begin{array}{c|c|c} J_{\frac{k}{2}} & \multirow{2}{*}{$P^\ell_{\frac{k}{2}}$} & O_{\frac{k}{2}} \\ O_{\frac{k}{2}} & & J_{\frac{k}{2}} \\ \hline O_{\frac{k}{2}} & \multirow{2}{*}{$Q^\ell_{\frac{k}{2}}$} & J_{\frac{k}{2}} \\ J_{\frac{k}{2}} & & O_{\frac{k}{2}} \end{array}\right)$$

be the matrix obtained from A after $\ell > 0$ iterations of the central-antichain algorithm, i.e. $P^\ell_{\frac{k}{2}}$ is obtained from $P_{\frac{k}{2}}$ after ℓ iterations of the Chain algorithm and $Q^\ell_{\frac{k}{2}}$ is obtained from $Q_{\frac{k}{2}}$ after ℓ iterations of the Rev–Chain algorithm.

Now consider $\ell < t$ and A^{ca}_ℓ and A^{ca}_t (i.e. A^{ca}_t is obtained from A^{ca}_ℓ after $t - \ell$ iterations of the central-antichain algorithm): obviously we have $P^\ell_{\frac{k}{2}} \precneqq P^t_{\frac{k}{2}}$ as elements of $\mathscr{A}(k, \frac{k}{2})$, i.e. there exist (u, v) with $1 \le u, v \le k$ such that $\sigma_{uv}\left(P^\ell_{\frac{k}{2}}\right) > \sigma_{uv}\left(P^t_{\frac{k}{2}}\right)$. Similarly, $Q^t_{\frac{k}{2}} \precneqq Q^\ell_{\frac{k}{2}}$ as elements of $\mathscr{A}(k, \frac{k}{2})$, i.e. there exist (w, z) with $1 \le w, z \le k$ such that $\sigma_{uv}\left(Q^t_{\frac{k}{2}}\right) > \sigma_{uv}\left(Q^\ell_{\frac{k}{2}}\right)$.

Therefore, considering in $\mathscr{A}(2k, k)$ the two matrices A^{ca}_ℓ and A^{ca}_t we have

$$\sigma_{u,\frac{k}{2}+v}\left(A_\ell^{ca}\right) = \sigma_{u,\frac{k}{2}}\begin{pmatrix} J_{\frac{k}{2}} \\ O_{\frac{k}{2}} \end{pmatrix} + \sigma_{uv}\left(P_{\frac{k}{2}}^{\ell}\right)$$

$$> \sigma_{u,\frac{k}{2}}\begin{pmatrix} J_{\frac{k}{2}} \\ O_{\frac{k}{2}} \end{pmatrix} + \sigma_{uv}\left(P_{\frac{k}{2}}^{t}\right) = \sigma_{u,\frac{k}{2}+v}\left(A_t^{ca}\right).$$

We now show that $\sigma_{k+w,\frac{k}{2}+z}\left(A_\ell^{ca}\right) < \sigma_{k+w,\frac{k}{2}+z}\left(A_t^{ca}\right)$, and therefore A_ℓ^{ca} and A_t^{ca} are incomparable in $\mathscr{A}(2k,k)$.

Note that since $P_{\frac{k}{2}}^{\ell}, P_{\frac{k}{2}}^{t} \in \mathscr{A}(k,\frac{k}{2})$, for any $1 \le j \le k$, we have

$$\sigma_{jk}\left(P_{\frac{k}{2}}^{\ell}\right) = \sigma_{jk}\left(P_{\frac{k}{2}}^{t}\right) = \sigma_{kj}\left(P_{\frac{k}{2}}^{\ell}\right) = \sigma_{kj}\left(P_{\frac{k}{2}}^{t}\right) = \frac{jk}{2},$$

hence

$$\sigma_{k+w,\frac{k}{2}+z}\left(A_\ell^{ca}\right) = \sigma_{k,\frac{k}{2}}\begin{pmatrix} J_{\frac{k}{2}} \\ O_{\frac{k}{2}} \end{pmatrix} + \sigma_{kz}\left(P_{\frac{k}{2}}^{\ell}\right) + \sigma_{w,\frac{k}{2}}\begin{pmatrix} O_{\frac{k}{2}} \\ J_{\frac{k}{2}} \end{pmatrix} + \sigma_{wz}\left(Q_{\frac{k}{2}}^{\ell}\right)$$

$$= \sigma_{k,\frac{k}{2}}\begin{pmatrix} J_{\frac{k}{2}} \\ O_{\frac{k}{2}} \end{pmatrix} + \frac{kz}{2} + \sigma_{w,\frac{k}{2}}\begin{pmatrix} O_{\frac{k}{2}} \\ J_{\frac{k}{2}} \end{pmatrix} + \sigma_{wz}\left(Q_{\frac{k}{2}}^{\ell}\right)$$

$$< \sigma_{k,\frac{k}{2}}\begin{pmatrix} J_{\frac{k}{2}} \\ O_{\frac{k}{2}} \end{pmatrix} + \frac{kz}{2} + \sigma_{w,\frac{k}{2}}\begin{pmatrix} O_{\frac{k}{2}} \\ J_{\frac{k}{2}} \end{pmatrix} + \sigma_{wz}\left(Q_{\frac{k}{2}}^{t}\right)$$

$$= \sigma_{k,\frac{k}{2}}\begin{pmatrix} J_{\frac{k}{2}} \\ O_{\frac{k}{2}} \end{pmatrix} + \sigma_{kz}\left(P_{\frac{k}{2}}^{t}\right) + \sigma_{w,\frac{k}{2}}\begin{pmatrix} O_{\frac{k}{2}} \\ J_{\frac{k}{2}} \end{pmatrix} + \sigma_{wz}\left(Q_{\frac{k}{2}}^{t}\right)$$

$$= \sigma_{k+w,\frac{k}{2}+z}\left(A_t^{ca}\right).$$

Analogously, we can apply simultaneously Chain and Rev–Chain algorithms to the submatrices $*$ and † in Eq. (5), denoting this operation by *lateral-antichain* algorithm, and we get $\left(\frac{k}{2}\right)^4 + 1$ incomparable elements, as well.

In fact, it is possible to apply independently both central-antichain and lateral-antichain algorithms to A in Eq. (5) and still get an antichain, namely $Z = \{A^{ij} \mid 0 \leqslant i,j \leqslant (\frac{k}{2})^4\}$ is an antichain, where A^{ij} is the matrix obtained from A applying i-times the central-antichain algorithm and j-times the lateral-antichain algorithm; thus we get an instance of an antichain having size

$$\left(\left(\frac{k}{2}\right)^4 + 1\right)^2.$$

It is easy to see that Z is an antichain because the upper half of the matrix A is the disjoint union of two submatrices $P_{\frac{k}{2}}$, whereas the lower half is the disjoint union of two submatrices $Q_{\frac{k}{2}}$, hence for any transformation we apply, the upper half goes up in the Bruhat order, and the lower half goes down, and therefore the resulting elements are incomparable.

For any integer $k \geqslant 3$, not necessary even, we obviously have

$$w(2(k-1), k-1) \leqslant w(2k, k),$$

and the desired result follows. □

Consider now the class $\mathscr{A}(n, 2)$. Recall that by the discussion after Lemma 4, we know that for $t \geq 0$ and $n \geq 2$, the set $\nu^{-1}(t)$ of all matrices $A \in \mathscr{A}(n, 2)$ such that $\nu(A) = t$ forms an antichain in the Bruhat order of $\mathscr{A}(n, 2)$.

Given a $m \times n$ matrix $A = [a_{ij}]$, let $A' = [b_{ij}]$ be the $m \times n$ matrix obtained from A by reversing the order of their columns, i.e. with $b_{ij} = a_{i,n-j+1}$ for all $1 \leq i \leq m$ and $1 \leq j \leq n$. We say that A' is the conjugate of A. Applying the inclusion-exclusion principle, we find that if $A \in \mathscr{A}(R, S)$, with $R = (r_1, \ldots, r_m)$ and $S = (s_1, \ldots, s_n)$, then

$$\nu(A) + \nu(A') = \binom{r_1 + \cdots + r_m}{2} - \sum_{i=1}^{m} \binom{r_i}{2} - \sum_{j=1}^{n} \binom{s_j}{2}.$$

In particular, if $A \in \mathscr{A}(n, 2)$ is such that $A = A'$, then we get $\nu(A) = n^2 - 3n/2$. From this equality we conclude that if n is odd there is no self-conjugate matrix in the class $\mathscr{A}(n, 2)$, but for n even self-conjugate matrices were used in [20] to construct antichains. In the next results we describe such constructions.

Theorem 7 *If $n \geq 2$ is an even integer, then there is an antichain of size $\frac{n!}{2^{n/2}}$ in the Bruhat order of $\mathscr{A}(n, 2)$.*

Proof Let $n = 2k$ be an even integer. The first $n/2$ columns of any self-conjugate matrix $A_C \in \mathscr{A}(n, 2)$ induces a matrix C of size $2k \times k$, with columns sums equal to 2 and row sums equal to 1, such that $A = [C\,C']$. Since $\nu(A_C) = n^2 - 3n/2$, the set of all self-conjugate matrices forms, an which we can identify with the set $\mathscr{A}(R, S)$, where $R = (1^n)$ and $S = (2^{n/2})$. In [18], the cardinal of such class $\mathscr{A}(R, S)$ was proved to be $n!/2^{n/2}$. Thus, the antichain formed by the self-conjugate matrices of $\mathscr{A}(n, 2)$ have$(2k)!/2^k$ elements. □

The construction of an antichain in the Bruhat order of $\mathscr{A}(n, 2)$ when n is odd follows the same lines of the even case.

Theorem 8 *If $n \geq 3$ is an odd integer, then there is an antichain of size $\frac{(n-1)!}{2^{(n-3)/2}}$ in the Bruhat order of $\mathscr{A}(n, 2)$.*

Proof Let $n = 2k + 1$ be an odd integer and let C be an $2k \times k$ $(0, 1)$-matrix with columns sums equal to 2 and row sums equal to 1. Let $A_C = [a_{ij}]$ be the $n \times n$ matrix such that its the restriction to the submatrix induced by rows $1, \ldots, 2k$ and columns $1, \ldots, k$ is C, the restriction to the submatrix induced by rows $2, \ldots, 2k + 1$ and columns $k + 1, \ldots, 2k$ is C, has ones in positions a_{1n} and a_{nn}, and zeros in the remaining positions. The number of inversions in both A_C and A'_C is given by

$$\nu(A_C) = \nu(A'_C) = k(4k-1).$$

Thus, the set of matrices of the form A_C and A'_C is an antichain in the Bruhat order of $\mathscr{A}(n,2)$. As mentioned in the proof of Theorem 7, there are $(2k)!/2^k$ such matrices C, and each induces two matrices A_C and A'_C. Thus, there are $(2k)!/2^{k-1}$ elements in the antichain. □

In the following theorem, whose proof we refer to M. Ghebleh [20], we present a construction of antichains in the Bruhat order of the class $\mathscr{A}(R,S)$ that are products of known antichains. This construction was used in [20] to improve the lower bound for the antichains of $\mathscr{A}(2k,k)$ obtained in Theorem 6. We use the notation $R \otimes S$ to denote the Kronecker product of the vectors R and $S = (s_1, \ldots, s_n)$, and let $t + S = (t + s_1, \ldots, t + s_n)$.

Theorem 9 *For positive integers a, b, m, n, let $R_1=(r_1, \ldots, r_a)$, $R_2=(r'_1, \ldots, r'_m)$, $R_3 = (r''_1, \ldots, r''_m)$, $S_1 = (s_1, \ldots, s_b)$, $S_2 = (s'_1, \ldots, s'_n)$, and $S_3 = (s''_1, \ldots, s''_m)$ be nonnegative integer vectors. Let $u = r_1 + r_2 + \cdots + r_a$, $u' = r'_1 + r'_2 + \cdots + r'_m$ and $u'' = r''_1 + r''_2 + \cdots + r''_m$, and suppose that $u' \neq u''$. If $\mathscr{D}_1$, $\mathscr{D}_2$ and $\mathscr{D}_3$ are antichains in the Bruhat order of the classes $\mathscr{A}(R_1,S_1)$, $\mathscr{A}(R_2,S_2)$ and $\mathscr{A}(R_3,S_3)$ respectively, then there is an antichain of size $|\mathscr{D}_1||\mathscr{D}_2|^u|\mathscr{D}_3|^{ab-u}$ in the Bruhat order of the class $\mathscr{A}(R,S)$, where $R = R_1 \otimes R_2 + (b - R_1) \otimes R_3$ and $S = S_1 \otimes S_2 + (a - S_1) \otimes S_3$.*

Corollary 1 *For any integer $k \geqslant 2$, let $w(2k,k)$ be the largest size of an antichain in the Bruhat order in $\mathscr{A}(2k,k)$. Then*

$$w(2k,k) \geq g(k) = \begin{cases} \frac{(k!)^4}{4^k} & \text{if } k \text{ is even} \\ \frac{((k-1)!)^4}{4^{k-3}} & \text{if } k \text{ is odd} \end{cases}.$$

Proof Consider the antichains $\mathscr{D}_1 = \{I_2\}$ in the Bruhat order of $\mathscr{A}(2,1)$, and $\mathscr{D}_2$ in the Bruhat order of $\mathscr{A}(k,2)$ given in Theorem 7 or 8, depending on the parity of k. Then, $\mathscr{D}_3 = \{J_2 - X : X \in \mathscr{D}_2\}$, formed by the complements of the matrices in $\mathscr{D}_2$, is also an antichain in the Bruhat order of $\mathscr{A}(k,2)$, since matrix complements reverses the Bruhat order. Applying Theorem 9 to the antichains $\mathscr{D}_1$, $\mathscr{D}_2$ and $\mathscr{D}_3$ we get an antichain of the desired length in the Bruhat order in $\mathscr{A}(2k,k)$. □

Acknowledgements This work was partially supported by the Centre for Mathematics of the University of Coimbra – UID/MAT/00324/2013, funded by the Portuguese Government through FCT/MEC and co-funded by the European Regional Development Fund through the Partnership Agreement PT2020.

References

1. Barvinok, A.: On the number of matrices and a random matrix with prescribed row and column sums and $0-1$ entries. Adv. Math. **224**(1), 316–339 (2010)
2. Borovnick, A.V., Gelfand, I.M., White, N.: Coxeter Matroids. Birkhäuser, Boston (2003)
3. Brualdi, R.A.: Matrices of zeros and ones with fixed row and column sum vectors. Linear Algebra Appl. **33**, 159–231 (1980)
4. Brualdi, R.A.: Short proofs of the Gale & Ryser and Ford & Fulkerson characterizations of the row and column sum vectors of (0, 1)-matrices. Math. Inequal. Appl. **4**, 157–159 (2001)
5. Brualdi, R.A.: Combinatorial matrix classes. Encyclopedia of Mathematics and its Applications, vol. 108. Cambridge University Press, Cambridge (2006)
6. Brualdi, R.A.: Algorithms for constructing (0, 1)-matrices with prescribed row and column sum vectors. Discrete Math. **306**(3), 3054–3062 (2006)
7. Brualdi, R.A., Deaett, L.: More on the Bruhat order for (0, 1)-matrices. Linear Algebra Appl. **421**(2–3), 219–232 (2007)
8. Brualdi, R.A., Hwang, S.-G.: A Bruhat order for the class of (0, 1)–matrices with row sum vector R and column sum vector S. Electron. J. Linear Algebra **12**, 6–16 (2004/05)
9. Brylawski, T.: The lattice of integer partitions. Discrete Math. **6**, 201–219 (1973)
10. Canfield, E.R., McKay, B.D.: Asymptotic enumeration of dense $0-1$ matrices with equal row sums and equal column sums. Electron. J. Combin. **12**, Research Paper 29, 31 pp. (2005)
11. Conflitti, A., da Fonseca, C.M., Mamede, R.: The maximal length of a chain in the Bruhat order for a class of binary matrices. Linear Algebra Appl. **436**(3), 753–757 (2012)
12. Conflitti, A., da Fonseca, C.M., Mamede, R.: On the largest size of an antichain in the Bruhat order for $A(2k, k)$. Order **30**(1), 255–260 (2013)
13. Dahl, G.: Network flows and combinatorial matrix theory, Lecture Notes (2010)
14. Dilworth, R.P.: A decomposition theorem for partially ordered sets. Ann. Math. **51**(2), 161–166 (1950)
15. Everett Jr., C.J., Stein, P.R.: The asymptotic number of integer stochastic matrices. Discrete Math. **1**, 33–72 (1971)
16. Fulkerson, D.R.: Zero-one matrices with zero trace. Pac. J. Math. **10**, 831–836 (1960)
17. Gale, D.: A theorem on flows in networks. Pac. J. Math. **7**, 1073–1082 (1957)
18. Gao, S., Matheis, K.: Closed formulas and integer sequences arising from the enumeration of (0, 1)-matrices with row sum two and some constant column sums. Congr. Numer. **202**, 45–53 (2010)
19. Ghebleh, M.: On maximum chains in the Bruhat order of $A(n, 2)$. Linear Algebra Appl. **446**, 377–387 (2014)
20. Ghebleh, M.: Antichains of (0, 1)-matrices through inversions. Linear Algebra Appl. **458**, 503–511 (2014)
21. Goldstein, D., Stong, R.: On the number of possible row and column sums of 0, 1–matrices. Electron. J. Combin. **13**(1), Note 8, 6 pp. (electronic) (2006)
22. Greenhill, C., McKay, B.D., Wang, X.: Asymptotic enumeration of sparse 0–1 matrices with irregular row and column sums. J. Combin. Theory Ser. A **113**(2), 291–324 (2006)
23. Greenhill, C., McKay, B.D.: Asymptotic enumeration of sparse nonnegative integer matrices with specified row and column sums. Adv. Appl. Math. **41**(4), 459–481 (2008)
24. Haber, R.M.: Term rank of 0, l matrices. Rend. Sem. Mat. Univ. Padova **30**, 24–51 (1960)
25. Krause, M.: A simple proof of the Gale-Ryser theorem. Am. Math. Mon. **103**, 335–337 (1996)
26. van Lint, J.H., Wilson, R.M.: A Course in Combinatorics. Cambridge University Press, Cambridge (2001)
27. McKay, B.D., Wang, X.: Asymptotic enumeration of $0-1$ matrices with equal row sums and equal column sums. Linear Algebra Appl. **373**, 273–287 (2003)
28. Mirsky, L.: A dual of Dilworth's decomposition theorem. Am. Math. Mon. **78**, 876–877 (1971)
29. O'Neil, P.E.: Asymptotics and random matrices with row-sum and column-sum restrictions. Bull. Am. Math. Soc. **75**, 1276–1282 (1969)

30. Pérez-Salvador, B.R., de-los Cobos-Silva, S., Gutiérrez-Andrade, M.A., Torres-Chazaro, A.: A reduced formula for the precise number of (0, 1)–matrices in $\mathscr{A}$ (**R**, **S**). Discrete Math. **256**(1–2), 361–372 (2002)
31. Ryser, H.J.: Combinatorial properties of matrices of zeros and ones. Canad. J. Math. **9**, 371–377 (1957)
32. Ryser, H.J.: Combinatorial mathematics. The Carus Mathematical Monographs, vol. 14, (Wiley, New York, 1963)
33. Wang, B.-Y.: Precise number of (0, 1)-matrices in $\mathfrak{A}(R, S)$. Sci. Sinica Ser. A **31**(1), 1–6 (1988)
34. Wang, B.-Y., Zhang, F.: On the precise number of (0, 1)-matrices in $\mathfrak{A}(R, S)$. Discrete Math. **187**(1–3), 211–220 (1998)

Iterative Method for Linear System with Coefficient Matrix as an $M_\vee$-matrix

Manideepa Saha

Abstract An $M_\vee$-matrix A has the form $A = sI - B$, with B an eventually nonnegative matrix and $s \geq \rho(B)$, the spectral radius of B. In this paper we study iterative procedures associated with a splitting of A, to solve the linear system $Ax = b$, with the coefficient matrix A an $M_\vee$-matrix. We generalize the concepts of regular and weak regular splitting of a matrix using the notion of eventually nonnegative matrix, and term them as E-regular and weak E-regular splitting, respectively. We obtain necessary and sufficient conditions for the convergence of these types of splittings. We also discuss the convergence of Jacobi and Gauss-Seidel splittings for $M_\vee$-matrices.

Keywords E-regular splitting · Weak E-regular splitting · Jacobi splitting · Gauss-Seidel splitting

1 Introduction

Consider the linear system

$$Ax = b \tag{1}$$

where $x, b \in \mathbb{R}^n$ and $A \in \mathbb{R}^{n,n}$, is an invertible matrix. An iterative technique to solve the linear system (1) involves an initial approximation x_0 to the solution x and determines a sequence $\{x_k\}$ that converges to the exact solution x. Most of these methods reduce to the iterative scheme $x^{k+1} = Hx^k + c$, with $k \geq 0$, where the matrix H is called an iteration matrix of the system (1). It is well known that the iterative scheme converges to the exact solution x of (1) if and only if $\rho(H) < 1$ for $\rho(H)$ the spectral radius of H.

As it is well known with a splitting $A = M - N$ of A, one may associate an iterative scheme

M. Saha (✉)
Department of Mathematics, National Institute of Technology Meghalaya,
Shillong, Meghalaya, India
e-mail: sahamanideepa@gmail.com

N. Bebiano (ed.), *Applied and Computational Matrix Analysis*,
Springer Proceedings in Mathematics & Statistics 192,
DOI 10.1007/978-3-319-49984-0_16

$$x^{k+1} = M^{-1}Nx^k + M^{-1}b \tag{2}$$

for solving the system (1) (see [2, 15]), and the convergence of such iterative scheme depends on the spectral radius of $M^{-1}N$. An M-matrix has the form $A = sI - B$ with B a nonnegative matrix and $s \geq \rho(B)$. To solve (1) with the coefficient matrix A an M-matrix, deserves attention due to its occurrence in a wide variety of areas including finite difference method for solving partial differential equations. In [8], the authors considered such system of linear equations with A an M-matrix, and the convergence of iterative scheme (2) was obtained via regular and weak regular splittings of A, concept introduced in [12, 15].

Initiated by Friedland [7], attempts were made to study generalized nonnegative matrices, called eventually nonnegative matrices (see [3, 4, 6, 9, 10]), and subsequently generalized M-matrices were studied (see [5, 11]). In [11], the authors introduced $M_{\vee}$-matrices, which have the form $A = sI - B$, where B is eventually nonnegative and $s \geq \rho(B)$. Thereafter, in [13, 14], researchers studied some combinatorial properties of this class of matrices. One of the reason that motivated researchers to study this class of matrices is due to its occurrence in engineering, biological and economic applications (see [1]).

Elhashash and Szyld in [5], generalized the concept of regular and weak regular splitting based on Perron-Frobenius property and studied the convergence of such splittings for another generalization of M-matrices, known as GM-matrices. In this paper, we are concerned with the system (1), where the coefficient matrix A is an $M_{\vee}$-matrix. We generalize regular and weak regular splitting using the notion of eventually nonnegative matrices, to study the convergence of the iterative scheme (2).

The paper proceeds as follows. In Sect. 2, we consider the basic definitions and some preliminary notations. In Sect. 3, we generalize the concept of regular and weak regular splitting and discuss the convergence of the iterative scheme (2), when the coefficient matrix A in (1) is a nonsingular $M_{\vee}$-matrix. In particular, we concern with the convergence of Jacobi and Gauss-Seidel methods for such type of linear systems. Lastly, in Sect. 4, we consider singular linear systems and derive a necessary and sufficient condition for semi-convergence of the linear system (1).

2 Notations and Preliminaries:

Let $\mathbb{R}^{m,n}$ denote the set of all $m \times n$ real matrices. We say a matrix $A \in \mathbb{R}^{m,n}$ is nonnegative (or positive) if $a_{ij} \geq (\text{ or } >)0$, for all i, j, and we denote it by $A \geq 0$ (or $A > 0$). For any matrix $A \in \mathbb{R}^{n,n}$, and for any negative integer k with $0 < |k| < n$, $\text{tril}(A, k)$ is the lower triangular part of A with $a_{ij} = 0$ for $i = j + r$, $r = 0, 1, 2, \ldots, |k| - 1$, and for any positive integer k with $0 < k < n$, $\text{triu}(A, k)$ is the upper triangular part of A with $a_{ij} = 0$ for $j = i + r$, $r = 0, 1, 2, \ldots, k - 1$.

The spectral radius of A is denoted by $\rho(A)$, and by $\sigma(A)$, we mean the spectrum of A. Let $\lambda \in \sigma(A)$, then $\text{index}_{\lambda}(A)$ defines the size of the largest Jordan block associated with λ. When A is singular, we simply write $\text{index}(A)$ for $\text{index}_0(A)$.

We begin with some preliminary definitions.

Definition 1 ([11]) A matrix B is said to be an eventually nonnegative matrix if there exists a positive integer k_0 such that $B^k \geq 0$ for all $k \geq k_0$. A matrix A which has the form $A = sI - B$, with eventually non-negative B and $s \geq \rho(B)$, is called an $M_\vee$-matrix.

Definition 2 ([9]) A matrix B is said to possess Perron-Frobenius property if there exists a nonnegative vector $y \neq 0$ such that $By = \rho(B)y$. By $WPFn$, we denote the collection of all matrices $B \in \mathbb{R}^{n,n}$ such that both B and B^T possess Perron-Frobenious property.

Definition 3 ([12, 15]) Recall that a splitting of a matrix A is of the form

$$A = M - N \tag{3}$$

with a nonsingular matrix M. Then the splitting (3) is called

(i) a nonnegative splitting if $M^{-1}N \geq 0$.
(ii) a regular splitting if $M^{-1} \geq 0$ and $N \geq 0$.
(iii) a weak regular splitting if $M^{-1}N \geq 0$ and $M^{-1} \geq 0$.

Lemma 1 ([2]) *Let $A = M - N \in \mathbb{R}^{n,n}$ with nonsingular matrices A and M. Then for $H = M^{-1}N$ and $c = M^{-1}b$, the iterative method (2) converges to the solution $A^{-1}b$ of (1) for each x^0 if and only if $\rho(H) < 1$.*

The following definition is due to Elhashash and Syzld, which generalized the above definition.

Definition 4 ([6]) A splitting $A = M - N$ is called a Perron-Frobenius splitting if $M^{-1}N$ is a nonnilpotent matrix having the Perron-Frobenius property.

3 Splitting of Nonsingular $M_\vee$-matrices

In this section we generalize the concepts of regular and weak regular splitting using the notion of eventually nonnegative matrices and call them as E-regular and weak E-regular splitting, respectively. We study the convergence of such types of splittings for nonsingular A. We also obtain sufficient conditions for the convergence of classical Jacobi and Gauss-Seidel iterative methods, in case the coefficient matrix A of (1) is a nonsingular $M_\vee$-matrix. We now define the new splittings introduced in this paper.

Definition 5 For $A \in \mathbb{R}^{n,n}$, a splitting of A is defined as $A = M - N$, with nonsingular M. The splitting $A = M - N$ is said to be an E-regular splitting if both M^{-1} and N are nonnilpotent eventually nonnegative matrices.

Definition 6 For $A \in \mathbb{R}^{n,n}$, a splitting $A = M - N$ is said to be a weak E-regular splitting if both $M^{-1}N$ and M^{-1} are nonnilpotent eventually nonnegative matrices.

We now consider the iterative schemes (2) starting with two different initial approximations and show that their convex combination approximates the exact solution $A^{-1}b$ of (1). We also give a sufficient condition for the existence such initial guess.

Theorem 1 *Let $A = M - N$ with nonsingular matrices A and M, and let the iterative matrix $H = M^{-1}N$ be a nonnilpotent eventually nonnegative matrix. Consider the system (1) and the iterative scheme (2).*

(i) If there exist vectors x^0 and y^0 such that $x^0 \le x^1,\ x^0 \le y^0,\ \ y^0 \le y^1$, where x^1 and y^1 are computed from the iterative scheme (2) with initial values x^0 and y^0, respectively, then there exists k_0 such that

$$x^{k_0} \le x^{k_0+1} \le \ldots \le x^k \le \ldots \le A^{-1}b \le \ldots \le y^k \le \ldots \le y^{k_0+1} \le y^{k_0} \tag{4}$$

and for any scalar λ

$$A^{-1}b = \lambda \lim_{k\to\infty} x^k + (1-\lambda) \lim_{k\to\infty} y^k. \tag{5}$$

(ii) If the iterative scheme (2) converges, then the existence of such x^0 and y^0 is ensured.

Proof (i) As H is eventually nonnegative, so there exists a positive integer k_0 such that $H^k \ge 0$, for all $k \ge k_0$. Equation (2) implies that for any $k \ge k_0$ we have

$$\begin{aligned} x^k &= H^k x^0 + H^{k-1}M^{-1}b + H^{k-2}M^{-1}b + \ldots + HM^{-1}b + M^{-1}b \\ \text{and } x^{k+1} &= H^k x^1 + H^{k-1}M^{-1}b + H^{k-2}M^{-1}b + \ldots + HM^{-1}b + M^{-1}b, \end{aligned}$$

so that $x^{k+1} - x^k = H^k(x^1 - x^0) \ge 0$. Thus $x^{k+1} \ge x^k$, for all $k \ge k_0$. Similarly it can be checked that for $k \ge k_0$, $y^{k+1} \le y^k$ and $x^k \le y^k$. Thus for any k we have

$$x^{k_0} \le x^{k_0+1} \le \ldots \le x^k \le y^k \le \ldots \le y^{k_0+1} \le y^{k_0},$$

so that both sequences $\{x^k\}$ and $\{y^k\}$ are bounded and so they converge. Hence both the iterative schemes (2) with initial values x^0 and y^0 converge to $A^{-1}b$.

(ii) Suppose that the iterative scheme (2) converges, say to x. Then it follows that $x = A^{-1}b$ and $\rho(H) < 1$. Since H is nonnilpotent eventually nonnegative, there exists $z \ge 0$ such that $Hz = \rho(H)z < z$ (see [3]). If we take $x^0 = A^{-1}b - z$ and $y^0 = A^{-1}b + z$, then $y^0 - x^0 = 2z \ge 0$ and $x^1 = Hx^0 + M^{-1}b = HA^{-1}b - \rho(H)z + M^{-1}b$. As $A^{-1} = (I-H)^{-1}M^{-1}$, which implies that $M^{-1} = (I-H)A^{-1}$, so $x^1 = A^{-1}b - \rho(H)z \ge A^{-1}b - z = x^0$. Similarly, it can be verified that $y^1 \le y^0$. □

Our next result contains a necessary and sufficient condition for the convergence of a weak E-regular splitting. We first state a theorem from [9], used to prove our result.

Theorem 2 ([9]) *If* (i) $A^T \in \mathbb{R}^{n,n}$ *possesses the Perron-Frobenius property and* $x \geq 0$ $(x \neq 0)$ *is such that* $Ax - \alpha x \leq 0$ *for a constant* $\alpha > 0$, *or,* (ii) $A \in \mathbb{R}^{n,n}$ *possesses the Perron-Frobenius property and* $x \geq 0$ $(x \neq 0)$ *is such that* $x^T A - \alpha x^T \leq 0$, *for a constant* $\alpha > 0$, *then* $\alpha \leq \rho(A)$.

Theorem 3 *Let* $A = sI - B \in \mathbb{R}^{n,n}$, *with* B *a nonnilpotent eventually nonnegative matrix, be a nonsingular matrix. Then* A *is an* $M_\vee$*-matrix if and only if every weak* E*-regular splitting* $A = M - N$ *with* $M \geq 0$ *is convergent.*

Proof Suppose that $\rho = \rho(M^{-1}N) \geq 1$. As $M^{-1}N$ is a nonnilpotent, eventually nonnegative matrix, there exists $x \geq 0$ $(x \neq 0)$ such that $M^{-1}Nx = \rho x$ which implies that $Nx = \rho Mx \geq Mx$, that is, $Ax \leq 0$ or, $sx \leq Bx$. Hence by Theorem 2 we have that $s \leq \rho(B)$, which is a contradiction. Hence the splitting $A = M - N$ converges.

Conversely let every weak E-regular splitting is convergent. As $A = sI - B$ is a weak E-regular splitting of A, hence $\rho(s^{-1}B) < 1$, that is $\rho(B) < s$. Thus A is an $M_\vee$-matrix. □

We now turn to the special splitting of $M_\vee$-matrices, namely Jacobi and Gauss-Seidel splittings and to their convergence.

Corollary 1 *Let* $A = sI - B$ *be an nonsingular* $M_\vee$*-matrix with positive diagonals. If the Jacobi iterative matrix* $J = D^{-1}(L + U)$, *with* $D = \operatorname{diag}(A)$ $L = -\operatorname{tril}(A, -1)$, $U = \operatorname{triu}(A, 1)$, *is a nonnilpotent eventually nonnegative matrix, then the Jacobi splitting converges.*

Similarly, if the Gauss-Seidel iterative matrix $G = (D - L)^{-1}U$ *is a nonnilpotent eventually nonnegative matrix and* $L \geq 0$, *then Gauss-Seidel method for solving the system (1) converges.*

In [2], the authors established that for nonsingular M-matrices, both Jacobi and SOR(and hence Gauss-Seidel) splittings converge. But the following example shows that neither Jacobi nor Gauss-Seidel methods may converge for $M_\vee$-matrices, if the associated iterative matrix is not a nonnilpotent eventually nonnegative matrix.

Example 1 Consider the nonsingular $M_\vee$-matrix $A = 12.5I - B$, with

$$B = \begin{bmatrix} 9.5 & 1 & 1.5 \\ -14.5 & 16 & 10.5 \\ 10.5 & -3 & 4.5 \end{bmatrix}.$$

Consider the Jacobi splitting $A = M - N$, with $M = \operatorname{diag}(A)$ and $N = M - A$. If $J = M^{-1}N$ is the Jacobi iteration matrix, $\rho(J) = 2.0454$ and hence the Jacobi splitting of A does not converge.

Again, if we consider the Gauss-Seidel iterative matrix $G = (D - L)^{-1}U$, with $L = -\operatorname{tril}(A, -1)$ and $U = -\operatorname{triu}(A, 1)$, $\rho(G) = 4.248$, the Gauss-Seidel splitting of A also diverges.

As both Jacobi and Gauss-Seidel methods converge for M-matrices, and M-matrices have nonnegative diagonals and off-digonals are nonpositive, so one may raise the question whether Jacobi and Gauss-Seidel methods converge for $M_\vee$-matrices if $D \geq 0$ or $-L - U \in WPFn$ or eventually nonnegative matrices. But this is not the case as the following example shows.

Example 2 Consider the $M_\vee$-matrix $A = 12I - B$, where

$$B = \begin{bmatrix} 9.5 & 1 & 1.5 \\ -14.5 & 11.9 & 10.5 \\ 10.5 & -3 & 4.5 \end{bmatrix}$$

Let $M = \text{diag}(A)$ and $N = -L - U$, where $L = \text{tril}(A, -1)$, $U = \text{triu}(A, 1)$. Note that $M \geq 0$, and the eigenvalues of N are -3.8763, $-1.9381 \pm 6.4435i$. The Jacobi iterative matrix $J = M^{-1}N$ has eigenvalues $-0.4678 \pm 9.9908i$ so that Jacobi method does not converge, because $\rho(J) = 10.0018 > 1$.

Let $M = \text{diag}(A) + L$ and $N = -U$, where $L = \text{tril}(A, -1)$, $U = \text{triu}(A, 1)$. Note that the Gauss iterative matrix $G = M^{-1}N$ has eigenvalues $0, -65.2610, 0.9010$ so that Jacobi method does not converge, because $\rho(G) = 65.2610 > 1$.

The following example shows that there are some $M_\vee$-matrices for which both Jacobi and Gauss-Seidel methods converge, whereas the corresponding iterative matrices are not eventually nonnegative matrices.

Example 3 Consider the $M_\vee$-matrix $A = 3I - B$ with

$$B = \begin{bmatrix} 0 & 1 & 1 & -1 \\ 1 & 0 & 1 & 1 \\ 0 & 0 & 1 & 1 \\ 0 & 0 & 1 & 1 \end{bmatrix}$$

Consider the Jacobi splitting $A = M - N$, with $M = \text{diag}(A)$ and $N = M - A$. If $J = M^{-1}N$ is the Jacobi iteration matrix, $\rho(J) = 0.5 < 1$ and hence Jacobi splitting of A converges. But note that the matrix

$$J = \begin{bmatrix} 0 & \frac{1}{3} & \frac{1}{3} & -\frac{1}{3} \\ \frac{1}{3} & 0 & \frac{1}{3} & \frac{1}{3} \\ 0 & 0 & 0 & 0.5 \\ 0 & 0 & 0.5 & 0 \end{bmatrix}$$

is not an eventually nonnegative matrix.

Again if we consider the Gauss-Seidel iterative matrix $G = (D - L)^{-1}U$, with $L = -\text{tril}(A, -1)$ and $U = -\text{triu}(A, 1)$, $\rho(G) = 0.8431$, the Gauss-Seidel splitting of A also converges, whereas the matrix

$$G = \begin{bmatrix} 0 & 0.3333 & 0.3333 & -0.3333 \\ 0.6667 & -0.1111 & 0.2222 & 0.4444 \\ 0 & 0 & 0 & 0.5 \\ 0 & 0 & 1 & -0.25 \end{bmatrix}$$

is not an eventually nonnegative matrix.

Remark 1 Jacobi splitting of the matrix A in Example 3 is not an E-regular splitting, but the splitting converges. Like with Theorem 3, it is not possible to characterize nonsingular $M_\vee$-matrices in terms of convergence of E-regular splittings.

The following theorem gives a sufficient condition for a matrix $A = sI - B$ with B an eventually nonnegative matrix to be an $M_\vee$-matrix and for the convergence of the Jacobi method for A. But the condition is not necessary. An example has been considered to illustrate the fact.

Lemma 2 *If $M \in \mathbb{R}^n$ and $D = \operatorname{diag}(d_i)$, is an nonsingular diagonal matrix, then* $\min\limits_i |d_i| \cdot \rho(M) \leq \rho(DM) \leq \max\limits_i |d_i| \cdot \rho(M)$.

Proof Let y be a nonzero vector such that $y^T DM = \lambda y^T$, where $|\lambda| = \rho(DM)$. Let x be a nonzero vector such that $Mx = \rho x$, where $|\rho| = \rho(M)$. Then $DMx = \rho Dx$ implies that $\lambda y^T x = \rho y^T Dx$. But,

$$|\rho| \cdot \min_i |d_i| \cdot |y^T x| \leq |\rho| \cdot |y^T Dx| \leq |\rho| \cdot \max_i |d_i| \cdot |y^T x|.$$

Hence $|\rho| \cdot \min\limits_i |d_i| \cdot |y^T x| \leq |\lambda| \cdot |y^T x| \leq |\rho| \cdot \max\limits_i |d_i| \cdot |y^T x|$ Thus, if $y^T x \neq 0$,

$$\rho(M) \cdot \min_i |d_i| \leq \rho(DM) \leq \rho(M) \cdot \max_i |d_i|. \tag{6}$$

If $y^T x = 0$, we consider a small perturbation of the matrices M and D such that the corresponding eigenvectors $\tilde{x}$ and $\tilde{y}$ of M and DM, respectively, satisfy $\tilde{y}^T \tilde{x} \neq 0$. Equation (6) holds for the new matrices and as the eigenvalues are continuous functions on the matrix entries, so (6) is true for the given M and DM. □

Theorem 4 *Let $A = sI - B = D + L + U$, where $D = \operatorname{diag}(A)$, $L = \operatorname{tril}(A, -1)$ and $U = \operatorname{triu}(A, 1)$, and let B be an eventually nonnegative matrix. If $(-L - U) \in WPFn$ and $\rho(L + U) < \min\limits_i |a_{ii}|$, then A is a nonsingular $M_\vee$-matrix and the Jacobi splitting of A converges.*

Proof If A is an $M_\vee$-matrix and $\rho(L + U) < \min\limits_i |a_{ii}|$, then from the righthand side inequality of Lemma 2, $\rho(-D^{-1}(L + U)) \leq \frac{\rho(L+U)}{\min\limits_i |a_{ii}|} < 1$, and hence the Jacobi splitting converges.

Let $\min\limits_i |a_{ii}| = d$, and let $\lambda = \rho(L + U)$. As $(-L - U) \in WPFn$ and B is an eventually nonnegative matrix, we choose nonnegative vectors x, y such that

$(L+U)x = -\lambda x$ and $y^T A = \lambda_n y^T$, where $\lambda_n = s - \rho(B)$. Now, $y^T Ax = y^T(D - \lambda I)x \geq (d-\lambda)y^T x$. Therefore $\lambda_n \geq (d-\lambda)$, if $y^T x \neq 0$. Otherwise, the statement is also true considering perturbed matrices and using the continuity of spectral radius on the entries of the matrix, as discussed in Lemma 2. Thus, in any case $\lambda_n \geq (d-\lambda) > 0$, and hence $s > B$, so that A is a nonsingular $M_\vee$-matrix. □

Example 4 Consider the $M_\vee$-matrix $A = 3I - B$ with

$$B = \begin{bmatrix} 0 & 0 & 1 & 1 \\ 0 & 0 & 1 & 1 \\ 1 & 1 & 1 & 1 \\ -1 & -1 & 1 & 1 \end{bmatrix}$$

Consider the Jacobi splitting $A = M - N$, with $M = \text{diag}(A)$ and $N = M - A$. If $J = M^{-1}N$ is the Jacobi iteration matrix, $\rho(J) = 0.5 < 1$ and hence Jacobi splitting of A converges. But note that the matrix $N = -L - U \notin WPFn$.

4 Splitting of Singular $M_\vee$-matrices

In this section we consider singular $M_\vee$-matrices and characterize an interesting subclass of these matrices A with $\text{index}(A) \leq 1$, with the convergence of weak E-regular splitting of A and with eventually monotone property.

Definition 7 ([2]) A matrix $A \in \mathbb{R}^{n,n}$ is said to be *semiconvergent* if $\lim_{j\to\infty} A^j$ exists.

Theorem 5 ([2]) *Let $A \in \mathbb{R}^{n,n}$. Then A is semiconvergent if and only if each of the following conditions hold.*

(i) $\rho(A) \leq 1$.
(ii) *if* $\rho(A) = 1$, *then* $\text{index}_1(A) = 1$.
(iii) *if* $\rho(A) = 1$, *then* $\lambda \in \sigma(A)$ *with* $|\lambda| = 1$, *implies that* $\lambda = 1$.

Definition 8 Let $A \in \mathbb{R}^{n,n}$ and $S \subseteq \mathbb{R}^n$. Then we say that A is *eventually monotone on S*, if there exists a positive integer k_0, such that for any $x \in S$, $A^k x \geq 0$, for all $k \geq k_0$, implies $x \geq 0$.

Theorem 6 *Let $A = \rho I - B$ be a singular $M_\vee$-matrix where $\rho(B) = \rho$, B is an irreducible, nonnilpotent, eventually nonnegative matrix with* $\text{index}(B) \leq 1$. *Then A is an $M_\vee$-matrix with* $\text{index}(A) \leq 1$ *if and only if every weak E-regular splitting $A = M - N$ with M^{-1} eventually monotone on* $\text{range}(M)$ *is semiconvergent.*

Proof Suppose that every weak E-regular splitting is semiconvergent. Note that $A = sI - B$ is an weak E-regular splitting of A and hence by the assumption $\rho(s^{-1}B) \leq 1$ so that A is an $M_\vee$-matrix. If $\rho(B) < s$, then A is nonsingular and hence $\text{index}(A) < 1$.

As the splitting $A = sI - B$ is semiconvergent, so Theorem 5 implies that $\text{index}(A) = 1$.

Conversely, suppose that A is an $M_\vee$-matrix with $\text{index}(A) \leq 1$ and choose $k_0 > 0$ such that for all $k \geq k_0$, $(M^{-1}N)^k \geq 0$, $M^{-k} \geq 0$. For $k \geq k_0$, consider the series $\sum_{i=0}^{\infty}(M^{-1}N)^i M^{-(k+1)}x$, where $x \geq 0$ and $x \in \text{range}(M^k A)$.

Let $S_p = \sum_{i=0}^{p-1}(M^{-1}N)^i M^{-(k+1)}$. Note that $\{S_p x\}$ is a monotonic increasing sequence. If we set $x = M^k Az$ and $z \geq 0$, then

$$S_p x = \sum_{i=0}^{p-1}(M^{-1}N)^i M^{-(k+1)}x = \sum_{i=0}^{p-1}(M^{-1}N)^i M^{-1}Az = z - (M^{-1}N)^p z$$

so that for a large value of p, $S_p x \leq z$. Thus the sequence $\{S_p x\}$ converges, and hence the series $\sum_{i=0}^{\infty}(M^{-1}N)^i M^{-(k+1)}x$ converges.

Assume that $\rho = \rho(M^{-1}N)$ and let $\rho > 1$. Let z be a nonzero nonnegative vector such that $M^{-1}Nz = \rho z$, so that $z = \left(\frac{1}{1-\rho}\right) M^{-1}Az$. Now, if we set $\alpha = \left(\frac{1}{1-\rho}\right)$, then

$$\sum_{i=0}^{\infty}(M^{-1}N)^i z = \alpha \sum_{i=0}^{\infty}(M^{-1}N)^i M^{-(k+1)}M^k Az = \sum_{i=0}^{\infty}(M^{-1}N)^i M^{-(k+1)}x,$$

where $x = \left(\frac{1}{1-\rho}\right) M^k Az \in \text{range}(M^k)$ for large k, which implies that $M^{-k}x = \left(\frac{1}{1-\rho}\right) Az = Mz$ so that $M^{-(k+1)}x \geq 0$, for sufficiently large k. As M^{-1} is eventually monotone on $\text{range}(M) = \bigcap_{k=1}^{\infty} \text{range}(M^k)$, then $x \geq 0$. Hence the series $\sum_{i=0}^{\infty}(M^{-1}N)^i z$ converges, which contradicts the fact that $\rho > 1$. Hence we have $\rho \leq 1$.

If $\rho < 1$, the Drazin inverse $(I - M^{-1}N)^{\#} = (I - M^{-1}N)^{-1}$ exists. Let $\rho = 1$ so that $M^{-1}A = I - M^{-1}N$ is an $M_\vee$-matrix. As $\text{index}(A) < 1$ and M is nonsingular, $\text{index}(M^{-1}A) < 1$ and hence $(I - M^{-1}N)^{\#}$ exists. □

The following example shows that the condition $\text{index}(B) \leq 1$ in Theorem 6 can not be relaxed.

Example 5 Consider an $M_\vee$-matrix $A = 2I - B$, with

$$B = \begin{bmatrix} 0 & 0 & 1 & 1 \\ 0 & 0 & 1 & 1 \\ 1 & 1 & 1 & 1 \\ -1 & -1 & 1 & 1 \end{bmatrix}$$

Consider the splitting $A = M - N$ of A, where

$$M = \text{tril}(A) = \begin{bmatrix} 2 & 0 & 0 & 0 \\ 0 & 2 & 0 & 0 \\ -1 & -1 & 1 & 0 \\ 1 & 1 & -1 & 1 \end{bmatrix} \text{ and } N = M - A.$$

As $M^{-1} = \begin{bmatrix} \frac{1}{2} & 0 & 0 & 0 \\ 0 & \frac{1}{2} & 0 & 0 \\ \frac{1}{2} & \frac{1}{2} & 1 & 0 \\ 0 & 0 & 1 & 1 \end{bmatrix}$ and $M^{-1}N = \begin{bmatrix} 0 & 0 & \frac{1}{2} & \frac{1}{2} \\ 0 & \frac{1}{2} & \frac{1}{2} & 0 \\ 0 & 0 & 1 & 2 \\ 0 & 0 & 0 & 1 \end{bmatrix}$ are both nonnegative matrices, the splitting $A = M - N$ is a weak E-regular splitting of A. But index$(I - M^{-1}N) = 2 > 1$, and hence $(I - M^{-1}N)^{\#}$ does not exist, which implies that the E-regular splitting $A = M - N$ is not semiconvergent. Note that index$(A) = 1$ and index$(B) = 2 > 1$ and thus the condition index$(B) < 1$ in Theorem 6 cannot be relaxed.

5 Conclusion

In this article, we considered splittings of $M_{\vee}$-matrices. We introduced two types of splittings of a matrix, named as E-regular and weak E-regular splittings. We characterized an important subclass of $M_{\vee}$-matrices in terms of convergence of weak E-regular splittings. We also discussed necessary conditions for the convergence of Jacobi and Gauss-Seidel methods for $M_{\vee}$-matrices, and examples are considered to illustrate that the conditions are not sufficient.

Theorems 6 and 3, respectively, characterize an important subclass of singular and nonsingular $M_{\vee}$-matrices in terms of weak E-regular splittings. As E-regular splittings generalize regular splittings using the notion of eventually nonnegative matrices, and M-matrices are characterized using regular splittings (see [8]), an interesting open problem in this context is to discuss the convergence of E-regular splittings, in particular to develop necessary and sufficient conditions for their convergence, or for the convergence of Jacobi and Gauss-Seidel splittings.

As in the entire work we use the Perron-Frobenius property of the matrix B, where $A = sI - B$, the results obtained in the paper are also true for GM-matrices which have the form $A = sI - B$, where $s \geq \rho(B)$ and $B \in WPFn$.

Acknowledgements The author would like to express her gratitude to an anonymous referee for valuable suggestions and helpful comments. The author was supported by National Institute of Technology Meghalaya under Start-up grant.

References

1. Berman, A., Neumann, M., Stern, R.J.: Nonnegative Matrices in Dynamic Systems. Wiley, New York (1989)
2. Berman, A., Plemmons, R.: Nonnegative Matrices in the Mathematical Sciences. SIAM, Philadelphia (1994)
3. Carnochan Naqvi, S., Mcdonald, J.J.: The combinatorial structure of eventually nonnegative matrices. Electron. J. Linear Algebra **9**, 255–269 (2002)
4. Carnochan Naqvi, S., Mcdonald, J.J.: Eventually nonnegative matrices are similar to seminonegative matrices. Linear Algebra Appl. **381**, 245–258 (2004)
5. Elhashash, A., Szyld, D.B.: Generalizations of M-matrices which may not have a nonnegative inverse. Linear Algebra Appl. **429**(10), 2435–2450 (2008)
6. Elhashash, A., Szyld, D.B.: On general matrices having the Perron-Frobenius property. Electron. J. Linear Algebra **17**, 389–413 (2008)
7. Friedland, S.: On an inverse problem for nonnegative and eventually nonnegative matrices. Israel J. Math. **29**, 43–60 (1978)
8. Neumann, M., Plemmons, R.J.: Convergence of nonnegative matrices and Iterative Methods for consistent linear systems. Numer. Math. **31**, 265–279 (1978)
9. Noutsos, D.: On Perron-Frobenius property of matrices having some negative entries. Linear Algebra Appl. **412**, 132–153 (2006)
10. Noutsos, D., Tsatsomeros, M.J.: Reachibility and holdibility of nonnegative states. SIAM J. Matrix Anal. Appl. **30**(2), 700–712 (2008)
11. Olesky, D.D., Tsatsomeros, M.J.: Driessche, P.van den.: $M_\vee$-matrices: a generalization of M-matrices based on eventually nonnegative matrices. Electron. J. Linear Algebra **18**, 339–351 (2009)
12. Ortega, M., Rheinboldt, W.C.: Monotone iterations for nonlinear equations with applications to Gauss-Seidel methods. SIAM J. Numer. Anal. **4**, 171–190 (1967)
13. Saha, M., Bandopadhyay, S.: Combinatorial properties of generalized M-matrices. Electron. J. Linear Algebra **30**, 550–576 (2015)
14. Saha, M., Bandopadhyay, S.: On some generalized inverses of $M_\vee$-matrices. Communicated
15. Varga, R.S.: Matrix Iterative Analysis. Prentice Hall, New Jersey (1962)

Symmetrized Tensors and Spherical Functions

Carlos Gamas

Abstract Let G be a subgroup of the symmetric group and φ a complex function on G. A longstanding question in Multilinear Algebra is to find conditions for the vanishing of the decomposable symmetrized tensor associated with G and φ (we recall the definition below). When φ is an irreducible complex character of G, the problem has been studied by several authors, see for example [1–3, 5]. In the present paper we study and solve the vanishing problem for the case when G is the full symmetric group and φ is a certain type of spherical function.

Keywords Symmetric group · Decomposable symmetrized tensor · Spherical function

1 Introduction

Let V a finite dimensional vector space over the complex numbers. Let N be a positive integer with $N \geq 2$. Let $\bigotimes^N V$ be the Nth tensor power of V, and $x_1 \otimes \cdots \otimes x_N$ the tensor product of the vectors $x_1, \ldots, x_N$. Let S_N be symmetric group of degree N. Let G be a subgroup of S_N. For each $\sigma \in G$ there exists a unique linear mapping $P(\sigma) : \bigotimes^N V \longrightarrow \bigotimes^N V$ such that

$$P(\sigma)(x_1 \otimes \cdots \otimes x_N) = x_{\sigma^{-1}(1)} \otimes \cdots \otimes x_{\sigma^{-1}(N)}$$

for all $x_i \in V$, $i = 1, \ldots, N$. If φ is a complex valued function of G we denote by $T(G, \varphi)$ the operator

$$T(G, \varphi) = \frac{\varphi(1)}{|G|} \sum_{\sigma \in G} \varphi(\sigma) P(\sigma).$$

C. Gamas (✉)
Department of Mathematics, University of Coimbra, P 3001-454, Coimbra, Portugal
e-mail: gamas@mat.uc.pt

N. Bebiano (ed.), *Applied and Computational Matrix Analysis*,
Springer Proceedings in Mathematics & Statistics 192,
DOI 10.1007/978-3-319-49984-0_17

The image of $x_1 \otimes \cdots \otimes x_N$ under $T(G, \varphi)$ is called decomposable symmetrized tensor associated with G and φ.

Let $\lambda = (\lambda_1, \ldots, \lambda_q)$ be a partition of N. We denote the partition and the character it induces in S_N by the same letter λ. We define in the set of all partitions of N the dominance order: If $\alpha = (\alpha_1, \ldots, \alpha_t)$, $\beta = (\beta_1, \ldots, \beta_s)$ are partitions of N then

$$\alpha \prec \beta \Leftrightarrow s \leq t \wedge \sum_{i=1}^{v} \alpha_i \leq \sum_{i=1}^{v} \beta_i \ , \quad \forall v = 1, \ldots, s. \tag{1}$$

Let m and p be positive integers with $m < p$. We identify S_m with the subgroup $\{\sigma \in S_p : \sigma(j) = j \ , \forall j = m+1, \ldots, p\}$ of S_p. Let λ (respectively χ) be an irreducible complex character of S_p (respectively S_m).

The spherical function $\varphi_{\lambda,\chi}$ is a complex valued function of S_p defined by

$$\varphi_{\lambda,\chi}(g) = \frac{\lambda(1)\chi(1)}{m!p!} \sum_{h \in S_m} \lambda(gh)\chi(h^{-1}) \ , \quad g \in S_p. \tag{2}$$

We denote by $(\lambda, \chi)_{S_m}$ the nonnegative integer

$$(\lambda, \chi)_{S_m} = \frac{1}{m!} \sum_{h \in S_m} \lambda(h)\chi(h^{-1})$$

and by A_λ the set

$$A_\lambda = \{\chi \in Irr(S_m) : (\lambda, \chi)_{S_m} \neq 0\}, \tag{3}$$

where $Irr(S_m)$ denotes the set of all irreducible characters of S_m. Note that if $\lambda = (\lambda_1, \ldots, \lambda_t)$ and $\chi = (\chi_1, \ldots, \chi_s)$ then, $\chi \in A_\lambda$ if and only if $s \leq t$ and $\chi_1 \leq \lambda_1, \ldots, \chi_s \leq \lambda_s$.

Let χ be a minimal element of A_λ relatively to the partial order $\prec$. A necessary and sufficient condition on the vectors $x_1 \otimes \cdots \otimes x_p$ is given for $T(S_p, \varphi_{\lambda,\chi})(x_1 \otimes \cdots \otimes x_p)$ to be zero (Theorem 2).

2 Definitions

Let N be a positive integer and $\lambda = (\lambda_1, \ldots, \lambda_q)$ a partition of N. We denote by F_λ the corresponding Young table. The Young diagram, $D_{\lambda,\rho}$, associated with the partition λ and $\rho \in S_N$ is the table F_λ whose boxes are occupied by the integers $1, \ldots, N$ in the following way: The box in the ith row and jth column, $i = 1, \ldots, q$, $j = 1, \ldots, \lambda_i$ is occupied by the integer

$$\rho(\lambda_1 + \cdots + \lambda_{i-1} + j).$$

A standard diagram, $D_{\lambda,\rho}$, is one in which the integers appearing in each row and each column increase. For a fixed λ, we arrange the diagrams $D_{\lambda,\rho}$ lexicrographically, according to the sequence $(\rho(1), \ldots, \rho(N))$.

It is well known that the number of standard diagrams is $\lambda(1)$ and if $D_{\lambda,\rho}$, $D_{\lambda,\sigma}$ are standard diagrams with $D_{\lambda,\rho} < D_{\lambda,\sigma}$, there are two integers in the same column of $D_{\lambda,\sigma}$ and in the same row of $D_{\lambda,\rho}$.

Let L_j, $1 \leq j \leq q$, denote the set of integers in the jth row of $D_{\lambda,\rho}$ and E_j, $1 \leq j \leq \lambda_1$, the set of integers in its jth column. We define $R(D_{\lambda,\rho})$, $C(D_{\lambda,\rho})$ and $\xi(D_{\lambda,\rho})$ as follows:

$$R(D_{\lambda,\rho}) = \{\sigma \in S_N : \sigma(L_j) = L_j \ , \ \ j = 1, \ldots, q\},$$

$$C(D_{\lambda,\rho}) = \{\sigma \in S_N : \sigma(E_j) = E_j \ , \ \ j = 1, \ldots, \lambda_1\},$$

$$\xi(D_{\lambda,\rho}) = \frac{\lambda(1)}{N!} \sum_{\tau \in R(D_{\lambda,\rho})} \sum_{\sigma \in C(D_{\lambda,\rho})} \varepsilon(\sigma)\tau\sigma,$$

where ε is the alternating character. As is well known $\xi(D_{\lambda,\rho})$ is a primitive idempotent element in the group algebra $\mathbb{C}S_N$.

Let $\chi = (\chi_1, \ldots, \chi_s)$ be an irreducible character of S_m and $\lambda = (\lambda_1, \ldots, \lambda_t)$ an irreducible character of S_p, $m < p$. Suppose $\chi \in A_\lambda$. Then $s \leq t$ and $\chi_1 \leq \lambda_1, \ldots, \chi_s \leq \lambda_s$. Let N be a positive integer. We denote by $< N >$ the set $\{1, \ldots, N\}$. Let $\sigma \in S_m$. We define $H_{\lambda,\chi} \subset S_p$ as follows:

$$H_{\lambda,\chi} = \{\rho \in S_p : \rho(\lambda_1 + \cdots + \lambda_{i-1} + j) \in < m >, \ \ i=1, \ldots, s, \ \ j=1, \ldots, \chi_i\}.$$

We define in $H_{\lambda,\chi}$ an equivalence relation $\sim$ putting for all $\rho, \gamma \in H_{\lambda,\chi}$

$$\begin{aligned} \rho \sim \gamma \Leftrightarrow \rho(\lambda_1 + \cdots + \lambda_{i-1} + j) \\ = \gamma(\lambda_1 + \ldots + \lambda_{i-1} + j), \ \ \forall i = 1, \ldots, s, \ \ j = 1, \ldots, \chi_i. \end{aligned}$$

Let $\rho \in H_{\lambda,\chi}$. We denote by ρ^χ the element of S_m defined as follows:

$$\rho^\chi(\chi_1 + \cdots + \chi_{i-1} + j) = \rho(\lambda_1 + \cdots + \lambda_{i-1} + j), \ \ \forall i=1, \ldots, s, \ \ j = 1, \ldots, \chi_i. \tag{4}$$

It is not difficult to see that if, $\rho, \gamma \in H_{\lambda,\chi}$ belong to the same equivalence class, then $\rho^\chi = \gamma^\chi$ and, for all $\sigma \in S_m$, there exists a $\rho \in S_p$ such that $\rho^\chi = \sigma$. Thus, there exists a bijective correspondence between S_m and the set of the $\sim$ - equivalence classes. We denote by $U_{\chi,\sigma}, \sigma \in S_m$, the $\sim$ - equivalence class such that for $\rho \in U_{\chi,\sigma}$ we have $\rho^\chi = \sigma$. We denote by $Z_{\lambda,\chi}$ the set of the elements ρ of $H_{\lambda,\chi}$ such that $D_{\lambda,\rho}$

is a standard table associated with λ and ρ. It is clear that if $\rho \in Z_{\lambda,\chi}$ then D_{χ,ρ^χ} is a standard table associated with χ and ρ^χ. It is also clear that if $D_{\chi,\sigma}$ is a standard table, $\sigma \in S_m$, there exists $\rho \in Z_{\lambda,\chi}$ such that $\rho^\chi = \sigma$.

3 Auxiliary Results

Lemma 1 ([4, Lemma 3.6]) *Let $D_{\chi,\sigma_1}, \ldots, D_{\chi,\sigma_{\chi(1)}}$ be the standard tables associated with χ. We have*

$$|U_{\chi,\sigma_1} \cap Z_{\lambda,\chi}| = \cdots = |U_{\chi,\sigma_{\chi(1)}} \cap Z_{\lambda,\chi}|.$$

Let $< \wp >$ denote the set $\{1, \ldots, \wp\}$. Let $i \in < \wp >$ and let

$$D_{\chi^i,\sigma_1^i} < \cdots < D_{\chi^i,\sigma_{w_i}^i}, w_i = \chi^i(1), \ \sigma_1^i = 1, \ldots, \sigma_{w_i}^i \in S_m \tag{5}$$

be the standard tables associated with χ^i. For all $j = 1, \ldots, w_i$ let

$$U_{\chi^i,\sigma_j^i} \cap Z_{\lambda,\chi^i} = \{\rho_{i,j,1}, \ldots, \rho_{i,j,g_i}\},$$

where $g_i = |U_{\chi^i,\sigma_1^i} \cap Z_{\lambda,\chi^i}|$. Let

$$D_{\lambda,\rho_{i,j,1}} < \cdots < D_{\lambda,\rho_{i,j,g_i}}, \quad j \in < w_i > \tag{6}$$

be the standard tables associated with λ and the elements of $U_{\chi^i,\sigma_j^i} \cap Z_{\lambda,\chi^i}$.

Lemma 2 ([4, Lemma 3.8]) *Let $\rho \in H_{\lambda,\chi^i}$, $i \in < \wp >$. Let $\pi \in S_p$ be such that*

$$\pi\rho_{i,w_i,g_i} = \rho. \tag{7}$$

We have

(a) $\pi \in S_m$.
(b) $\pi\rho_{i,f,l} \in H_{\lambda,\chi^i}$ *for all* $f = 1, \ldots, w_i$, $l = 1, \ldots, g_i$.

Definition 1 For all $i \in < \wp - 1 >$, $f \in < \omega_i >$ and for all $l \in < g_i >$, let ϕ_f^i and $\delta_{i,f,l}$ denote $\xi(D_{\chi^i,\sigma_f^i})$ and $\xi(D_{\lambda,\rho_{i,f,l}})$, respectively. For all $f \in < \wp >$, $l \in < g_\wp >$, let $\phi_f^\wp$ and $\delta_{\wp,f,l}$ denote $\xi(D_{\chi^\wp,(\pi|<m>)\sigma_f^\wp})$ and $\xi(D_{\lambda,\pi\rho_{\wp,f,l}})$, respectively.

For all $i \in < \wp >$ let

$$\psi_1^i = \phi_1^i$$

and for $f = 2, \ldots, \omega_i$,

$$\psi_f^i = (1 - \psi_1^i - \cdots - \psi_{f-1}^i)\phi_f^i.$$

Let $\Theta_{1,1,1}$ denote $\delta_{1,1,1}$. For $i \in <\wp>$, $f \in <w_i>$, $l = 2, \ldots, g_i$, let

$$\Theta_{i,f,l} = \left(1 - \sum_{x=1}^{i} \sum_{y=1}^{f} \sum_{z=1}^{l-1} \Theta_{x,y,z}\right) \delta_{i,f,l}.$$

For $i \in <\wp>$, $f = 2, \ldots, w_i$, let

$$\Theta_{i,f,1} = \left(1 - \sum_{x=1}^{i} \sum_{y=1}^{f-1} \sum_{z=1}^{g_i} \Theta_{x,y,z}\right) \delta_{i,f,1}.$$

For $i = 2, \ldots, \wp$, let

$$\Theta_{i,1,1} = \left(1 - \sum_{x=1}^{i-1} \sum_{y=1}^{w_{i-1}} \sum_{z=1}^{g_{i-1}} \Theta_{x,y,z}\right) \delta_{i,1,1}.$$

Lemma 3 ([4, Lemma 3.10]) *We have*

(a) $\psi^i_f, i = 1, \ldots, \wp, f = 1, \ldots, w_i$, *are orthogonal idempotents in the group algebra* $\mathbb{C}S_m$.

(b) $\Theta_{i,f,l}$, $i = 1, \ldots, \wp$, $f = 1, \ldots, w_i$, $l = 1, \ldots, g_i$, *are ortogonal idempotents in the group algebra* $\mathbb{C}S_p$.

(c) For all $i = 1, \ldots, \wp$ *and* $\sigma \in S_m$, *we have*

$$\chi^i(\sigma) = \frac{m!}{\chi^i(1)}((\psi^i_1 + \cdots + \psi^i_{w_i})(\sigma).$$

(d) For all $\sigma \in S_p$ *we have*

$$\lambda(\sigma) = \lambda(1)\left(\sum_{u=1}^{\wp} \sum_{v=1}^{w_u} \sum_{h=1}^{g_u} \Theta_{u,v,h}\right)(\sigma).$$

(e) $\phi^i_j \psi^i_j = \phi^i_j$ *for all* $i = 1, \ldots, \wp$ *and* $j = 1, \ldots, w_i$.

(f) Let $u \in <\wp>$, $v \in <w_u>$, $h, l \in <g_u>$ *with* $l > h$. *We have*

$$\delta_{u,v,l}\phi^u_v \Theta_{u,v,h} = 0,$$

$$\delta_{u,v,h}\phi^u_v \Theta_{u,v,h} = M\delta_{u,v,h},$$

where M is a positive rational integer.

Lemma 4 ([4, Theorem 3.2]) *For all* $u \in < \wp >$, $v \in < w_u >$, $h \in < g_u >$, *we have*

$$\Theta_{u,v,h} = \sum_{i=1}^{u-1} \sum_{f=1}^{w_i} \sum_{l=1}^{g_i} \Upsilon_{i,f,l}^{u,v,h} \psi_f^i \rho_{i,f,l} (\rho_{u,v,h})^{-1} \delta_{u,v,h} +$$

$$+ \sum_{f=1}^{v} \sum_{l=1}^{h} \Upsilon_{u,f,l}^{u,v,h} \psi_f^u \rho_{u,f,l} (\rho_{u,v,h})^{-1} \delta_{u,v,h}, \tag{8}$$

with $\Upsilon_{i,f,l}^{u,v,h} \in \mathbb{C}$ *and* $\Upsilon_{u,v,h}^{u,v,h} = 1$.

If $\alpha = \sum_{\sigma \in S_n} \alpha(\sigma)\sigma \in \mathbb{C}S_n$ let $\overline{\alpha}$ denote $\dfrac{\alpha(1)}{n!}\alpha$.

4 Results

Definition 2 Let $\lambda = (\lambda_1, \ldots, \lambda_t)$ be a partition of p and $(x_1, \ldots, x_p)$ be a family of nonzero vectors of V. The collection of subfamilies of $(x_1, \ldots, x_p)$, $(\Re_1 = (x_i)_{i \in \Lambda_1}, \ldots, \Re_t = (x_i)_{i \in \Lambda_t})$, is said to be a λ-coloring of $(x_1, \ldots, x_p)$ if the following conditions hold:

(a) $\Re_i$ is a set of linearly independent vectors, $i = 1, \ldots, t$.
(b) $\Lambda_i \cap \Lambda_j = \emptyset, i \neq j, i, j = 1, \ldots, t$.
(c) $|\Lambda_i| = \lambda_i, i = 1, \ldots, t$.

The collection $(\Lambda_1, \ldots, \Lambda_t)$ is called *support* of the coloring $(\Re_1, \ldots, \Re_t)$.

Definition 3 Let $\lambda = (\lambda_1, \ldots, \lambda_t)$ be a partition of p and let $(x_1, \ldots, x_p)$ be a family of nonzero vectors of V. Let $\chi = (\chi_1, \ldots, \chi_s)$ be a partition of m with $m < p$ and $s \leq t$. We say that a collection $(\Re_1 = (x_i)_{i \in \Lambda_1}, \ldots, \Re_t = (x_i)_{i \in \Lambda_t})$ of subfamilies of $(x_1, \ldots, x_p)$ is a (λ, χ)-coloring of $(x_1, \ldots, x_p)$ if the following conditions hold:

(a) $(\Re_1, \ldots, \Re_t)$ is λ-coloring of $(x_1, \ldots, x_p)$.
(b) $((x_i)_{i \in \Lambda_1 \cap <m>}, \ldots, (x_i)_{i \in \Lambda_s \cap <m>})$ is a χ-coloring of $(x_1, \ldots, x_m)$.

Let $a = \sum_{\sigma \in S_p} a(\sigma)\sigma$ be an element of the group algebra $\mathbb{C}S_p$. We denote by $P(a)$ the linear mapping $\sum_{\sigma \in S_p} a(\sigma)P(\sigma)$ where $P(\sigma)$ is the linear mapping defined above. Note that if $a, b \in \mathbb{C}S_p$ then $P(ab) = P(a)P(b)$ and $P(a + b) = P(a) + P(b)$.

Lemma 5 *Let* $\lambda = (\lambda_1, \ldots, \lambda_t)$ *be a partition of* p *and let* $(x_1, \ldots, x_p)$ *be a family of nonzero vectors of* V. *Let* $D_{\lambda,\rho}$ *be a Young table associated with* λ *and* $\rho \in S_p$. *Let* Λ_j, $j = 1, \ldots, \lambda_1$ *denote the set of integers in its* j*th column. We have*

$$P(\xi(D_{\lambda,\rho}))(x_1 \otimes \cdots \otimes x_p) \neq 0,$$

if and only if $(\Lambda_1, \ldots, \Lambda_{\lambda_1})$ *is the support of a* λ'*-coloring of* $(x_1, \ldots, x_p)$*, where* λ' *denotes the conjugate partition of* λ.

Theorem 1 *Let* $\lambda = (\lambda_1, \ldots, \lambda_t)$ *be a partition of p and let* $(x_1, \ldots, x_p)$ *be a family of nonzero vectors of V. Let m be a positive integer with* $m < p$ *and let* $A_\lambda = \{\chi^1, \ldots, \chi^\wp\}$ *with* $\chi^r = (\chi_1^r, \ldots, \chi_{q_r}^r)$*, for* $r = 1, \ldots, \wp$*. Let* χ^r *be an element of* A_λ*. If*

$$T(S_p, \varphi_{\lambda,\chi^r})(x_1 \otimes \cdots \otimes x_p) \neq 0,$$

then there exists a $(\lambda', (\chi^r)')$*-coloring of* $(x_1, \ldots, x_p)$*.*

Proof Suppose

$$T(S_p, \varphi_{\lambda,\chi^r})(x_1 \otimes \cdots \otimes x_p) \neq 0.$$

From this inequality we get

$$P(\varphi_{\lambda,\chi^r})(x_1 \otimes \cdots \otimes x_p) \neq 0.$$

As $\varphi_{\lambda,\chi^r} = \overline{\chi^r}\ \overline{\lambda}$, from this inequality, (a), (b), (c) and (d) of Lemma 3 we obtain

$$P\left((\psi_1^r + \cdots + \psi_{w_r}^r)\left(\sum_{u=1}^{\wp}\sum_{v=1}^{w_u}\sum_{h=1}^{g_u} \Theta_{u,v,h}\right)\right)(x_1 \otimes \cdots \otimes x_p) \neq 0.$$

From this inequality we can conclude that there exist $f \in< w_r >$, $u \in< \wp >$, $v \in< w_u >$, $h \in< g_u >$ such that

$$P(\psi_f^r \Theta_{u,v,h})(x_1 \otimes \cdots \otimes x_p) \neq 0. \tag{9}$$

From this inequality and Lemma 4 we have $r < u$ or $r = u$ and $f \leq v$ and: if $r < u$ then

$$P(\psi_f^r \Theta_{u,v,h}) = P\left(\sum_{l=1}^{g_r} \Upsilon_{r,f,l}^{u,v,h} \psi_f^r \rho_{r,f,l}(\rho_{u,v,h})^{-1}\delta_{u,v,h}\right),$$

if $r = u$, $f \leq v$ then

$$P(\psi_f^r \Theta_{u,v,h}) = P\left(\sum_{l=1}^{h} \Upsilon_{r,f,l}^{u,v,h} \psi_f^r \rho_{r,f,l}(\rho_{u,v,h})^{-1}\delta_{u,v,h}\right).$$

From these two last equalities and (9) we can conclude that

$$P(\psi_f^r \rho_{r,f,l}(\rho_{u,v,h})^{-1}\delta_{u,v,h})(x_1 \otimes \cdots \otimes x_p) \neq 0,$$

with $l \in < g_r >$ if $r < u$ or $l \in < h >$ if $r = u$, $f \leq v$. From this inequality we derive (note that $\delta_{u,v,h} = \rho_{u,v,h}\xi(D_{\lambda,1})(\rho_{u,v,h})^{-1}$)

$$P(\psi^r_f \delta_{r,f,l} \rho_{r,f,l} (\rho_{u,v,h})^{-1})(x_1 \otimes \cdots \otimes x_p) \neq 0.$$

Putting $\sigma^{-1} = \rho_{r,f,l}(\rho_{u,v,h})^{-1}$ this inequality becomes

$$P(\psi^r_f \delta_{r,f,l})(x_{\sigma(1)} \otimes \cdots \otimes x_{\sigma(p)}) \neq 0.$$

This inequality leads to

$$P(\delta_{r,f,l})(x_{\sigma(1)} \otimes \cdots \otimes x_{\sigma(p)}) \neq 0.$$

This inequality and the definition of $\delta_{r,f,l}$ imply

$$P(\xi(D_{\lambda,\rho_{r,f,l}}))(x_{\sigma(1)} \otimes \cdots \otimes x_{\sigma(p)}) \neq 0. \tag{10}$$

Let Λ_j, $j = 1, \ldots, \lambda_1$ denote the set of integers in the jth column of $D_{\lambda,\rho_{r,f,l}}$ and let $\Re_j = (x_i)_{i \in \Lambda_j}$, $j = 1, \ldots, \lambda_1$. From (10) and Lemma 5

$$(\Re_1, \ldots, \Re_{\lambda_1}) \tag{11}$$

is a λ'-coloring of $(x_1, \ldots, x_p)$. By definition we have

$$\rho_{r,f,l} \in H_{\lambda,\chi^r}.$$

From this relation we derive

$$|\Lambda_j \cap < m > | = (\chi^r)'_j, \quad j = 1, \ldots, \chi^r_1.$$

From this last relation and (11) we can conclude that

$$((x_i)_{i \in \Lambda_1 \cap <m>}, \ldots, (x_i)_{i \in \Lambda_{\chi^r_1} \cap <m>})$$

is a $(\chi^r)'$-coloring of $(x_1, \ldots, x_m)$. □

Theorem 2 *Let $\lambda = (\lambda_1, \ldots, \lambda_t)$ be a partition of p and let $(x_1, \ldots, x_p)$ be a family of nonzero vectors of V. Let m be a positive integer wih $m < p$ and let $\chi = (\chi_1, \ldots, \chi_s)$ be a minimal element of A_λ. We have*

$$T(S_p, \varphi_{\lambda,\chi})(x_1 \otimes \cdots \otimes x_p) \neq 0,$$

if and only if there exists a (λ', χ')-coloring of $(x_1, \ldots, x_p)$.

Proof From Theorem 1 we have only to prove the sufficiency. As χ is a minimal element of A_λ we can assume, without lost of generality, that $\chi = \chi^{\wp}$. Suppose

$$(\Re_1, \ldots, \Re_{\lambda_1}) \tag{12}$$

is a $(\lambda', (\chi^{\wp})')$-coloring of $(x_1, \ldots, x_p)$ with $\Re_j = (x_i)_{i\in\Lambda_j},\ j = 1, \ldots, \lambda_1$. Let

$$\Lambda_j = \{a_j^1, \ldots, a_j^{l_j}\},\ \ j = 1, \ldots, \lambda_1, \tag{13}$$

with $a_j^1 < \cdots < a_j^{l_j}$. We define $\rho \in S_p$ as follows:

$$\rho(\lambda_1 + \cdots + \lambda_{i-1} + j) = a_j^i,\ \ i = 1, \ldots, t,\ \ j = 1, \ldots, \lambda_i. \tag{14}$$

From (12) to (14) we have

$$\rho \in H_{\lambda,\chi^{\wp}}. \tag{15}$$

Let $\pi \in S_p$ such that

$$\pi\rho_{\wp,w_\wp,g_\wp} = \rho. \tag{16}$$

From (15), (16), (a)–(d) of Lemma 3 we have

$$\begin{aligned} &T(S_p, \varphi_{\lambda,\chi})(x_1 \otimes \cdots \otimes x_p) \\ &= P\left((\psi_1^{\wp} + \cdots + \psi_{w_\wp}^{\wp})\left(\sum_{u=1}^{\wp}\sum_{v=1}^{w_u}\sum_{h=1}^{g_u} \Theta_{u,v,h}\right)\right)(x_1 \otimes \cdots \otimes x_p). \end{aligned} \tag{17}$$

As $\psi_1^{\wp}, \ldots, \psi_{w_\wp}^{\wp}$ are ortogonal idempotents we have

$$P(\psi_1^{\wp} + \cdots + \psi_{w_\wp}^{\wp})\left(\bigotimes^p V\right) = P(\psi_1^{\wp})\left(\bigotimes^p V\right) \bigoplus \cdots \bigoplus P(\psi_{w_\wp}^{\wp})\left(\bigotimes^p V\right).$$

From this equality we can conclude that if

$$P\left(\psi_{w_\wp}^{\wp}\left(\sum_{u=1}^{\wp}\sum_{v=1}^{w_u}\sum_{h=1}^{g_u} \Theta_{u,v,h}\right)\right)(x_1 \otimes \cdots \otimes x_p) \neq 0, \tag{18}$$

then (17) is a nonzero element of $\bigotimes^p V$.

We prove (18) by contradiction. Suppose

$$P\left(\psi_{w_\wp}^{\wp}\left(\sum_{u=1}^{\wp}\sum_{v=1}^{w_u}\sum_{h=1}^{g_u} \Theta_{u,v,h}\right)\right)(x_1 \otimes \cdots \otimes x_p) = 0.$$

From this equality and Lemma 4 we obtain

$$P\left(\psi_{w_\wp}^{\wp}\left(\sum_{h=1}^{g_\wp}\Theta_{\wp,w_\wp,h}\right)\right)(x_1\otimes\cdots\otimes x_p)=0.$$

From this equality we get

$$P\left(\delta_{\wp,w_\wp,g_\wp}\phi_{w_\wp}^{\wp}\psi_{w_\wp}^{\wp}\left(\sum_{h=1}^{g_\wp}\Theta_{\wp,w_\wp,h}\right)\right)(x_1\otimes\cdots\otimes x_p)=0.$$

This equality, (f) of Lemma 3 leads to

$$MP(\delta_{\wp,w_\wp,g_\wp})(x_1\otimes\cdots\otimes x_p)=0,$$

where M is a positive rational number. Thus we have

$$P(\delta_{\wp,w_\wp,g_\wp})(x_1\otimes\cdots\otimes x_p)=0.$$

This equality and the definition of ρ lead to

$$P(\xi(D_{\lambda,\rho}))(x_1\otimes\cdots\otimes x_p)=0. \tag{19}$$

From (12), (14), Definition 3, and Lemma 5 we obtain

$$P(\xi(D_{\lambda,\rho}))(x_1\otimes\cdots\otimes x_p)\neq 0,$$

which contradicts (19). □

References

1. Dias da Silva, J.A.: On $\mu-$colorings of a matroid. Linear Multilinear Algebra **27**, 25–32 (1990)
2. Dias da Silva, J.A., Fonseca, A.: Nonzero star products. Linear Multilinear Algebra **27**, 49–55 (1990)
3. Gamas, C.: Conditions for a symmetrized decomposable tensor to be zero. Linear Algebra Appl. **108**, 83–119 (1988)
4. Gamas, C.: Symmetrized Tensors and spherical functions. Submitted
5. Pate, T.H.: Immanants and decomposable tensors that symmetrized to zero. L. M. A. **28**, 175–184 (1990)

Testing Independence via Spectral Moments

Jolanta Pielaszkiewicz, Dietrich von Rosen and Martin Singull

Abstract Assume that a matrix $X : p \times n$ is matrix normally distributed and that the Kolmogorov condition, i.e., $\lim_{n,p\to\infty} \frac{n}{p} = c > 0$, holds. We propose a test for identity of the covariance matrix using a goodness-of-fit approach. Calculations are based on a recursive formula derived by Pielaszkiewicz et al. [19]. The test performs well regarding the power compared to presented alternatives, for both $c < 1$ or $c \geq 1$.

Keywords Test of independence · Goodness of fit test · Covariance matrix · Wishart matrix · Spectral moments

1 Introduction

Nowadays a large amount of empirical problems generate high-dimensional data sets. We are interested in discussing an independence test for the covariance matrix that works in the case where the dimension p exceeds, is equal or is smaller than the sample size n, i.e., $p > n$, $p = n$ or $p < n$.

J. Pielaszkiewicz (✉) · D. von Rosen · M. Singull
Linköping University, Linköping, Sweden
e-mail: jolanta.pielaszkiewicz@liu.se

D. von Rosen
e-mail: dietrich.von.rosen@slu.se

M. Singull
e-mail: martin.singull@liu.se

J. Pielaszkiewicz
Linnaeus University, Växjö, Sweden

D. von Rosen
Swedish University of Agricultural Sciences, Uppsala, Sweden

N. Bebiano (ed.), *Applied and Computational Matrix Analysis*,
Springer Proceedings in Mathematics & Statistics 192,
DOI 10.1007/978-3-319-49984-0_18

1.1 Notation and Assumptions

The data matrix $X \in \mathbb{R}^{p\times n}$ follows the central matrix normal distribution, denoted $X \sim \mathcal{N}_{p,n}(0, \Sigma, I_n)$, where the dispersion matrix Σ is assumed to be positive definite and I_n is the identity matrix of size $n \times n$. Alternatively, one can think of a set of n independently distributed p-dimensional column vectors X_i, $i = 1, \ldots, n$, each distributed according to a multivariate normal distribution, $\mathcal{N}_p(0, \Sigma)$. Then, $W = XX' = \sum_{i=1}^{n} X_i X_i'$, where $X = (X_1, \ldots, X_n)$ and X' denotes the transpose of X, follows a Wishart distribution, $W \sim \mathcal{W}_p(\Sigma, n)$.

We assume that the Kolmogorov condition holds, so that both p and n increase with the same speed, i.e., $\lim_{n,p\to\infty} \frac{n}{p} = c \in (0, \infty)$.

Note also that for an arbitrary matrix A the matrix A^k denotes $AA\cdots A$, where usual matrix multiplication is applied k times. $\mathbb{E}[\cdot]$ denotes expectation and the trace $\mathrm{Tr}\{\cdot\}$ is defined as the sum of the diagonal elements of a square matrix.

1.2 Stating Hypothesis and Brief Review of Historical Results

The hypothesis for testing identity of the covariance matrix is given by

$$H_0 : \Sigma = I_p \;\text{ against }\; H_1 : \Sigma \neq I_p \tag{1}$$

which of course is the same as

$$H_0 : \Sigma = \Sigma_0 \;\text{ against }\; H_1 : \Sigma \neq \Sigma_0,$$

where the matrix Σ_0 is a given positive definite matrix. Equivalence of both formulations holds since we can consider the transformation $\Sigma_0^{-\frac{1}{2}} X$ instead of the data matrix X. Given the equivalence, that case will not be discussed further.

Stated in this way the hypothesis (1) was tested for the very first time by Mauchly in [15], using a likelihood ratio approach. Tests based on the likelihood ratio test statistics were, for a long time, commonly applied method. As the likelihood ratio approach is only suitable in the case $p < n$, see [1, 16], further results were derived.

Nagao [17] introduced a statistic based on $a_1 = \frac{1}{p}\sum_{i=1}^{p} \lambda_i$ and $a_2 = \frac{1}{p}\sum_{i=1}^{p} \lambda_i^2$, where λ_i are eigenvalues of $\frac{1}{n}XX'$. Furthermore, in the paper by Ledoit and Wolf [14] a modification of Nagao's test statistics was suggested and given as

$$T_W = \frac{1}{p}\mathrm{Tr}\left[\left(\frac{1}{n}XX' - I_p\right)^2\right] - \frac{p}{n}\left(\frac{1}{p}\mathrm{Tr}\left[\frac{1}{n}XX'\right]\right)^2 + \frac{p}{n}.$$

The reason for the improvement was the lack of consistency of Nagao's result for $p > n$, with that which was obtained for T_W. The result of [14] has been further analyzed for $\lim_{n,p\to\infty}\frac{n}{p}=c>0$ in [4, 8]. Another modification of the result presented in [17] is a paper by Chen et al. [5] in which the normality assumption is relaxed. Srivastava's publication [24] also follows the paper of Nagao [17] and proposes a test statistic of the form

$$T_S=\frac{n}{2}(\hat{a}_2-2\hat{a}_1+1),$$

where

$$\hat{a}_1=\frac{1}{p}\operatorname{Tr}\left\{\frac{1}{n}XX'\right\},$$
$$\hat{a}_2=\frac{n^2}{(n-1)(n+2)p}\left[\operatorname{Tr}\left\{\frac{1}{n^2}(XX')^2\right\}-\frac{1}{n}\left(\operatorname{Tr}\left\{\frac{1}{n}XX'\right\}\right)^2\right]$$

are unbiased and consistent, under Kolmogorov condition, estimators of a_1 and a_2.

Natural continuation of Srivastava's research is given in [6, 7], where asymptotically normally distributed test statistics

$$T_1=\frac{n}{c\sqrt{8}}(\hat{a}_4-4\hat{a}_3+6\hat{a}_2-4\hat{a}_1+1),$$
$$T_2=\frac{n}{\sqrt{8(c^2+12c+8)}}(\hat{a}_4-2\hat{a}_2+1)$$

are based on the unbiased and consistent, under Kolmogorov condition, estimators of $a_j=\frac{1}{p}\sum_{i=1}^{p}\lambda_i^j$, for $j=1,2,3,4$.

The results proposed in [6, 7, 24] are based on the idea that the null hypothesis $\Sigma=I_p$ implies, that all the eigenvalues are equal to 1. Then,

$$\frac{1}{p}\sum_{i=1}^{p}(\lambda_i-1)^{2k}=\frac{1}{p}\sum_{i=1}^{p}\sum_{j=0}^{2k}(-1)^j\binom{2k}{j}\lambda_i^{2k-j}=\sum_{j=0}^{2k}(-1)^j\binom{2k}{j}a_{2k-j}\geq 0$$

as it is a sum of even powers of λ_i-1. Moreover, keeping notation $a_{2k-j}=\frac{1}{p}\sum_{i=1}^{p}\lambda_i^{2k-j}$, we have

$$\frac{1}{p}\sum_{i=1}^{p}(\lambda_i-1)^{2k}=\sum_{j=0}^{2k}(-1)^j\binom{2k}{j}a_{2k-j}=0\quad\text{under}\quad H_0.$$

Furthermore, other methods to test the hypothesis (1) allowing for large p are given in, among others, [3, 9–11, 20–22].

1.3 Outline

The paper is organized as follows. After the introduction and brief review of the historical results in Sect. 1, Sect. 2 states and discusses a new test statistic for testing (1) based on Jonsson's result, see [12], on joint asymptotic normality of some, specified later, m-dimensional vector Y. The presented test statistics follows a χ^2-distribution under H_0 in contrast to a number of normally distributed results. Simulations and comparison to alternative test statistics are carried out in Sect. 3.

2 Test

It is well known that under the assumptions given in Sect. 1.1 and under H_0 the matrix $W = XX'$ of size $p \times p$ follows a Wishart distribution, i.e., $W = XX' \sim \mathcal{W}_p(I, n)$. Then, an asymptotic distribution of $\frac{1}{p}\operatorname{Tr}\{(\frac{1}{n}XX')^t\}$, when $\frac{n}{p} \overset{p,n\to\infty}{\to} c$, is degenerated (with variance converging to zero with increasing n and p) normal for any $t \in \mathbb{N}$ as proven e.g. in [18].

We present recursive formula (see [19])

$$\mathbb{E}\Big[\prod_{i=0}^{k}\operatorname{Tr}\{W^{m_i}\}\Big] = (n-p+m_k-1)\mathbb{E}\Big[\operatorname{Tr}\{W^{m_k-1}\}\prod_{i=0}^{k-1}\operatorname{Tr}\{W^{m_i}\}\Big]$$
$$+2\sum_{i=1}^{k-1} m_i\mathbb{E}\Big[\operatorname{Tr}\{W^{m_k+m_i-1}\}\prod_{\substack{j=0\\ j\neq i}}^{k-1}\operatorname{Tr}\{W^{m_j}\}\Big]$$
$$+\sum_{i=0}^{m_k-1}\mathbb{E}\Big[\operatorname{Tr}\{W^{i}\}\operatorname{Tr}\{W^{m_k-1-i}\}\prod_{j=0}^{k-1}\operatorname{Tr}\{W^{m_j}\}\Big], \tag{2}$$

where $k \in \mathbb{N}$, $m_0 = 0$, $m_k \in \mathbb{N}$ and $m_i \in \mathbb{N}_0$ for $i = 1, \ldots, k-1$. Let us denote expectation $\mathbb{E}[\frac{1}{p}\operatorname{Tr}\{(\frac{1}{n}XX')^t\}]$ by $m_1^{(t)}(n, p)$. Then, using (2) for each $t \in \mathbb{N}$ expectation $m_1^{(t)}(n, p)$ can be computed as a function of n and p. In particular

$$m_1^{(1)}(n,p) = \mathbb{E}\Big[\tfrac{1}{p}\operatorname{Tr}\Big\{\tfrac{1}{n}W\Big\}\Big] = 1,$$
$$m_1^{(2)}(n,p) = \mathbb{E}\Big[\tfrac{1}{p}\operatorname{Tr}\Big\{\Big(\tfrac{1}{n}W\Big)^2\Big\}\Big] = 1 + \tfrac{p}{n} + \tfrac{1}{n},$$
$$m_1^{(3)}(n,p) = \mathbb{E}\Big[\tfrac{1}{p}\operatorname{Tr}\Big\{\Big(\tfrac{1}{n}W\Big)^3\Big\}\Big] = \Big(1 + \tfrac{p}{n} + \tfrac{2}{n}\Big)\Big(1 + \tfrac{p}{n} + \tfrac{1}{n}\Big) + \tfrac{p}{n} + \tfrac{2}{n^2},$$
$$m_1^{(4)}(n,p) = \mathbb{E}\Big[\tfrac{1}{p}\operatorname{Tr}\Big\{\Big(\tfrac{1}{n}W\Big)^4\Big\}\Big] = \Big(1 + \tfrac{p}{n} + \tfrac{3}{n}\Big)\Big(\Big(1 + \tfrac{p}{n} + \tfrac{2}{n}\Big)\Big(1 + \tfrac{p}{n} + \tfrac{1}{n}\Big) + \tfrac{p}{n} + \tfrac{2}{n^2}\Big)$$
$$+ 2\Big(1 + \tfrac{p}{n} + \tfrac{1}{n}\Big)\Big(\tfrac{p}{n} + \tfrac{4}{n^2}\Big).$$

Moreover, $m_1^{(t)}(n, p)$ represents the tth spectral moment of matrix $\frac{1}{n}XX'$ and under Kolmogorov condition converges to moments of Marchenko–Pastur distribution.

Furthermore, by the application of the result given in [12] to

$$Y_t = \sqrt{np}\left(\frac{1}{p}\operatorname{Tr}\left\{\left(\tfrac{1}{n}XX'\right)^t\right\} - m_1^{(t)}(n, p)\right) \tag{3}$$

we claim the joint asymptotic multivariate normality of the vector $Y = (Y_1, \ldots, Y_m)$ under Kolmogorov condition. Moreover, the random vector Y has a mean equal to zero and covariance matrix

$$\begin{aligned}\Sigma_Y &= (Cov(Y_i, Y_j))_{i,j=1}^m \\ &= \begin{pmatrix} 2 & 4(1+\frac{1}{c}) & 6((1+\frac{1}{c})^2+\frac{1}{c}) & \cdots & Cov(Y_1, Y_m) \\ 4(1+\frac{1}{c}) & \frac{4}{c}+8(1+\frac{1}{c})^2 & 12(1+\frac{1}{c})((1+\frac{1}{c})^2+\frac{2}{c}) & \cdots & Cov(Y_2, Y_m) \\ 6((1+\frac{1}{c})^2+\frac{1}{c}) & 12(1+\frac{1}{c})((1+\frac{1}{c})^2+\frac{2}{c}) & 24((1+\frac{1}{c})^2+\frac{1}{c})^2+\frac{42}{c}(1+\frac{1}{c})^2 & \cdots & Cov(Y_3, Y_m) \\ \vdots & \vdots & \ddots & \vdots & \vdots \\ Cov(Y_m, Y_1) & Cov(Y_m, Y_2) & \cdots & \cdots & Var(Y_m) \end{pmatrix},\end{aligned}$$

where c stands for constant from the Kolmogorov condition, i.e., $\lim_{n,p\to\infty}\frac{n}{p} = c$. The result of Jonsson, mentioned above, was inspired by [2].

Elements of the covariance matrix can be calculated analytical using (2). For illustration purpose the calculations of the upper left element of the matrix are presented below:

$$\begin{aligned}Var[Y_1] &= Var\left[\sqrt{np}\left(\frac{1}{p}\operatorname{Tr}\left\{\tfrac{1}{n}XX'\right\} - m_1^{(1)}(n, p)\right)\right] = npVar\left[\frac{1}{p}\operatorname{Tr}\left\{\tfrac{1}{n}XX'\right\}\right] \\ &= np\left(\mathbb{E}\left[\left(\frac{1}{p}\operatorname{Tr}\left\{\tfrac{1}{n}XX'\right\}\right)^2\right] - \left(\mathbb{E}\left[\frac{1}{p}\operatorname{Tr}\left\{\tfrac{1}{n}XX'\right\}\right]\right)^2\right) \\ &= np\left(\mathbb{E}\left[\tfrac{1}{p^2n^2}\operatorname{Tr}\{XX'\}\operatorname{Tr}\{XX'\}\right] - 1\right) \\ &= np\left(\tfrac{1}{p^2n^2}(np(np+2)) - 1\right) = 2.\end{aligned}$$

Finally, we suggest new test for (1) through the goodness-of-fit approach that is based on the result regarding multivariate normality of the vector Y. We define a test statistic by

$$T_{Jm} = Y^T\Sigma_Y^{-1}Y \sim \chi^2(m), \tag{4}$$

where the distribution under H_0 is asymptotically, $\lim_{n,p\to\infty}\frac{n}{p} = c$, χ^2 with m degrees of freedom. We reject the hypothesis for large values of T_{Jm}, since our test statistics, that is by construction non negative, tends to zero under H_0.

3 Simulation Studies

Let $Y = (Y_1, Y_2, Y_3, Y_4)$, where $Y_t = \sqrt{np}\Big(\frac{1}{p}\operatorname{Tr}\Big\{\Big(\frac{1}{n}W\Big)^t\Big\} - m_1^{(t)}(n, p)\Big)$, for $t = 1, 2, 3, 4$. We are going to analyze the results of 2000 simulated matrices X of size $p = n$, $p > n$ and $p < n$ with a) $p = 125$, $n = 125$, b) $p = 250$, $n = 125$ and c) $p = 125$, $n = 250$.

3.1 On the Distribution of Y_t

In the Sect. 2 we claim the marginal and joint normal distribution of the vector $Y = (Y_1, \ldots, Y_m)$ following [12] and give the recursive formula (2) for calculating the variances and covariances of Y_i, $i = 1, \ldots, m$.

In this section the test statistics and p-value of Shapiro–Wilk test of normality will be given. Classical Shapiro–Wilk test have been introduced by [23] with test statistics obtained by dividing the square of the linear combination of the sample order statistics by the estimate of variance. Moreover, the comparison of empirical and theoretical density functions, and QQ-plots are provided.

By Table 1 we cannot reject normal distribution of Y_i, $i = 1, 2, 3, 4$ on a significance level of 2.5%. We see that normality is much stronger in the classical case when the sample size is bigger than the dimension of the problem ($p < n$). Nevertheless, normality holds even in the case when $p > n$.

Normality is also illustrated in Figs. 1 and 2 by the comparison of empirical and theoretical distribution functions and in Figs. 3 and 4 using QQ-plots.

3.2 On the Distribution of the Test Statistics

Data simulated in Sect. 3.1 is used to analyze the distribution of test statistics T_{Jm}, for $m = 2, 3, 4$. The Kolmogorov–Smirnov test (see [13]) is used to verify the $\chi^2(m)$ distribution and have been introduced by Kolmogorov in 1933.

Table 1 The results of the Shapiro–Wilk normality test for the distribution function of Y_1, Y_2, Y_3 and Y_4, where Y_i is given as (3). Values of Shapiro–Wilk test statistics and p-values for rejection under H_0 are given for particular choices of p and n in the two cases: $p < n$ and $p > n$

	Shapiro–Wilk test for normality			
	W	p-value	W	p-value
Y_1	0.99952	0.9213	0.99892	0.2666
Y_2	0.99949	0.8967	0.9986	0.09905
Y_3	0.99931	0.6909	0.99846	0.06208
Y_4	0.99907	0.3966	0.99833	0.04067
	$p = 125, n = 250$		$p = 250, n = 125$	

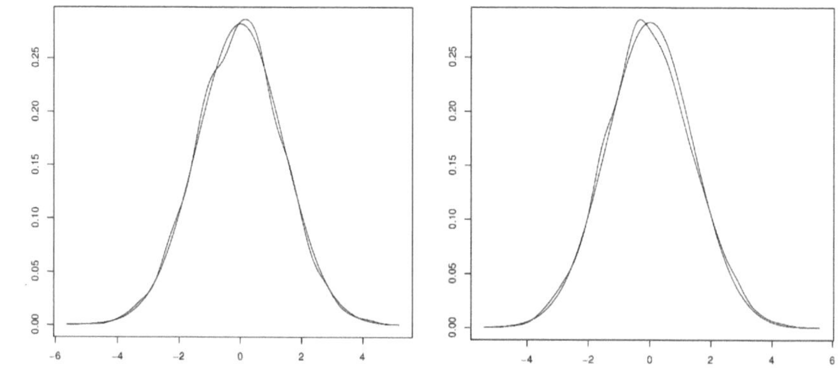

(a) Y_1 for $p = 125, n = 250$ (b) Y_1 for $p = 250, n = 125$

Fig. 1 Comparison of the empirical density function and theoretical asymptotic density function, i.e., the normal distribution, of Y_1, which is defined in (3)

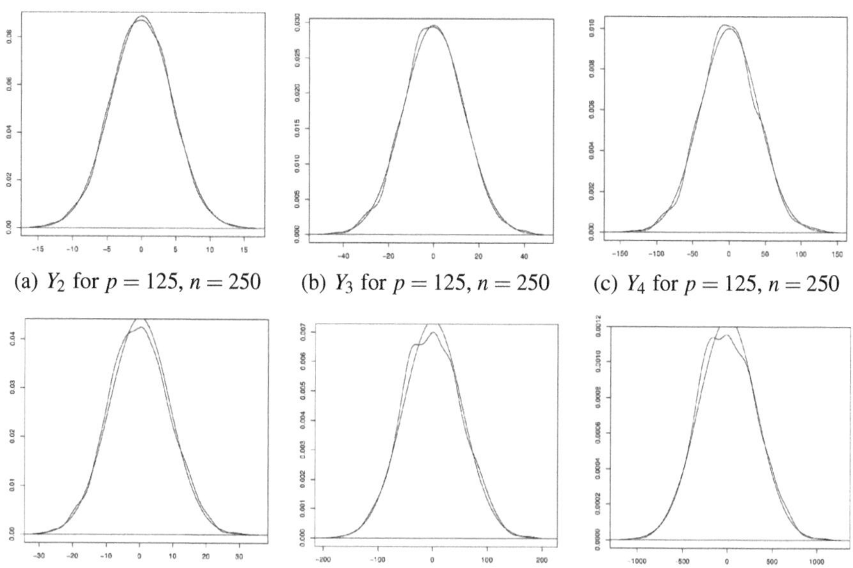

(a) Y_2 for $p = 125, n = 250$ (b) Y_3 for $p = 125, n = 250$ (c) Y_4 for $p = 125, n = 250$

(d) Y_2 for $p = 250, n = 125$ (e) Y_3 for $p = 250, n = 125$ (f) Y_4 for $p = 250, n = 125$

Fig. 2 Comparison of the empirical density function and theoretical asymptotic density function, i.e., the normal distribution, of Y_2, Y_3, Y_4, where Y_i is defined as in (3)

Following the p-values of the Kolmogorov–Smirnov test in Table 2 we cannot reject χ^2-distribution of test statistics for all considered values of the parameter m. Visual illustration of the comparison between the theoretical and empirical density function is given in Fig. 5.

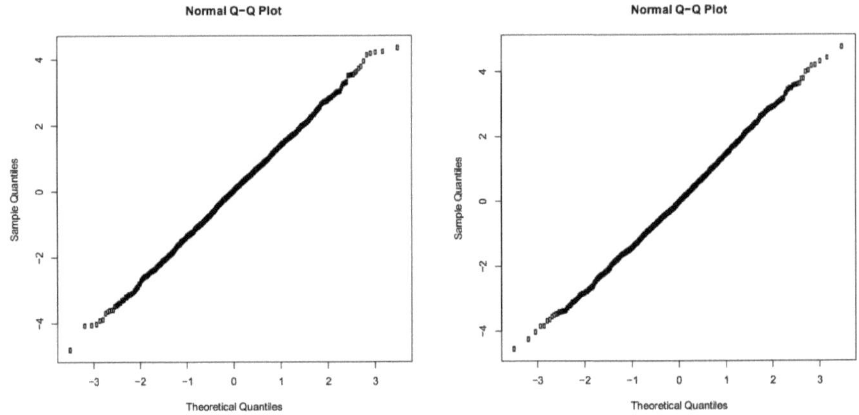

(a) Y_1 for $p = 125, n = 250$ (b) Y_1 for $p = 250, n = 125$

Fig. 3 Normal QQ-plots for the empirical distribution of Y_1, which is defined in (3)

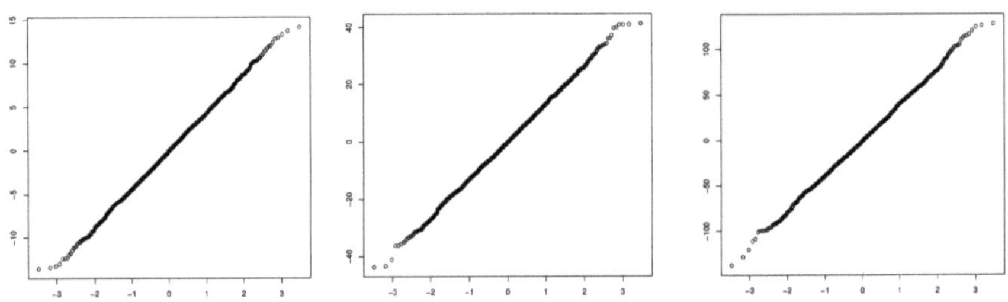

(a) Y_2 for $p = 125, n = 250$ (b) Y_3 for $p = 125, n = 250$ (c) Y_4 for $p = 125, n = 250$

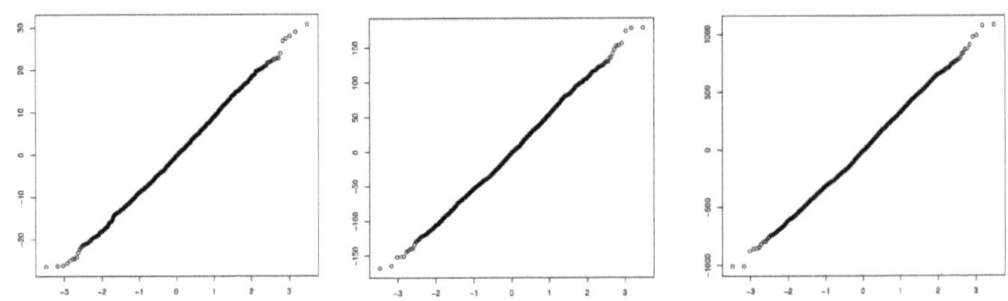

(d) Y_2 for $p = 250, n = 125$ (e) Y_3 for $p = 250, n = 125$ (f) Y_4 for $p = 250, n = 125$

Fig. 4 Normal QQ-plots for the empirical distribution of Y_2, Y_3, Y_4, where Y_i is defined as in (3)

3.3 *Attained Significance Level and Empirical Power*

To check how well the proposed test statistics T_{J2}, T_{J3} and T_{J4} perform we present a comparison with the tests obtained by Ledoit and Wolf in [14] and Srivastava in [24], as well as by Fisher et al. in [6, 7]. In Table 3 the significance levels are given, while Table 4 gives empirical statistical power.

Table 2 The results of the Kolmogorov–Smirnov test for testing χ^2-distribution of, defined by (4), test statistics T_{J2}, T_{J3} and T_{J4} for the particular choices of p and n in the two cases: $p < n$ and $p > n$

	Kolmogorov–Smirnov test			
	D	p-value	D	p-value
T_{J2}	0.024805	0.1706	0.015531	0.7203
T_{J3}	0.01566	0.7107	0.021177	0.3311
T_{J4}	0.015463	0.6403	0.010643	0.9773
	$p = 125, n = 250$		$p = 250, n = 125$	

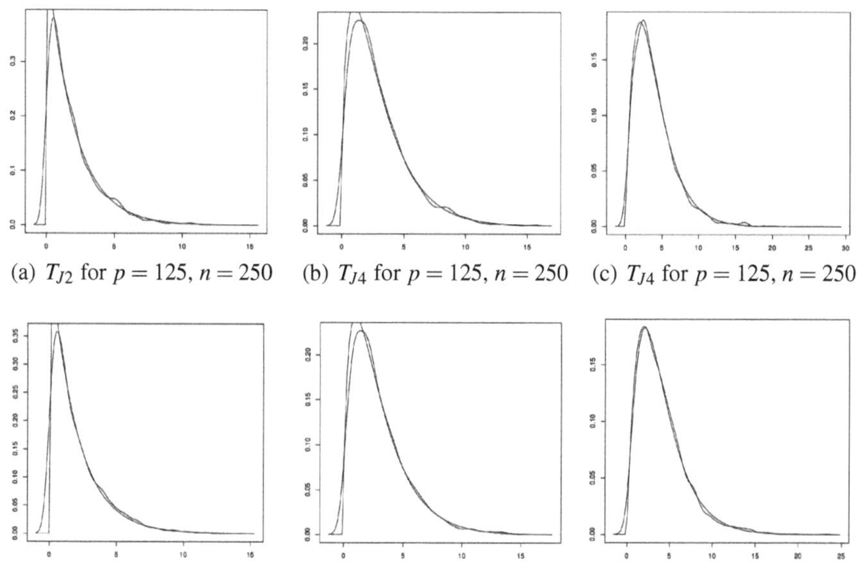

(a) T_{J2} for $p = 125, n = 250$ (b) T_{J4} for $p = 125, n = 250$ (c) T_{J4} for $p = 125, n = 250$

(d) T_{J2} for $p = 250, n = 125$ (e) T_{J4} for $p = 250, n = 125$ (f) T_{J4} for $p = 250, n = 125$

Fig. 5 Comparison of the empirical density function of test statistics T_{J2}, T_{J3} and T_{J4}, defined by (4), and theoretical asymptotic density function, i.e., $\chi^2(2)$-, $\chi^2(3)$- and $\chi^2(4)$-distribution, respectively

Power studies are performed for the alternative hypothesis that the data comes from a distribution with covariance matrix $\Sigma = aI$ for fixed a close to 1, for a values following uniform distribution on the interval symmetric around 1, i.e., $a \sim U(1-\varepsilon, 1+\varepsilon)$, for $\Sigma = \begin{pmatrix} aI_k & 0 \\ 0 & I_{p-k} \end{pmatrix}$ as well as $\Sigma = \begin{pmatrix} I_{p/2} & aJ_{p/2} \\ aJ_{p/2} & I_{p/2} \end{pmatrix}$, where $J_{p/2}$ stands for matrix of ones of the size $\frac{p}{2} \times \frac{p}{2}$.

In Tables 3 and 4, we see that this paper proposes tests which provides better empirical power than alternative methods and keeps a similar performance with respect to the size of the test. Already with $\Sigma = 1.03^2 I$ we reject $H_0 : \Sigma = I_p$

Table 3 Comparison of empirical significance levels. T_{J2}, T_{J3}, T_{J4} are defined in (4). Test statistics introduced by Ledoit and Wolf is denoted by T_W, T_S stands for Srivastava's test statistics, T_1 and T_2 for the two test statistics introduced by Fisher, see Sect. 1.2

	α	T_J			Alternative tests			
		T_{J2}	T_{J3}	T_{J4}	T_W	T_S	T_1	T_2
$p = 125$ $n = 250$	0.1	0.097	0.093	0.0935	0.1045	0.104	0.097	0.0995
	0.05	0.0405	0.051	0.0515	0.049	0.048	0.058	0.0565
	0.025	0.021	0.0225	0.028	0.024	0.023	0.0305	0.0325
	0.01	0.009	0.0075	0.0135	0.0095	0.0095	0.0155	0.0175
$p = 250$ $n = 125$	0.1	0.1065	0.098	0.0955	0.107	0.106	0.107	0.103
	0.05	0.0505	0.0455	0.0495	0.053	0.0515	0.058	0.056
	0.025	0.024	0.023	0.028	0.0285	0.028	0.0305	0.0315
	0.01	0.0085	0.01	0.0135	0.013	0.012	0.012	0.0135

Table 4 Comparison of empirical powers of tests for $\alpha = 0.05$. The test statistics T_{J2}, T_{J3} are T_{J4} are defined in (4), T_W stands for the Ledoit and Wolf test statistic, T_S for Srivastava's test statistic, T_1 and T_2 for the two test statistics introduced by Fisher, see Sect. 1.2. The highest power is marked with black

		T_J			Alternative tests			
		T_{J2}	T_{J3}	T_{J4}	T_W	T_S	T_1	T_2
$p = 125$ $n = 125$	$\Sigma = 1.005^2 I$	0.1195	0.121	**0.13**	0.0585	0.0585	0.0685	0.0805
	$\Sigma = 1.01^2 I$	0.2855	**0.296**	0.2855	0.054	0.0525	0.0705	0.0715
	$\Sigma = 1.03^2 I$	**0.9985**	0.9975	0.9935	0.124	0.1135	0.102	0.1295
$p = 125$ $n = 250$	$\Sigma = 1.005^2 I$	**0.19**	0.16	0.152	0.0475	0.0465	0.0595	0.061
	$\Sigma = 1.01^2 I$	**0.5995**	0.534	0.504	0.0705	0.069	0.07	0.0795
	$\Sigma = 1.03^2 I$	**1**	**1**	**1**	0.175	0.159	0.111	0.1745
$p = 250$ $n = 125$	$\Sigma = 1.005^2 I$	**0.194**	0.171	0.163	0.052	0.049	0.0655	0.0695
	$\Sigma = 1.01^2 I$	**0.6155**	0.551	0.512	0.068	0.064	0.069	0.067
	$\Sigma = 1.03^2 I$	**1**	**1**	**1**	0.117	0.1045	0.102	0.115
$p = 125$ $n = 125$	$\Sigma = diag(U)$, $U \sim U(1 \pm .01)^2$	**0.1405**	0.127	0.118	0.059	0.058	0.0575	0.063
	$\Sigma = diag(U)$, $U \sim U(1 \pm .1)^2$	**0.885**	0.875	0.87	0.2585	0.265	0.0815	0.142
	$\Sigma = diag(U)$, $U \sim U(1 \pm .3)^2$	**0.961**	0.9575	0.956	0.748	0.747	0.2715	0.659

(continued)

Table 4 (continued)

		T_J			Alternative tests			
		T_{J2}	T_{J3}	T_{J4}	T_W	T_S	T_1	T_2
$p = 128$ $n = 128$	$\Sigma = \begin{pmatrix} 0.95I_{64} & 0 \\ 0 & I_{64} \end{pmatrix}$	**0.508**	0.4215	0.3615	0.046	0.047	0.0405	0.048
	$\Sigma = \begin{pmatrix} 0.9I_{64} & 0 \\ 0 & I_{64} \end{pmatrix}$	**0.99**	0.9825	0.968	0.065	0.071	0.0315	0.0465
	$\Sigma = \begin{pmatrix} 0.5I_{64} & 0 \\ 0 & I_{64} \end{pmatrix}$	1	1	1	1	1	0.321	0.996
	$\Sigma = \begin{pmatrix} 0.9I_{32} & 0 \\ 0 & I_{96} \end{pmatrix}$	**0.531**	0.4565	0.3895	0.065	0.069	0.0475	0.054
	$\Sigma = \begin{pmatrix} 0.9I_{16} & 0 \\ 0 & I_{112} \end{pmatrix}$	**0.1505**	0.1255	0.1185	0.069	0.0695	0.0575	0.0545
	$\Sigma = \begin{pmatrix} I_{64} & 0.01J_{64} \\ 0.01J_{64} & I_{64} \end{pmatrix}$	0.1215	0.1875	0.2465	0.2085	0.2085	0.249	**0.358**
	$\Sigma = \begin{pmatrix} I_{64} & 0.02J_{64} \\ 0.02J_{64} & I_{64} \end{pmatrix}$	0.7305	0.944	0.9735	0.85	0.8495	0.971	**0.983**
	$\Sigma = \begin{pmatrix} I_{64} & 0.1J_{64} \\ 0.1J_{64} & I_{64} \end{pmatrix}$	1	1	1	1	1	1	1

with probability 1, while the other tests reach a maximum power of 17.5%. Test performance remains good while the elements of the diagonal covariance matrix come from the uniform distribution on the interval surrounding 1.

References

1. Anderson, T.W.: An Introduction to Multivariate Statistical Analysis, 3rd edn. Wiley, New York (2003)
2. Arharov, L.V.: Limits theorems for the characteristical roots of a sample covariance matrix. Sov. Math. Dokl. **12**, 1206–1209 (1971)
3. Bai, Z., Jiang, D., Yao, J., Zheng, S.: Corrections to LRT on large-dimensional covariance matrix by RMT. Ann. Stat. **37**, 3822–3840 (2009)
4. Birke, M., Dette, H.A.: A note on testing the covariance matrix for large dimension. Stat. Probab. Lett. **74**, 281–289 (2005)
5. Chen, S., Zhang, L., Zhong, P.: Tests for high dimensional covariance matrices. J. Am. Stat. Assoc. **105**, 810–819 (2010)
6. Fisher, T.J., Sun, X., Gallagher, C.M.: A new test for sphericity of the covariance matrix for high dimensional data. J. Multivar. Anal. **101**(10), 2554–2570 (2010)
7. Fisher, T.J.: On testing for an identity covariance matrix when the dimensionality equals or exceeds the sample size. J. Stat. Plan. Infer. **142**(1), 312–326 (2012)

8. Fujikoshi, Y., Ulyanov, V.V., Shimizu, R.: Multivariate Statistics: High-Dimensional and Large-Sample Approximations. Wiley, New York (2010)
9. Jiang, D., Jiang, T., Yang, F.: Likelihood ratio tests for covariance matrices of high-dimensional normal distributions. J. Stat. Plan. Inference **142**, 2241–2256 (2012)
10. Jiang, T., Qi, Y.: Likelihood ratio tests for high-dimensional normal distributions. Scand. J. Stat. (2015). doi:10.1111/sjos.12147
11. Jiang, T., Yang, F.: Central limit theorems for classical likelihood ratio tests for high-dimensional normal distributions. Ann. Stat. **41**, 2029–2074 (2013)
12. Jonsson, D.: Some limit theorems for the eigenvalues of a sample covariance matrix. J. Multivar. Anal. **12**, 1–38 (1982)
13. Kolmogorov, A.N.: Sulla determinazione empirica di una legge di distribuzione (On the empirical determination of a distribution law). Giorn. Inst. Ital. Attuari. **4**, 83–91 (1933)
14. Ledoit, O., Wolf, M.: Same hypothesis tests for the covariance matrix when the dimension is large compared to the sample size. Ann. Stat. **30**(4), 1081–1102 (2002)
15. Mauchly, J.W.: Significance test for sphericity of a normal N-variate distribution. Ann. Math. Stat. **11**, 204–209 (1940)
16. Muirhead, R.J.: Aspects of Multivariate Statistical Theory. Wiley, New York (1982)
17. Nagao, H.: On same test criteria for covariance matrix. Ann. Stat. **1**, 700–709 (1973)
18. Pielaszkiewicz, J.: Contributions to High-Dimensional Analysis under Kolmogorov Condition. Linkopings Studies in Science and Technology. Dissertation No. 1724 (2015)
19. Pielaszkiewicz, J., von Rosen, D., Singull, M.: On $\mathbb{E}\big[\prod_{i=0}^{k}\mathrm{Tr}\{W^{m_i}\}\big]$, where $W \sim \mathcal{W}_p(I, n)$ to appear in Commun. Stat. Theory. doi:10.1080/03610926.2015.1053942
20. Schott, J.R.: Some tests for the equality of covariance matrices. J. Stat. Plan. Inference **94**, 25–36 (2001)
21. Schott, J.R.: Testing of complete independence in high dimensions. Biometrika **92**(4), 951–956 (2005)
22. Schott, J.R.: A test for the equality of covariance matrices when the dimension is large relative to the sampel size. Comput. Stat. Data Anal. **51**(12), 6535–6542 (2007)
23. Shapiro, S.S., Wilk, M.B.: An analysis of variance test for normality (complete samples). Biometrika **52**(3/4), 591–611 (1965)
24. Srivastava, M.S.: Some tests concerning the covariance matrix in high dimensional data. J. Japan Stat. Soc. **35**(2), 251–272 (2005)

Some Further Remarks on the Linear Sufficiency in the Linear Model

Radosław Kala, Augustyn Markiewicz and Simo Puntanen

Abstract In this article we consider the linear sufficiency of statistic $\mathbf{F}\mathbf{y}$ when estimating the estimable parametric function of $\boldsymbol{\beta}$ under the linear model $\mathscr{A} = \{\mathbf{y}, \mathbf{X}\boldsymbol{\beta}, \mathbf{V}\}$. We review some properties that have not been received much attention in the literature and provide some new results and insight into the meaning of the linear sufficiency. In particular, we consider the best linear unbiased estimation (BLUE) under the transformed model $\mathscr{A}_t = \{\mathbf{F}\mathbf{y}, \mathbf{F}\mathbf{X}\boldsymbol{\beta}, \mathbf{F}\mathbf{V}\mathbf{F}'\}$ and study the possibilities to measure the relative linear sufficiency of $\mathbf{F}\mathbf{y}$ by comparing the BLUEs under $\mathscr{A}$ and $\mathscr{A}_t$. We also consider some new properties of the Euclidean norm of the distance of the BLUEs under $\mathscr{A}$ and $\mathscr{A}_t$. The concept of linear sufficiency was essentially introduced in early 1980s by Baksalary, Kala and Drygas, but to our knowledge the concept of relative linear sufficiency nor the Euclidean norm of the difference between the BLUEs under $\mathscr{A}$ and $\mathscr{A}_t$ have not appeared in the literature. To make the article more self-readable we go through some basic concepts related to linear sufficiency. We also provide a rather extensive list of relevant references.

Keywords Best linear unbiased estimator · generalized inverse · linear model · linear sufficiency · orthogonal projector · transformed linear model

1 Introduction

In this paper we consider the linear model defined by

R. Kala · A. Markiewicz
Department of Mathematical and Statistical Methods,
Poznań University of Life Sciences, Wojska Polskiego 28, 60637 Poznań, Poland
e-mail: kalar@up.poznan.pl

A. Markiewicz
e-mail: amark@up.poznan.pl

S. Puntanen (✉)
School of Information Sciences, University of Tampere, FI-33014 Tampere, Finland
e-mail: simo.puntanen@uta.fi

N. Bebiano (ed.), *Applied and Computational Matrix Analysis*,
Springer Proceedings in Mathematics & Statistics 192,
DOI 10.1007/978-3-319-49984-0_19

$$\mathbf{y} = \mathbf{X}\boldsymbol{\beta} + \boldsymbol{\varepsilon}, \quad \text{or shortly notated} \quad \mathscr{A} = \{\mathbf{y}, \mathbf{X}\boldsymbol{\beta}, \mathbf{V}\}, \tag{1}$$

where $\mathbf{y}$ is an n-dimensional observable response variable, $\mathbf{X}$ is a known $n \times p$ matrix, i.e., $\mathbf{X} \in \mathbb{R}^{n\times p}$, $\boldsymbol{\beta} \in \mathbb{R}^p$ is a vector of fixed (but unknown) parameters, and $\boldsymbol{\varepsilon}$ is an unobservable random error with a known covariance matrix $\operatorname{cov}(\boldsymbol{\varepsilon}) = \mathbf{V} = \operatorname{cov}(\mathbf{y})$ and expectation $\operatorname{E}(\boldsymbol{\varepsilon}) = \mathbf{0}$.

Under the model $\mathscr{A} = \{\mathbf{y}, \mathbf{X}\boldsymbol{\beta}, \mathbf{V}\}$, the statistic $\mathbf{Gy}$, where $\mathbf{G} \in \mathbb{R}^{n\times n}$, is the best linear unbiased estimator, BLUE, of $\mathbf{X}\boldsymbol{\beta}$ whenever $\mathbf{Gy}$ is unbiased, i.e., $\mathbf{GX} = \mathbf{X}$, and it has the minimal covariance matrix in the Löwner sense among all unbiased linear estimators of $\mathbf{X}\boldsymbol{\beta}$. The BLUE of an estimable parametric function $\mathbf{K}\boldsymbol{\beta}$, where $\mathbf{K} \in \mathbb{R}^{k\times p}$, is defined in the corresponding way. Recall that $\mathbf{K}\boldsymbol{\beta}$ is said to be estimable under $\mathscr{A}$ if it has a linear unbiased estimator $\mathbf{Ly}$, say, so that $\operatorname{E}(\mathbf{Ly}) = \mathbf{LX}\boldsymbol{\beta} = \mathbf{K}\boldsymbol{\beta}$ for all $\boldsymbol{\beta} \in \mathbb{R}^p$, which happens if and only if

$$\mathscr{C}(\mathbf{K}') \subset \mathscr{C}(\mathbf{X}'), \tag{2}$$

where $\mathscr{C}(\cdot)$ stands for the column space (range) of the matrix argument.

In what follows, we frequently refer to the following lemma; see, e.g., [18, p. 55], [38, p. 282], and [3].

Lemma 1 *Consider the general linear model* $\mathscr{A} = \{\mathbf{y}, \mathbf{X}\boldsymbol{\beta}, \mathbf{V}\}$. *Then the statistic* $\mathbf{Gy}$ *is the* BLUE *for* $\mathbf{X}\boldsymbol{\beta}$ *if and only if* $\mathbf{G}$ *satisfies the equation*

$$\mathbf{G}(\mathbf{X} : \mathbf{V}\mathbf{X}^{\perp}) = (\mathbf{X} : \mathbf{0}). \tag{3}$$

The corresponding condition for $\mathbf{By}$ *to be the* BLUE *of an estimable parametric function* $\mathbf{K}\boldsymbol{\beta}$ *is*

$$\mathbf{B}(\mathbf{X} : \mathbf{V}\mathbf{X}^{\perp}) = (\mathbf{K} : \mathbf{0}). \tag{4}$$

The notation $(\mathbf{X} : \mathbf{V}\mathbf{X}^{\perp})$ refers to a columnwise partitioned matrix by juxtaposing matrices $\mathbf{X}$ and $\mathbf{V}\mathbf{X}^{\perp}$. The matrix $\mathbf{X}^{\perp}$ refers to a matrix spanning the orthocomplement of the column space $\mathscr{C}(\mathbf{X})$. One convenient choice for $\mathbf{X}^{\perp}$ is $\mathbf{M} := \mathbf{I}_n - \mathbf{P}_{\mathbf{X}} = \mathbf{I}_n - \mathbf{H}$, with $\mathbf{P}_{\mathbf{X}} = \mathbf{X}\mathbf{X}^+ =: \mathbf{H}$ denoting the orthogonal projector onto $\mathscr{C}(\mathbf{X})$ and $\mathbf{X}^+$ referring to the Moore–Penrose inverse of $\mathbf{X}$. Of course, $\mathscr{C}(\mathbf{X}^{\perp}) = \mathscr{C}(\mathbf{M}) = \mathscr{N}(\mathbf{X}')$, where $\mathscr{N}(\cdot)$ stands for the null space.

The solution $\mathbf{G}$ for (3) always exists but is unique if and only if $\mathscr{C}(\mathbf{X} : \mathbf{V}) = \mathbb{R}^n$. However, the observed value of $\mathbf{Gy}$ is unique (with probability 1) once the random vector $\mathbf{y}$ has realized its value in the space

$$\mathscr{C}(\mathbf{X} : \mathbf{V}) = \mathscr{C}(\mathbf{X}) \oplus \mathscr{C}(\mathbf{VM}). \tag{5}$$

In (5) the symbol $\oplus$ stands for the direct sum. Two estimators $\mathbf{G}_1\mathbf{y}$ and $\mathbf{G}_2\mathbf{y}$ are said to be equal (with probability 1) whenever $\mathbf{G}_1\mathbf{y} = \mathbf{G}_2\mathbf{y}$ for all $\mathbf{y} \in \mathscr{C}(\mathbf{X} : \mathbf{V})$. When talking about the equality of estimators we sometimes may drop the phrase "with probability 1". The consistency of the model $\mathscr{A}$ means that the observed $\mathbf{y}$ lies in

$\mathscr{C}(\mathbf{X} : \mathbf{V})$ which is assumed to hold whatever model we have. For the consistency concept, see, e.g., [13].

In this paper we use the notation $\mathscr{W}$ for the set of nonnegative definite matrices defined as

$$\mathscr{W} = \{\mathbf{W} \in \mathbb{R}^{n\times n} : \mathbf{W} = \mathbf{V} + \mathbf{XUU'X'},\ \mathscr{C}(\mathbf{W}) = \mathscr{C}(\mathbf{X} : \mathbf{V})\}. \tag{6}$$

In (6) $\mathbf{U}$ can be any $p \times m$ matrix as long as $\mathscr{C}(\mathbf{W}) = \mathscr{C}(\mathbf{X} : \mathbf{V})$ is satisfied. One obvious choice is of course $\mathbf{U} = \mathbf{I}_p$. In particular, if $\mathscr{C}(\mathbf{X}) \subset \mathscr{C}(\mathbf{V})$, we can choose $\mathbf{U} = \mathbf{0}$. The set $\mathscr{W}$ appears to be a very useful class of matrices and it has numerous applications related to linear models. For example, it is easy to confirm the following lemma.

Lemma 2 *Let* $\mathbf{W} \in \mathscr{W}$*. Then* $\mathbf{Gy}$ *is the* BLUE *for* $\mathbf{X}\boldsymbol{\beta}$ *under* $\mathscr{A} = \{\mathbf{y}, \mathbf{X}\boldsymbol{\beta}, \mathbf{V}\}$ *if and only if* $\mathbf{Gy}$ *is the* BLUE *for* $\mathbf{X}\boldsymbol{\beta}$ *under* $\mathscr{A}_{\mathbf{W}} = \{\mathbf{y}, \mathbf{X}\boldsymbol{\beta}, \mathbf{W}\}$.

We will later consider some interesting properties of $\mathscr{W}$ and the corresponding extended set

$$\mathscr{W}_* = \{\mathbf{W} \in \mathbb{R}^{n\times n} : \mathbf{W} = \mathbf{V} + \mathbf{XTX'},\ \mathscr{C}(\mathbf{W}) = \mathscr{C}(\mathbf{X} : \mathbf{V})\}. \tag{7}$$

Notice that $\mathbf{W}$ that belongs to $\mathscr{W}_*$ is not necessarily nonnegative definite and it can be nonsymmetric. For example, the following statements concerning $\mathbf{W}$ belonging to $\mathscr{W}_*$ are equivalent:

$$\mathscr{C}(\mathbf{X} : \mathbf{V}) = \mathscr{C}(\mathbf{W}), \tag{8a}$$
$$\mathscr{C}(\mathbf{X}) \subset \mathscr{C}(\mathbf{W}), \tag{8b}$$
$$\mathbf{X'W^-X} \text{ is invariant for any choice of } \mathbf{W}^-, \tag{8c}$$
$$\mathscr{C}(\mathbf{X'W^-X}) = \mathscr{C}(\mathbf{X'}) \text{ for any choice of } \mathbf{W}^-, \tag{8d}$$
$$\mathbf{X(X'W^-X)^-X'W^-X} = \mathbf{X} \text{ for any choices of } \mathbf{W}^- \text{ and } (\mathbf{X'W^-X})^-. \tag{8e}$$

Moreover, each of these statements is equivalent also to $\mathscr{C}(\mathbf{X} : \mathbf{V}) = \mathscr{C}(\mathbf{W}')$, and hence to the statements (8b)–(8e) by replacing $\mathbf{W}$ with $\mathbf{W}'$. Notice that obviously $\mathscr{C}(\mathbf{W}) = \mathscr{C}(\mathbf{W}')$ and that the invariance properties in (8d) and (8e) concern also the choice of $\mathbf{W} \in \mathscr{W}_*$. For further properties of $\mathscr{W}_*$, see, e.g., [11, Theorem 1], [12, Theorem 2], [10, Theorem 2], and [37, Sect. 12.3].

The usefulness of $\mathscr{W}_*$ appears, e.g., from the following well-known representation of the BLUE of $\mathbf{X}\boldsymbol{\beta}$:

$$\mathrm{BLUE}(\mathbf{X}\boldsymbol{\beta} \mid \mathscr{A}) = \mathbf{X(X'W^-X)^-X'W^-y} =: \mathbf{Cy}, \tag{9}$$

where $\mathbf{W} \in \mathscr{W}_*$. The *general* representation for the BLUE can be written as $\mathbf{Ay}$, where

$$\mathbf{A} = \mathbf{C} + \mathbf{N}(\mathbf{I}_n - \mathbf{P}_{\mathbf{W}}), \tag{10}$$

with $\mathbf{N} \in \mathbb{R}^{n\times n}$ being free to vary. In this context we might mention also the following expression:

$$\mathrm{BLUE}(\mathbf{X}\boldsymbol{\beta} \mid \mathscr{A}) = [\mathbf{I}_n - \mathbf{VM}(\mathbf{MVM})^{-}\mathbf{M}]\mathbf{y}. \tag{11}$$

For further expressions, see, e.g., [37, Sect. 10.4].

Recall that the multipliers of the random vector $\mathbf{y}$ in (9) and (11) are not necessarily the same but the following holds:

$$\mathbf{X}(\mathbf{X}'\mathbf{W}^{-}\mathbf{X})^{-}\mathbf{X}'\mathbf{W}^{-}\mathbf{y} = [\mathbf{I}_n - \mathbf{VM}(\mathbf{MVM})^{-}\mathbf{M}]\mathbf{y} \quad \text{for all } \mathbf{y} \in \mathscr{C}(\mathbf{W}). \tag{12}$$

One more property requiring attention before proceeding into the concept of linear sufficiency is the invariance of the matrix product $\mathbf{AB}^{-}\mathbf{C}$. According to [39, Lemma 2.2.4], for any nonnull $\mathbf{A}$ and $\mathbf{C}$ the following holds:

$$\mathbf{AB}^{-}\mathbf{C} = \mathbf{AB}^{+}\mathbf{C} \text{ for all } \mathbf{B}^{-} \iff \mathscr{C}(\mathbf{C}) \subset \mathscr{C}(\mathbf{B}) \text{ and } \mathscr{C}(\mathbf{A}') \subset \mathscr{C}(\mathbf{B}'). \tag{13}$$

We shall frequently need the invariance property (13). For example, we immediately see that for $\mathbf{W} \in \mathscr{W}_{*}$, the matrices $\mathbf{X}'\mathbf{W}^{-}\mathbf{X}$ and $\mathbf{X}(\mathbf{X}'\mathbf{W}^{-}\mathbf{X})^{-}\mathbf{X}'$ are invariant for any choice of $\mathbf{W}^{-}$. Similarly in (12) we can use any generalized inverses involved.

2 Definition of the Linear Sufficiency

Now we can formally define the concept of linear sufficiency as done by [7]. Actually they talked about "linear transformations preserving best linear unbiased estimators" and it was [19] who adopted the term "linear sufficiency".

Definition 1 A linear statistic $\mathbf{Fy}$, where $\mathbf{F} \in \mathbb{R}^{f\times n}$, is called linearly sufficient for $\mathbf{X}\boldsymbol{\beta}$ under the model $\mathscr{A} = \{\mathbf{y}, \mathbf{X}\boldsymbol{\beta}, \mathbf{V}\}$, if there exists a matrix $\mathbf{A} \in \mathbb{R}^{n\times f}$ such that $\mathbf{AFy}$ is the BLUE for $\mathbf{X}\boldsymbol{\beta}$. Correspondingly, $\mathbf{Fy}$ is linearly sufficient for estimable $\mathbf{K}\boldsymbol{\beta}$, where $\mathbf{K} \in \mathbb{R}^{k\times p}$, if there exists a matrix $\mathbf{A} \in \mathbb{R}^{k\times f}$ such that $\mathbf{AFy}$ is the BLUE for $\mathbf{K}\boldsymbol{\beta}$.

Sometimes we may use the short notations

$$\mathbf{Fy} \in \mathrm{S}(\mathbf{X}\boldsymbol{\beta}), \quad \mathbf{Fy} \in \mathrm{S}(\mathbf{K}\boldsymbol{\beta}) \tag{14}$$

to indicate that $\mathbf{Fy}$ is linearly sufficient for $\mathbf{X}\boldsymbol{\beta}$ or for $\mathbf{K}\boldsymbol{\beta}$, respectively.

By definition, $\mathbf{Fy}$ is linearly sufficient for $\mathbf{X}\boldsymbol{\beta}$ if and only if the equation

$$\mathbf{AF}(\mathbf{X} : \mathbf{VM}) = (\mathbf{X} : \mathbf{0}) \tag{15}$$

has a solution for $\mathbf{A}$, which happens if and only if

$$\mathscr{C}\begin{pmatrix}\mathbf{X}'\\ \mathbf{0}\end{pmatrix} \subset \mathscr{C}\begin{pmatrix}\mathbf{X}'\mathbf{F}'\\ \mathbf{MVF}'\end{pmatrix}. \tag{16}$$

The concept of linear minimal sufficiency, introduced by [19], is defined as follows.

Definition 2 A linear statistic $\mathbf{Fy}$ is called linearly minimal sufficient if for any other linearly sufficient statistics $\mathbf{Sy}$, there exists a matrix $\mathbf{A}$ such that $\mathbf{Fy} = \mathbf{ASy}$ almost surely.

In Lemma 3 we collect some well-known equivalent conditions for $\mathbf{Fy}$ being linearly sufficient for $\mathbf{X}\boldsymbol{\beta}$. For the proofs of parts (c) and (d), see [7]; part (e), see [8, Corollary 2]; and part (f), [32, Proposition 3.1a]. For further related references, see [4, 9, 19, 20, 26–28, 30].

Lemma 3 *The statistic* $\mathbf{Fy}$ *is linearly sufficient for* $\mathbf{X}\boldsymbol{\beta}$ *under the linear model* $\mathscr{A} = \{\mathbf{y}, \mathbf{X}\boldsymbol{\beta}, \mathbf{V}\}$ *if and only if any of the following equivalent statements holds:*

(a) $\mathscr{C}\begin{pmatrix}\mathbf{X}'\\ \mathbf{0}\end{pmatrix} \subset \mathscr{C}\begin{pmatrix}\mathbf{X}'\mathbf{F}'\\ \mathbf{MVF}'\end{pmatrix}$,
(b) $\mathscr{N}(\mathbf{FX} : \mathbf{FVX}^{\perp}) \subset \mathscr{N}(\mathbf{X} : \mathbf{0})$,
(c) $\mathscr{C}(\mathbf{X}) \subset \mathscr{C}(\mathbf{WF}')$, *where* $\mathbf{W} \in \mathscr{W}$,
(d) $\operatorname{rank}(\mathbf{X} : \mathbf{VF}') = \operatorname{rank}(\mathbf{WF}')$, *where* $\mathbf{W} \in \mathscr{W}$,
(e) $\mathscr{C}(\mathbf{X}'\mathbf{F}') = \mathscr{C}(\mathbf{X}')$ *and* $\mathscr{C}(\mathbf{FX}) \cap \mathscr{C}(\mathbf{FVX}^{\perp}) = \{\mathbf{0}\}$,
(f) $\mathscr{N}(\mathbf{F}) \cap \mathscr{C}(\mathbf{X} : \mathbf{V}) \subset \mathscr{C}(\mathbf{VX}^{\perp})$,
(g) *there exists a matrix* $\mathbf{A}$ *such that* $\mathbf{AF}(\mathbf{X} : \mathbf{VX}^{\perp}) = (\mathbf{X} : \mathbf{0})$.

Moreover, $\mathbf{Fy}$ *is linearly minimal sufficient for* $\mathbf{X}\boldsymbol{\beta}$ *if and only if* $\mathscr{C}(\mathbf{X}) = \mathscr{C}(\mathbf{WF}')$, *or equivalently, the equality holds in (a), (b) or (f).*

Baksalary and Kala [8] proved the following:

Lemma 4 *Let* $\mathbf{K}\boldsymbol{\beta}$ *be an estimable parametric function under* $\mathscr{A} = \{\mathbf{y}, \mathbf{X}\boldsymbol{\beta}, \mathbf{V}\}$, *i.e.,* $\mathscr{C}(\mathbf{K}') \subset \mathscr{C}(\mathbf{X}')$. *Then* $\mathbf{Fy}$ *is linearly sufficient for* $\mathbf{K}\boldsymbol{\beta}$ *under* $\mathscr{A}$ *if and only if any of the following equivalent statements holds:*

(a) $\mathscr{C}\begin{pmatrix}\mathbf{K}'\\ \mathbf{0}\end{pmatrix} \subset \mathscr{C}\begin{pmatrix}\mathbf{X}'\mathbf{F}'\\ \mathbf{MVF}'\end{pmatrix}$,
(b) $\mathscr{N}(\mathbf{FX} : \mathbf{FVX}^{\perp}) \subset \mathscr{N}(\mathbf{K} : \mathbf{0})$,
(c) $\mathscr{C}[\mathbf{X}(\mathbf{X}'\mathbf{W}^{-}\mathbf{X})^{-}\mathbf{K}'] \subset \mathscr{C}(\mathbf{WF}')$, *where* $\mathbf{W} \in \mathscr{W}$,
(d) there exists a matrix $\mathbf{A}$ *such that* $\mathbf{AF}(\mathbf{X} : \mathbf{VX}^{\perp}) = (\mathbf{K} : \mathbf{0})$.

Moreover, $\mathbf{Fy}$ *is linearly minimal sufficient for* $\mathbf{K}\boldsymbol{\beta}$ *if and only if equality (instead of subspace inclusion) holds in (a), (b) or equivalently (c).*

Suppose that $\mathbf{Fy}$ is linearly sufficient for $\mathbf{X}\boldsymbol{\beta}$ under the model $\mathscr{A}$, and $\mathbf{F}_1$ is some arbitrary matrix with n columns. Then it is interesting to observe that the extended statistic

$$\mathbf{F}_0\mathbf{y} := \begin{pmatrix}\mathbf{F}\\ \mathbf{F}_1\end{pmatrix}\mathbf{y} \tag{17}$$

is also linearly sufficient for $\mathbf{X}\boldsymbol{\beta}$. This is so because

$$\mathscr{C}(\mathbf{X}) \subset \mathscr{C}(\mathbf{WF}') \subset \mathscr{C}[\mathbf{W}(\mathbf{F}' : \mathbf{F}_1')] = \mathscr{C}(\mathbf{WF}_0'). \tag{18}$$

Similarly

$$\mathbf{Fy} \in \mathrm{S}(\mathbf{X}\boldsymbol{\beta}) \implies \mathbf{F}_*\mathbf{y} \in \mathrm{S}(\mathbf{X}\boldsymbol{\beta}), \quad \text{if } \mathscr{C}(\mathbf{F}') = \mathscr{C}(\mathbf{F}_*'). \tag{19}$$

Thus if $\operatorname{rank}(\mathbf{F}) = r$ we can replace $\mathbf{F} \in \mathbb{R}^{f\times n}$ with $\mathbf{F}_* \in \mathbb{R}^{r\times n}$, where $r \leq f$, i.e., the columns of $\mathbf{F}_*'$ provide a spanning basis for $\mathscr{C}(\mathbf{F}')$.

Notice also that the linear sufficiency condition $\mathscr{C}(\mathbf{X}) \subset \mathscr{C}(\mathbf{WF}')$ implies that we necessarily must have

$$\operatorname{rank}(\mathbf{X}_{n\times p}) \leq p \leq \operatorname{rank}(\mathbf{F}_{f\times n}) \leq f. \tag{20}$$

In passing we note that $\mathbf{X}'\mathbf{W}^-\mathbf{y}$ is linearly minimal sufficient for $\mathbf{X}\boldsymbol{\beta}$ under the model $\mathscr{A}$; this follows from $\mathscr{C}(\mathbf{X}) = \mathscr{C}[\mathbf{W}(\mathbf{W}^-)'\mathbf{X}]$.

3 The Transformed Model $\mathscr{A}_t$

Consider the model $\mathscr{A} = \{\mathbf{y}, \mathbf{X}\boldsymbol{\beta}, \mathbf{V}\}$ and let $\mathbf{F} \in \mathbb{R}^{f\times n}$ be such a matrix that $\mathbf{Fy}$ is linearly sufficient for $\mathbf{X}\boldsymbol{\beta}$. Then the transformation $\mathbf{F}$ applied to $\mathbf{y}$ induces the transformed model

$$\mathscr{A}_t = \{\mathbf{Fy}, \mathbf{FX}\boldsymbol{\beta}, \mathbf{FVF}'\}. \tag{21}$$

Now, as the statistic $\mathbf{Fy}$ is linearly sufficient for $\mathbf{X}\boldsymbol{\beta}$, it sounds intuitively believable that both models provide the same starting point for obtaining the BLUE of $\mathbf{X}\boldsymbol{\beta}$. Indeed this appears to be true as proved by [7, 8]. Moreover, [40, Theorem 2.8] and [29, Theorem 2] showed the following:

Lemma 5 *Consider the model* $\mathscr{A} = \{\mathbf{y}, \mathbf{X}\boldsymbol{\beta}, \mathbf{V}\}$ *and its transformed version*

$$\mathscr{A}_t = \{\mathbf{Fy}, \mathbf{FX}\boldsymbol{\beta}, \mathbf{FVF}'\}, \tag{22}$$

and let $\mathbf{K}\boldsymbol{\beta}$ *be estimable under* $\mathscr{A}$. *Then the following statements are equivalent:*

(a) $\mathbf{Fy}$ *is linearly sufficient for* $\mathbf{K}\boldsymbol{\beta}$.
(b) $\mathrm{BLUE}(\mathbf{K}\boldsymbol{\beta} \mid \mathscr{A}) = \mathrm{BLUE}(\mathbf{K}\boldsymbol{\beta} \mid \mathscr{A}_t)$ *with probability* 1.
(c) *There exists at least one representation of* BLUE *of* $\mathbf{K}\boldsymbol{\beta}$ *under* $\mathscr{A}$ *which is the* BLUE *also under the transformed model* $\mathscr{A}_t$.

It is noteworthy that if $\mathbf{Fy}$ is linearly sufficient for $\mathbf{X}\boldsymbol{\beta}$, then, in view of (16), we have

$$\mathscr{C}(\mathbf{X}') = \mathscr{C}(\mathbf{X}'\mathbf{F}'), \quad \text{i.e.,} \quad \operatorname{rank}(\mathbf{FX}) = \operatorname{rank}(\mathbf{X}). \tag{23}$$

On the other hand, on account of (2), $\mathbf{X}\boldsymbol{\beta}$ is estimable under the transformed model $\mathscr{A}_t = \{\mathbf{Fy}, \mathbf{FX}\boldsymbol{\beta}, \mathbf{FVF}'\}$ if and only if

$$\mathscr{C}(\mathbf{X}') \subset \mathscr{C}(\mathbf{X}'\mathbf{F}'), \tag{24}$$

i.e., $\mathscr{C}(\mathbf{X}') = \mathscr{C}(\mathbf{X}'\mathbf{F}')$, which is (23). This confirms the following:

$$\mathbf{Fy} \in \mathrm{S}(\mathbf{X}\boldsymbol{\beta}) \implies \mathbf{X}\boldsymbol{\beta} \text{ is estimable under } \mathscr{A}_t. \tag{25}$$

However, the reverse implication in (25) does not hold. In view of part (e) of Lemma 3, we need the following *two* conditions for $\mathbf{Fy} \in \mathrm{S}(\mathbf{X}\boldsymbol{\beta})$:

$$\mathscr{C}(\mathbf{X}'\mathbf{F}') = \mathscr{C}(\mathbf{X}') \quad \text{and} \quad \mathscr{C}(\mathbf{FX}) \cap \mathscr{C}(\mathbf{FVX}^{\perp}) = \{\mathbf{0}\}, \tag{26}$$

which can be expressed equivalently as

$$\mathbf{X}\boldsymbol{\beta} \text{ is estimable under } \mathscr{A}_t \quad \text{and} \quad \mathscr{C}(\mathbf{FX}) \cap \mathscr{C}(\mathbf{FVX}^{\perp}) = \{\mathbf{0}\}. \tag{27}$$

Let us consider some special choices of $\mathbf{F}$. For example, if $\mathbf{F}$ has the property $\mathscr{C}(\mathbf{F}') = \mathbb{R}^n$ (implying that the number of the rows in $\mathbf{F} \in \mathbb{R}^{f \times n}$ is at least n), then

$$\mathscr{C}(\mathbf{X}) \subset \mathscr{C}(\mathbf{W}) = \mathscr{C}(\mathbf{WF}'), \tag{28}$$

and thereby $\mathbf{Fy}$ is linearly sufficient for $\mathbf{X}\boldsymbol{\beta}$. In particular, for a nonsingular $\mathbf{F} \in \mathbb{R}^{n \times n}$, the statistic $\mathbf{Fy}$ is linearly sufficient. For a positive definite $\mathbf{V}$ the linear sufficiency condition becomes simply

$$\mathscr{C}(\mathbf{X}) \subset \mathscr{C}(\mathbf{VF}'). \tag{29}$$

Supposing that $\mathbf{V}^{1/2}$ is the positive definite square root of $\mathbf{V}$ we observe that $\mathbf{V}^{-1/2}\mathbf{y}$ is linearly sufficient and thus the BLUE of $\mathbf{X}\boldsymbol{\beta}$ under the transformed model

$$\mathscr{A}_t = \{\mathbf{V}^{-1/2}\mathbf{y},\ \mathbf{V}^{-1/2}\mathbf{X}\boldsymbol{\beta},\ \mathbf{I}_n\} \tag{30}$$

is the same as in the original model $\mathscr{A} = \{\mathbf{y}, \mathbf{X}\boldsymbol{\beta}, \mathbf{V}\}$, i.e., the $\mathrm{BLUE}(\mathbf{X}\boldsymbol{\beta})$ under $\mathscr{A}$ equals the ordinary least squares estimator of $\mathbf{X}\boldsymbol{\beta}$, $\mathrm{OLSE}(\mathbf{X}\boldsymbol{\beta})$, under $\mathscr{A}_t$:

$$\mathrm{BLUE}(\mathbf{X}\boldsymbol{\beta} \mid \mathscr{A}) = \mathrm{OLSE}(\mathbf{X}\boldsymbol{\beta} \mid \mathscr{A}_t). \tag{31}$$

This technique, sometimes referred to as the Aitken-approach, see [1], is well known in statistical textbooks. However, usually these textbooks do not mention anything about linear sufficiency feature of this transformation.

Consider then a more general case. By Lemma 2 we know that the BLUEs under $\mathscr{A} = \{\mathbf{y}, \mathbf{X}\boldsymbol{\beta}, \mathbf{V}\}$ and $\mathscr{A}_{\mathbf{W}} = \{\mathbf{y}, \mathbf{X}\boldsymbol{\beta}, \mathbf{W}\}$ are equal. Suppose that $\mathrm{rank}(\mathbf{W}) = w$ and that $\mathbf{W}$ has the eigenvalue decomposition $\mathbf{W} = \mathbf{Z}\boldsymbol{\Lambda}\mathbf{Z}'$, where the columns of $\mathbf{Z} \in$

$\mathbb{R}^{n\times w}$ are orthonormal eigenvectors of $\mathbf{W}$ with respect to nonzero eigenvalues $\lambda_1 \geq \cdots \geq \lambda_w > 0$ of $\mathbf{W}$, and $\boldsymbol{\Lambda} = \operatorname{diag}(\lambda_1, \ldots, \lambda_w)$. Choosing

$$\mathbf{F} = \boldsymbol{\Lambda}^{-1/2}\mathbf{Z}' \in \mathbb{R}^{w\times n}, \tag{32}$$

we observe that

$$\mathscr{C}(\mathbf{WF}') = \mathscr{C}(\mathbf{WZ}\boldsymbol{\Lambda}^{-1/2}) = \mathscr{C}(\mathbf{W}) \tag{33}$$

and hence $\mathscr{C}(\mathbf{X}) \subset \mathscr{C}(\mathbf{WF}')$ and thereby $\mathbf{Fy}$ is linearly sufficient in $\mathscr{A}_{\mathbf{W}}$. Thus the BLUE of $\mathbf{X}\boldsymbol{\beta}$ under the original model $\mathscr{A}$ is the same as under $\mathscr{A}_{\mathbf{W}}$ and further the same as under the transformed model

$$\mathscr{A}_t = \{\, \boldsymbol{\Lambda}^{-1/2}\mathbf{Z}'\mathbf{y},\ \boldsymbol{\Lambda}^{-1/2}\mathbf{Z}'\mathbf{X}\boldsymbol{\beta},\ \mathbf{I}_w\}. \tag{34}$$

Because $\mathbf{F}'\mathbf{F} = \mathbf{Z}\boldsymbol{\Lambda}^{-1}\mathbf{Z}' = \mathbf{W}^+$, we have

$$\begin{aligned} \operatorname{BLUE}(\mathbf{X}\boldsymbol{\beta} \mid \mathscr{A}) &= \operatorname{BLUE}(\mathbf{X}\boldsymbol{\beta} \mid \mathscr{A}_t) = \operatorname{OLSE}(\mathbf{X}\boldsymbol{\beta} \mid \mathscr{A}_t) \\ &= \mathbf{X}(\mathbf{X}'\mathbf{W}^+\mathbf{X})^-\mathbf{X}'\mathbf{W}^+\mathbf{y}, \end{aligned} \tag{35}$$

where we actually can use any generalized inverses involved.

We may note that [17, p. 239] uses the transformation matrix $\boldsymbol{\Lambda}^{-1/2}\mathbf{Z}'$ when considering the so-called weakly singular linear model, i.e., when $\mathscr{C}(\mathbf{X}) \subset \mathscr{C}(\mathbf{V})$, and [25, Sect. 4] while comparing the BLUEs under two linear models with different covariance matrices.

We complete this section by considering a partitioned linear model

$$\mathscr{A}_{12} = \{\mathbf{y},\ \mathbf{X}_1\boldsymbol{\beta}_1 + \mathbf{X}_2\boldsymbol{\beta}_2,\ \mathbf{V}\}. \tag{36}$$

Let us assume that $\mathscr{C}(\mathbf{X}_1) \cap \mathscr{C}(\mathbf{X}_2) = \{\mathbf{0}\}$ implying that $\mathbf{X}_1\boldsymbol{\beta}_1$ is estimable. Premultiplying the model $\mathscr{A}_{12}$ by $\mathbf{M}_2 = \mathbf{I}_n - \mathbf{P}_{\mathbf{X}_2}$ yields the reduced model

$$\mathscr{A}_{12\cdot 2} = \{\mathbf{M}_2\mathbf{y},\ \mathbf{M}_2\mathbf{X}_1\boldsymbol{\beta}_1,\ \mathbf{M}_2\mathbf{V}\mathbf{M}_2\}\,. \tag{37}$$

Now the well-known Frisch–Waugh–Lovell theorem, see, e.g., [22, 23] and [2, Theorem 1], states that the BLUEs of $\mathbf{X}_1\boldsymbol{\beta}_1$ under $\mathscr{A}_{12}$ and $\mathscr{A}_{12\cdot 2}$ coincide. Hence, in view of Lemma 5, the statistic $\mathbf{M}_2\mathbf{y}$ is linearly sufficient for $\mathbf{X}_1\boldsymbol{\beta}_1$. One expression for the BLUE of $\mathbf{X}_1\boldsymbol{\beta}_1$, obtainable from the reduced model $\mathscr{A}_{12\cdot 2}$, is

$$\mathbf{Ay} := \mathbf{X}_1(\mathbf{X}_1'\dot{\mathbf{M}}_2\mathbf{X}_1)^-\mathbf{X}_1'\dot{\mathbf{M}}_2\mathbf{y}, \tag{38}$$

where $\dot{\mathbf{M}}_2 = \mathbf{M}_2(\mathbf{M}_2\mathbf{W}_1\mathbf{M}_2)^-\mathbf{M}_2$ and $\mathbf{W}_1 = \mathbf{V} + \mathbf{X}_1\mathbf{U}_1\mathbf{U}_1'\mathbf{X}_1'$ is such that $\mathscr{C}(\mathbf{W}_1) = \mathscr{C}(\mathbf{X}_1 : \mathbf{V})$. Notice that of course the BLUE of $\mathbf{X}_1\boldsymbol{\beta}_1$ can be written also as

$$\mathbf{By} := (\mathbf{X}_1 : \mathbf{0})(\mathbf{X}'\mathbf{W}^-\mathbf{X})^-\mathbf{X}'\mathbf{W}^-\mathbf{y} = \mathbf{K}(\mathbf{X}'\mathbf{W}^-\mathbf{X})^-\mathbf{X}'\mathbf{W}^-\mathbf{y}, \tag{39}$$

where $\mathbf{K} = (\mathbf{X}_1 : \mathbf{0}) \in \mathbb{R}^{n\times p}$ and $\mathbf{W} \in \mathscr{W}$. The equality $\mathbf{AW} = \mathbf{BW}$ implies

$$\mathbf{W}\dot{\mathbf{M}}_2\mathbf{X}_1(\mathbf{X}_1'\dot{\mathbf{M}}_2\mathbf{X}_1)^-\mathbf{X}_1' = \mathbf{X}(\mathbf{X}'\mathbf{W}^-\mathbf{X})^-\mathbf{K}', \tag{40}$$

and it is easy to confirm that $\mathscr{C}[\mathbf{W}\dot{\mathbf{M}}_2\mathbf{X}_1(\mathbf{X}_1'\dot{\mathbf{M}}_2\mathbf{X}_1)^-\mathbf{X}_1'] = \mathscr{C}(\mathbf{W}\dot{\mathbf{M}}_2\mathbf{X}_1)$. Thus, in view of part (c) of Lemma 4, the statistic $\mathbf{Fy}$ is linearly sufficient for $\mathbf{X}_1\boldsymbol{\beta}_1$ if and only if

$$\mathscr{C}(\mathbf{W}\dot{\mathbf{M}}_2\mathbf{X}_1) \subset \mathscr{C}(\mathbf{WF}'). \tag{41}$$

From (41) we immediately see that $\mathbf{X}_1'\dot{\mathbf{M}}_2\mathbf{y}$ is linearly minimal sufficient for $\mathbf{X}_1\boldsymbol{\beta}_1$, as observed by [26, Theorem 2].

4 Properties of $\mathscr{C}(\mathbf{WF}')$

Consider the linear sufficiency condition

$$\mathscr{C}(\mathbf{X}) \subset \mathscr{C}(\mathbf{WF}'), \quad \text{where } \mathbf{W} \in \mathscr{W}. \tag{42}$$

One question: is the column space $\mathscr{C}(\mathbf{WF}')$ unique, i.e., does it remain invariant for any choice of $\mathbf{W} \in \mathscr{W}$? In statistical literature, the invariance of $\mathscr{C}(\mathbf{WF}')$ is not discussed. It might be somewhat tempting to conjecture that for a given $\mathbf{F}$, the column space $\mathscr{C}(\mathbf{WF}')$ would be invariant. However, our counterexample below shows that this is not the case. In any event, it is of interest to study the mathematical properties of the possible invariance.

Before our counterexample, we will take a quick look at the rank of $\mathbf{WF}'$ by allowing $\mathbf{W}$ to belong to set $\mathscr{W}_*$, defined as in (7),

$$\mathscr{W}_* = \{\mathbf{W} \in \mathbb{R}^{n\times n} : \mathbf{W} = \mathbf{V} + \mathbf{XTX}',\ \mathscr{C}(\mathbf{W}) = \mathscr{C}(\mathbf{X} : \mathbf{V})\}. \tag{43}$$

Now, on account of (5) and the equality $\mathscr{C}(\mathbf{W}) = \mathscr{C}(\mathbf{W}') = \mathscr{C}(\mathbf{X} : \mathbf{V})$, we have $\mathscr{C}(\mathbf{FW}') = \mathscr{C}(\mathbf{FW}) = \mathscr{C}[\mathbf{F}(\mathbf{X} : \mathbf{VM})]$. Using the rank rule for the partitioned matrix: $\operatorname{rank}(\mathbf{A} : \mathbf{B}) = \operatorname{rank}(\mathbf{A}) + \operatorname{rank}[(\mathbf{I} - \mathbf{P}_\mathbf{A})\mathbf{B}]$, see, e.g., [31, Theorem 19], we get

$$\operatorname{rank}(\mathbf{WF}') = \operatorname{rank}(\mathbf{FW}') = \operatorname{rank}(\mathbf{FW}) = \operatorname{rank}(\mathbf{FX}) + \operatorname{rank}(\mathbf{Q}_{\mathbf{FX}}\mathbf{FVM}), \tag{44}$$

where $\mathbf{Q}_{\mathbf{FX}} = \mathbf{I} - \mathbf{P}_{\mathbf{FX}}$. Now (44) means that $\operatorname{rank}(\mathbf{WF}')$ is invariant with respect to $\mathbf{W} \in \mathscr{W}_*$. In particular, if $\mathscr{C}(\mathbf{X}) \subset \mathscr{C}(\mathbf{WF}')$, we obtain

$$\begin{aligned}\operatorname{rank}(\mathbf{WF}') &= \operatorname{rank}(\mathbf{X} : \mathbf{WF}') = \operatorname{rank}(\mathbf{X}) + \operatorname{rank}(\mathbf{MWF}')\\ &= \operatorname{rank}(\mathbf{X}) + \operatorname{rank}(\mathbf{MVF}')\\ &= \operatorname{rank}(\mathbf{X} : \mathbf{VF}'). \end{aligned} \tag{45}$$

We can summarise our observations as follows:

Theorem 1 *Consider the linear model* $\mathscr{A} = \{\mathbf{y}, \mathbf{X}\boldsymbol{\beta}, \mathbf{V}\}$. *Then:*

(a) *The rank of* $\mathbf{W}\mathbf{F}'$ *is invariant for any* $\mathbf{W} \in \mathscr{W}_*$ *and it can be expressed as*

$$\operatorname{rank}(\mathbf{W}\mathbf{F}') = \operatorname{rank}(\mathbf{F}\mathbf{X}) + \operatorname{rank}(\mathbf{Q}_{\mathbf{F}\mathbf{X}}\mathbf{F}\mathbf{V}\mathbf{M}). \tag{46}$$

(b) *For any* $\mathbf{W} \in \mathscr{W}_*$, *the inclusion* $\mathscr{C}(\mathbf{X}) \subset \mathscr{C}(\mathbf{W}\mathbf{F}')$ *holds if and only if*

$$\operatorname{rank}(\mathbf{W}\mathbf{F}') = \operatorname{rank}(\mathbf{X}) + \operatorname{rank}(\mathbf{F}\mathbf{V}\mathbf{M}) = \operatorname{rank}(\mathbf{X} : \mathbf{V}\mathbf{F}'). \tag{47}$$

(c) *For any* $\mathbf{W} \in \mathscr{W}_*$, *we have* $\operatorname{rank}(\mathbf{W}'\mathbf{F}') = \operatorname{rank}(\mathbf{W}\mathbf{F}')$.

Example 1 Our purpose is to confirm that the following statement is not correct:
Let $\mathbf{W}_1, \mathbf{W}_2 \in \mathscr{W}$. Then for any matrix $\mathbf{F}$,

$$\mathscr{C}(\mathbf{W}_1\mathbf{F}') = \mathscr{C}(\mathbf{W}_2\mathbf{F}'). \tag{48}$$

Consider the model where

$$\mathbf{V} = \begin{pmatrix} 1 & 0 & 0 \\ 0 & 1 & 0 \\ 0 & 0 & 0 \end{pmatrix}, \quad \mathbf{X} = \begin{pmatrix} 1 & 0 \\ 0 & 0 \\ 0 & 1 \end{pmatrix}, \quad \mathbf{F}' = \begin{pmatrix} 1 \\ 0 \\ 0 \end{pmatrix}, \tag{49}$$

and let $\mathbf{U}_1\mathbf{U}_1' = \mathbf{I}_2,\ \mathbf{U}_2\mathbf{U}_2' = \left(\begin{smallmatrix} 2 & 1 \\ 1 & 1 \end{smallmatrix}\right)$. Denoting $\mathbf{W}_i = \mathbf{V} + \mathbf{X}\mathbf{U}_i\mathbf{U}_i'\mathbf{X}'$, we have

$$\mathscr{C}(\mathbf{W}_1\mathbf{F}') = \mathscr{C}\begin{pmatrix} 1 \\ 0 \\ 0 \end{pmatrix} \neq \mathscr{C}(\mathbf{W}_2\mathbf{F}') = \mathscr{C}\begin{pmatrix} 3 \\ 0 \\ 1 \end{pmatrix}, \tag{50}$$

and hence the statement (48) is not correct. □

It is interesting to observe that in the above Example 1 the linear sufficiency condition $\mathscr{C}(\mathbf{X}) \subset \mathscr{C}(\mathbf{W}\mathbf{F}')$ does not hold. Actually $\mathbf{X}\boldsymbol{\beta}$ is not even estimable under the transformed model $\mathscr{A}_t$ since $\operatorname{rank}(\mathbf{X}'\mathbf{F}') \neq \operatorname{rank}(\mathbf{X})$. For $\mathbf{F}\mathbf{y}$ to be linearly sufficient it is necessary that $\operatorname{rank}(\mathbf{X}) \leq \operatorname{rank}(\mathbf{F})$, which in this case would mean $\operatorname{rank}(\mathbf{F}) \geq 2$. Consider the Example 1 by extending the matrix $\mathbf{F}'$ by one column:

$$\mathbf{F}' = \begin{pmatrix} 1 & 0 \\ 0 & 0 \\ 0 & 1 \end{pmatrix} = \mathbf{X}. \tag{51}$$

Then we immediately observe that $\mathscr{C}(\mathbf{W}_1\mathbf{F}') = \mathscr{C}(\mathbf{W}_2\mathbf{F}')$. Actually,

$$\mathscr{C}(\mathbf{X}) = \mathscr{C}(\mathbf{W}_i\mathbf{F}') = \mathscr{C}(\mathbf{W}_i\mathbf{X}),\ i = 1, 2, \tag{52}$$

implying that in this situation $\mathbf{Fy} = \mathbf{X}'\mathbf{y}$ is linearly minimal sufficient for $\mathbf{X}\boldsymbol{\beta}$. This provokes the following questions:

(A) When is $\mathbf{X}'\mathbf{y}$ linearly sufficient for $\mathbf{X}\boldsymbol{\beta}$?
(B) What can be said about $\mathscr{C}(\mathbf{WF}')$ in such a case when $\text{rank}(\mathbf{X}'\mathbf{F}') = \text{rank}(\mathbf{X})$, i.e., $\mathbf{X}\boldsymbol{\beta}$ is estimable under $\mathscr{A}_t$?
(C) Is $\mathscr{C}(\mathbf{WF}')$ invariant for any choice of $\mathbf{W}$ if $\mathbf{Fy} \in \text{S}(\mathbf{X}\boldsymbol{\beta})$?

Let us first take a look at the problem (A). Now $\mathbf{X}'\mathbf{y}$ is linearly sufficient for $\mathbf{X}\boldsymbol{\beta}$ if and only if $\mathscr{C}(\mathbf{X}) \subset \mathscr{C}(\mathbf{WX})$, which, in light of $\text{rank}(\mathbf{WX}) = \text{rank}(\mathbf{X})$, becomes equality

$$\mathscr{C}(\mathbf{X}) = \mathscr{C}(\mathbf{WX}). \tag{53}$$

The column space equality (53) holds if and only if

$$\mathbf{HWX} = \mathbf{WX}, \tag{54}$$

where $\mathbf{H} = \mathbf{P}_{\mathbf{X}}$. Now (54) can be equivalently expressed as

$$\mathbf{HV} = \mathbf{VH}, \tag{55}$$

which is the well-known condition for the equality of the $\text{OLSE}(\mathbf{X}\boldsymbol{\beta}) = \mathbf{Hy}$ and $\text{BLUE}(\mathbf{X}\boldsymbol{\beta})$ under the model $\mathscr{A}$; see, e.g., [36] and [37, Chap. 10]. We can express our conclusion as follows:

Theorem 2 *The statistic* $\mathbf{X}'\mathbf{y}$ *is linearly sufficient for* $\mathbf{X}\boldsymbol{\beta}$ *under the model* $\mathscr{A} = \{\mathbf{y}, \mathbf{X}\boldsymbol{\beta}, \mathbf{V}\}$ *if and only if*

$$\text{OLSE}(\mathbf{X}\boldsymbol{\beta}) = \text{BLUE}(\mathbf{X}\boldsymbol{\beta}). \tag{56}$$

In this situation $\mathbf{X}'\mathbf{y}$ *is linearly minimal sufficient.*

The corresponding result as in Theorem 2, for a positive definite $\mathbf{V}$, appears also in [7, p. 913]. We recall that expression (56) is supposed to hold with probability 1, just like any other equality between estimators.

Example 2 As a reply to question (B) above, let us consider the situation where

$$\mathbf{V} = \begin{pmatrix} 1 & 0 & 0 \\ 0 & 0 & 0 \\ 0 & 0 & 1 \end{pmatrix}, \quad \mathbf{X} = \begin{pmatrix} 1 & 0 \\ 0 & 1 \\ 0 & 0 \end{pmatrix}, \quad \mathbf{F}' = \begin{pmatrix} 1 & 0 \\ 0 & 1 \\ 1 & 0 \end{pmatrix}. \tag{57}$$

In this situation the estimability condition $\text{rank}(\mathbf{FX}) = \text{rank}(\mathbf{X})$ holds but $\mathbf{Fy}$ is not linearly sufficient for $\mathbf{X}\boldsymbol{\beta}$. Choosing $\mathbf{U}_1\mathbf{U}_1' = \mathbf{I}_2$, $\mathbf{U}_2\mathbf{U}_2' = \left(\begin{smallmatrix} 0 & 0 \\ 0 & 1 \end{smallmatrix}\right)$, and denoting $\mathbf{W}_i = \mathbf{V} + \mathbf{XU}_i\mathbf{U}_i'\mathbf{X}'$, we have

$$\mathscr{C}(\mathbf{W}_1\mathbf{F}') = \mathscr{C}\begin{pmatrix} 2 & 0 \\ 0 & 1 \\ 1 & 0 \end{pmatrix} \neq \mathscr{C}(\mathbf{W}_2\mathbf{F}') = \mathscr{C}\begin{pmatrix} 1 & 0 \\ 0 & 1 \\ 1 & 0 \end{pmatrix}. \tag{58}$$

Thus the estimability condition is not enough for the invariance of $\mathscr{C}(\mathbf{WF}')$. □

The following theorem is a reply to question (C) above. However, we formulate it in a more general setup by using the set $\mathscr{W}_*$ of $\mathbf{W}$-matrices defined by (7) instead of $\mathscr{W}$.

Theorem 3 *Consider the linear model $\mathscr{A} = \{\mathbf{y}, \mathbf{X}\boldsymbol{\beta}, \mathbf{V}\}$, let $\mathbf{W} \in \mathscr{W}_*$ and suppose that $\mathscr{C}(\mathbf{X}) \subset \mathscr{C}(\mathbf{WF}')$. Then the column space $\mathscr{C}(\mathbf{WF}')$ is invariant for any choice of $\mathbf{W} \in \mathscr{W}_*$ and*

$$\mathscr{C}(\mathbf{WF}') = \mathscr{C}(\mathbf{X}) \oplus \mathscr{C}(\mathbf{MVF}') = \mathscr{C}(\mathbf{W}'\mathbf{F}'). \tag{59}$$

Proof Suppose that $\mathscr{C}(\mathbf{X}) \subset \mathscr{C}(\mathbf{WF}')$. Then

$$\mathscr{C}(\mathbf{WF}') = \mathscr{C}(\mathbf{X} : \mathbf{WF}') = \mathscr{C}(\mathbf{X}) \oplus \mathscr{C}(\mathbf{MVF}'), \tag{60}$$

and the proof is completed. □

Next we present the following extended version of Lemma 3:

Theorem 4 *Let $\mathbf{W} \in \mathscr{W}_*$. Then the statistic $\mathbf{Fy}$ is linearly sufficient for $\mathbf{X}\boldsymbol{\beta}$ under the linear model $\mathscr{A} = \{\mathbf{y}, \mathbf{X}\boldsymbol{\beta}, \mathbf{V}\}$ if and only if*

$$\mathscr{C}(\mathbf{X}) \subset \mathscr{C}(\mathbf{WF}'), \tag{61}$$

or, equivalently,

$$\mathscr{C}(\mathbf{X}) \subset \mathscr{C}(\mathbf{W}'\mathbf{F}'). \tag{62}$$

Proof The proof is parallel to that of [7, p. 914] who utilize the fact that $\mathbf{By}$ is a BLUE of estimable $\mathbf{K}\boldsymbol{\beta}$ if and only if

$$\mathbf{BW} = \mathbf{K}(\mathbf{X}'\mathbf{W}^{+}\mathbf{X})^{+}\mathbf{X}', \quad \text{where } \mathbf{W} \in \mathscr{W}. \tag{63}$$

However, it is easy to confirm, using (8a)–(8e), that in this condition the set $\mathscr{W}$ can be replaced with $\mathscr{W}_*$. Moreover, if $\mathbf{W} \in \mathscr{W}_*$, then also $\mathbf{W}' \in \mathscr{W}_*$ and (63) can be replaced with

$$\mathbf{BW}' = \mathbf{K}[\mathbf{X}'(\mathbf{W}')^{+}\mathbf{X}]^{+}\mathbf{X}'. \tag{64}$$

Proceeding then along the same lines as [7], we observe that $\mathbf{AFy}$ is the BLUE for $\mathbf{X}\boldsymbol{\beta}$ under $\mathscr{A} = \{\mathbf{y}, \mathbf{X}\boldsymbol{\beta}, \mathbf{V}\}$ if and only if

$$\mathbf{AFW}' = \mathbf{X}[\mathbf{X}'(\mathbf{W}')^{+}\mathbf{X}]^{+}\mathbf{X}'. \tag{65}$$

Now (65) has a solution for $\mathbf{A}$, i.e., $\mathbf{Fy}$ is linearly sufficient for $\mathbf{X}\boldsymbol{\beta}$, if and only if

$$\mathscr{C}[\mathbf{X}(\mathbf{X}'\mathbf{W}^{+}\mathbf{X})^{+}\mathbf{X}'] \subset \mathscr{C}(\mathbf{WF}'). \tag{66}$$

Using (8a)–(8e), we observe that $\mathscr{C}[\mathbf{X}(\mathbf{X}'\mathbf{W}^{+}\mathbf{X})^{+}\mathbf{X}'] = \mathscr{C}(\mathbf{X})$ and so we have obtained (61). Notice also that in light of Theorem 3, the statements (61) and (62) are equivalent. □

According to our knowledge, in all linear sufficiency considerations appearing in literature, it is assumed that $\mathbf{W}$ is nonnegative definite. However, this is not necessary, and $\mathbf{W}$ can also be nonsymmetric. Of course, sometimes it is simpler to have $\mathbf{W}$ from set $\mathscr{W}$.

Remark 1 There is one feature in the paper of [7] that is worth special attention. Namely in their considerations they need the "$\mathbf{W}$-matrix" in the transformed model $\mathscr{A}_t = \{\mathbf{Fy}, \mathbf{FX}\boldsymbol{\beta}, \mathbf{FVF}'\}$. The appropriate set is the following:

$$\mathscr{W}_t = \{\mathbf{W}_t : \mathbf{W}_t = \mathbf{F}(\mathbf{V} + \mathbf{XSX}')\mathbf{F}',\ \mathscr{C}(\mathbf{W}_t) = \mathscr{C}[\mathbf{F}(\mathbf{X} : \mathbf{V})]\}. \tag{67}$$

Let $\mathbf{W} = \mathbf{V} + \mathbf{XSX}'$ be some matrix from $\mathscr{W}_*$, and so $\mathbf{W}$ may not be nonnegative definite. We then have

$$\mathscr{C}(\mathbf{W}_t) = \mathscr{C}(\mathbf{FWF}') \subset \mathscr{C}(\mathbf{FW}) = \mathscr{C}[\mathbf{F}(\mathbf{X} : \mathbf{V})]. \tag{68}$$

If $\mathbf{W}$ is nonnegative definite, as [7] have, then we have equality in (68). However, if $\mathbf{W}$ belongs to $\mathscr{W}_*$ and is not nonnegative definite, then we must add the condition

$$\operatorname{rank}(\mathbf{FWF}') = \operatorname{rank}(\mathbf{FW}) \tag{69}$$

if we want to have $\mathbf{FWF}' \in \mathscr{W}_t$. Thus one representation for the BLUE of $\mathbf{FX}\boldsymbol{\beta}$ under $\mathscr{A}_t$ is

$$\mathbf{FX}[\mathbf{X}'\mathbf{F}'(\mathbf{FWF}')^{-}\mathbf{FX}]^{-}\mathbf{X}'\mathbf{F}'(\mathbf{FWF}')^{-}\mathbf{Fy}, \tag{70}$$

where $\mathbf{W} \in \mathscr{W}_*$ and $\mathbf{W}$ satisfies (69). □

5 Comments on the Relative Linear Sufficiency

When studying the relative efficiency of OLSE versus BLUE of $\boldsymbol{\beta}$ we are dealing with two linear models

$$\mathscr{A} = \{\mathbf{y}, \mathbf{X}\boldsymbol{\beta}, \mathbf{V}\}, \quad \mathscr{A}_{\mathrm{I}} = \{\mathbf{y}, \mathbf{X}\boldsymbol{\beta}, \mathbf{I}_n\}, \tag{71}$$

where the corresponding BLUEs are

$$\tilde{\boldsymbol{\beta}} = (\mathbf{X}'\mathbf{V}^{-1}\mathbf{X})^{-1}\mathbf{X}'\mathbf{V}^{-1}\mathbf{y}, \quad \hat{\boldsymbol{\beta}} = (\mathbf{X}'\mathbf{X})^{-1}\mathbf{X}'\mathbf{y}. \tag{72}$$

Then it is assumed that model $\{\mathbf{y}, \mathbf{X}\boldsymbol{\beta}, \mathbf{V}\}$ is correct and then the relative goodness of $\hat{\boldsymbol{\beta}}$ with respect to $\tilde{\boldsymbol{\beta}}$ is measured by various means. The most common measure is the Watson efficiency, see [16, 41],

$$\phi = \frac{|\mathrm{cov}(\tilde{\boldsymbol{\beta}})|}{|\mathrm{cov}(\hat{\boldsymbol{\beta}})|} = \frac{|\mathbf{X}'\mathbf{X}|^2}{|\mathbf{X}'\mathbf{V}\mathbf{X}| \cdot |\mathbf{X}'\mathbf{V}^{-1}\mathbf{X}|}, \tag{73}$$

where $|\cdot|$ refers to the determinant. Obviously $0 < \phi \leq 1$ and the upper bound is attained when $\tilde{\boldsymbol{\beta}} = \hat{\boldsymbol{\beta}}$.

Let us consider the models

$$\mathscr{A} = \{\mathbf{y}, \mathbf{X}\boldsymbol{\beta}, \mathbf{V}\}, \quad \mathscr{A}_t = \{\mathbf{F}\mathbf{y}, \mathbf{F}\mathbf{X}\boldsymbol{\beta}, \mathbf{F}\mathbf{V}\mathbf{F}'\}, \tag{74}$$

and try to do something similar with

$$\mathrm{BLUE}(\boldsymbol{\beta} \mid \mathscr{A}) = \tilde{\boldsymbol{\beta}} = (\mathbf{X}'\mathbf{V}^{-1}\mathbf{X})^{-1}\mathbf{X}'\mathbf{V}^{-1}\mathbf{y}, \tag{75}$$

$$\mathrm{BLUE}(\boldsymbol{\beta} \mid \mathscr{A}_t) = \tilde{\boldsymbol{\beta}}_t = [\mathbf{X}'\mathbf{F}'(\mathbf{F}\mathbf{V}\mathbf{F}')^{-}\mathbf{F}\mathbf{X}]^{-1}\mathbf{X}'\mathbf{F}'(\mathbf{F}\mathbf{V}\mathbf{F}')^{-}\mathbf{F}\mathbf{y}. \tag{76}$$

Above we have some rank problems. To simplify the considerations, we have assumed that $\mathbf{V}$ is positive definite. The model matrix $\mathbf{X}$ has to have full column rank so that $\boldsymbol{\beta}$ would be estimable under $\mathscr{A}$. Similarly, $\mathbf{FX}$ has to have full column rank for $\boldsymbol{\beta}$ to be estimable under $\mathscr{A}_t$; using the rank rule of [31, Corollary 6.2] for the matrix product, we must have

$$p = \mathrm{rank}(\mathbf{X}) = \mathrm{rank}(\mathbf{FX}) = \mathrm{rank}(\mathbf{X}) - \dim \mathscr{C}(\mathbf{X}) \cap \mathscr{C}(\mathbf{F}')^{\perp}, \tag{77}$$

so that

$$\mathscr{C}(\mathbf{X}) \cap \mathscr{C}(\mathbf{F}')^{\perp} = \{\mathbf{0}\}. \tag{78}$$

It is noteworthy that in view of $\mathscr{C}(\mathbf{FX}) \subset \mathscr{C}(\mathbf{FVF}') = \mathscr{C}(\mathbf{F})$ the model $\mathscr{A}_t = \{\mathbf{Fy}, \mathbf{FX}\boldsymbol{\beta}, \mathbf{FVF}'\}$ is so-called weakly singular linear, or Zyskind–Martin model, see [42], and hence the representation (76) indeed is valid for any $(\mathbf{FVF}')^{-}$. Moreover, it is easy to confirm that $\mathbf{X}'\mathbf{F}'(\mathbf{FVF}')^{-}\mathbf{FX}$ is positive definite.

Notice that $\mathrm{E}(\tilde{\boldsymbol{\beta}}) = \mathrm{E}(\tilde{\boldsymbol{\beta}}_t) = \boldsymbol{\beta}$ and

$$\mathrm{cov}(\tilde{\boldsymbol{\beta}}_t) = [\mathbf{X}'\mathbf{F}'(\mathbf{FVF}')^{-}\mathbf{FX}]^{-1}, \quad \mathrm{cov}(\tilde{\boldsymbol{\beta}}) = (\mathbf{X}'\mathbf{V}^{-1}\mathbf{X})^{-1}. \tag{79}$$

Remark 2 The following Löwner ordering obviously holds:

$$\mathrm{cov}(\tilde{\boldsymbol{\beta}}) \leq_{\mathrm{L}} \mathrm{cov}(\tilde{\boldsymbol{\beta}}_t), \tag{80}$$

i.e.,

$$(\mathbf{X}'\mathbf{V}^{-1}\mathbf{X})^{-1} \leq_{\mathrm{L}} [\mathbf{X}'\mathbf{F}'(\mathbf{FVF}')^{-}\mathbf{FX}]^{-1}. \tag{81}$$

We can rewrite (81) as

$$\mathbf{X}'\mathbf{V}^{-1/2}\mathbf{P}_{\mathbf{V}^{1/2}\mathbf{F}'}\mathbf{V}^{-1/2}\mathbf{X} \leq_{\mathrm{L}} \mathbf{X}'\mathbf{V}^{-1/2}\mathbf{V}^{-1/2}\mathbf{X}, \tag{82}$$

where the equality is obtained if and only if $\mathscr{C}(\mathbf{V}^{-1/2}\mathbf{X}) \subset \mathscr{C}(\mathbf{V}^{1/2}\mathbf{F}')$, i.e., $\mathscr{C}(\mathbf{X}) \subset \mathscr{C}(\mathbf{V}\mathbf{F}')$, which is precisely the condition for linear sufficiency (when $\mathbf{V}$ is positive definite). □

Corresponding to Watson efficiency, we could consider the ratio

$$\begin{aligned}\gamma = \frac{|\mathrm{cov}(\tilde{\boldsymbol{\beta}})|}{|\mathrm{cov}(\tilde{\boldsymbol{\beta}}_t)|} &= \frac{|(\mathbf{X}'\mathbf{V}^{-1}\mathbf{X})^{-1}|}{|[\mathbf{X}'\mathbf{F}'(\mathbf{F}\mathbf{V}\mathbf{F}')^{-}\mathbf{F}\mathbf{X}]^{-1}|} \\ &= \frac{|\mathbf{X}'\mathbf{F}'(\mathbf{F}\mathbf{V}\mathbf{F}')^{-}\mathbf{F}\mathbf{X}|}{|\mathbf{X}'\mathbf{V}^{-1}\mathbf{X}|} \\ &= \frac{|\mathbf{X}'\mathbf{V}^{-1/2}\mathbf{P}_{\mathbf{V}^{1/2}\mathbf{F}'}\mathbf{V}^{-1/2}\mathbf{X}|}{|\mathbf{X}'\mathbf{V}^{-1/2}\mathbf{V}^{-1/2}\mathbf{X}|}.\end{aligned} \tag{83}$$

Clearly

$$0 < \gamma \leq 1, \tag{84}$$

where the upper bound is attained if and only if $\mathbf{F}\mathbf{y}$ is linearly sufficient for $\boldsymbol{\beta}$. What might be the lower bound? Here we now keep $\mathbf{X}$ and $\mathbf{V}$ given and try to figure out which $\mathbf{F}$ yields the minimum of γ subject to the condition $\mathrm{rank}(\mathbf{X}) = \mathrm{rank}(\mathbf{F}\mathbf{X})$. The lower bound for the Watson efficiency was found by [16] (actually it appeared already in [41] but there was a flaw in the proof). However, it seems to be nontrivial to find the lower bound for γ. The (attainable) lower bound zero does not make sense, of course.

Remark 3 Consider matrices $\mathbf{F}_1$ and $\mathbf{F}_2$ and the corresponding transformed models

$$\mathscr{A}_{ti} = \{\mathbf{F}_i\mathbf{y}, \mathbf{F}_i\mathbf{X}\boldsymbol{\beta}, \mathbf{F}_i\mathbf{V}\mathbf{F}_i'\},\ i = 1, 2, \tag{85}$$

and suppose that $\mathrm{rank}(\mathbf{F}_1\mathbf{X}) = \mathrm{rank}(\mathbf{F}_2\mathbf{X}) = \mathrm{rank}(\mathbf{X}) = p$, so that β is estimable under both models. We observe that the Löwner ordering

$$\mathrm{cov}(\tilde{\boldsymbol{\beta}}_{t1}) \leq_{\mathrm{L}} \mathrm{cov}(\tilde{\boldsymbol{\beta}}_{t2}) \tag{86}$$

holds if and only if

$$\mathbf{X}'\mathbf{V}^{-1/2}\mathbf{P}_{\mathbf{V}^{1/2}\mathbf{F}_2'}\mathbf{V}^{-1/2}\mathbf{X} \leq_{\mathrm{L}} \mathbf{X}'\mathbf{V}^{-1/2}\mathbf{P}_{\mathbf{V}^{1/2}\mathbf{F}_1'}\mathbf{V}^{-1/2}\mathbf{X}, \tag{87}$$

i.e.,

$$\mathbf{X}'\mathbf{V}^{-1/2}(\mathbf{P}_{\mathbf{V}^{1/2}\mathbf{F}_1'} - \mathbf{P}_{\mathbf{V}^{1/2}\mathbf{F}_2'})\mathbf{V}^{-1/2}\mathbf{X} \geq_{\mathrm{L}} \mathbf{0}. \tag{88}$$

The matrix $\mathbf{P}_{\mathbf{V}^{1/2}\mathbf{F}_1'} - \mathbf{P}_{\mathbf{V}^{1/2}\mathbf{F}_2'}$ is nonnegative definite if and only if

$$\mathscr{C}(\mathbf{F}_2') \subset \mathscr{C}(\mathbf{F}_1'). \tag{89}$$

Hence we can conclude that (86) holds if $\mathscr{C}(\mathbf{F}_2') \subset \mathscr{C}(\mathbf{F}_1')$. In this case we can say that in a sense $\mathbf{F}_1\mathbf{y}$ is "more than or equally linearly sufficient" than $\mathbf{F}_2\mathbf{y}$ even though neither of them need to be "fully linearly sufficient". Notice that if $\mathscr{C}(\mathbf{F}_1') = \mathbb{R}^n$, i.e., $\mathbf{F}_1$ is a nonsingular $n \times n$ matrix, then $\mathrm{cov}(\tilde{\boldsymbol{\beta}}_{t1})$ is the smallest in the Löwner sense in the set of $\mathrm{cov}(\tilde{\boldsymbol{\beta}}_t)$: it is $\mathrm{cov}(\tilde{\boldsymbol{\beta}})$.

However, it may well be that there is no Löwner ordering between the covariance matrices $\mathrm{cov}(\tilde{\boldsymbol{\beta}}_{t1})$ and $\mathrm{cov}(\tilde{\boldsymbol{\beta}}_{t2})$. Then some other criteria should be used to compare the "linear sufficiency" of $\mathbf{F}_1\mathbf{y}$ and $\mathbf{F}_2\mathbf{y}$. □

Bloomfield, Watson [16] introduced also another measure of efficiency of the OLSE, based on the Frobenius norm of the commutator $\mathbf{HV} - \mathbf{VH}$:

$$\delta = \tfrac{1}{2}\,\|\mathbf{HV} - \mathbf{VH}\|_F^2 = \|\mathbf{HVM}\|_F^2 = \mathrm{tr}(\mathbf{HVMVH}), \tag{90}$$

where $\mathrm{tr}(\cdot)$ refers to the trace. They showed that the maximum of δ is attained in the same situation as the minimum of the Watson efficiency ϕ. Of course, $\delta = 0$ if and only if $\mathrm{OLSE}(\mathbf{X}\boldsymbol{\beta})$ equals $\mathrm{BLUE}(\mathbf{X}\boldsymbol{\beta})$.

We can now try to develop something similar as the commutator criterion for the linear sufficiency condition $\mathscr{C}(\mathbf{X}) \subset \mathscr{C}(\mathbf{WF}')$ which is equivalent to

$$\mathbf{P}_{\mathbf{WF}'}\mathbf{X} = \mathbf{X}. \tag{91}$$

Hence one can wonder how "badly" (42) is satisfied by considering the difference

$$\mathbf{D} := \mathbf{X} - \mathbf{P}_{\mathbf{WF}'}\mathbf{X}. \tag{92}$$

The "size" of $\mathbf{D}$ could be measured by the Frobenius norm as

$$\|\mathbf{D}\|_F^2 = \mathrm{tr}(\mathbf{D}'\mathbf{D}) = \mathrm{tr}(\mathbf{X}'\mathbf{X}) - \mathrm{tr}(\mathbf{X}'\mathbf{P}_{\mathbf{WF}'}\mathbf{X}). \tag{93}$$

Hence the relative linear sufficiency of $\mathbf{Fy}$ could be defined as

$$\psi = \frac{\mathrm{tr}(\mathbf{X}'\mathbf{P}_{\mathbf{WF}'}\mathbf{X})}{\mathrm{tr}(\mathbf{X}'\mathbf{X})}. \tag{94}$$

Now

$$0 \le \psi \le 1, \tag{95}$$

where the lower bound is attained when $\mathscr{C}(\mathbf{X}) \subset \mathscr{C}(\mathbf{WF}')^{\perp}$ and the upper bound is attained when $\mathscr{C}(\mathbf{X}) \subset \mathscr{C}(\mathbf{WF}')$, i.e., when $\mathbf{Fy}$ is linearly sufficient for $\mathbf{X}\boldsymbol{\beta}$.

6 Euclidean Norm of the Difference Between the BLUEs Under $\mathscr{A}$ and $\mathscr{A}_t$

In this section we will study the properties of the Euclidean norm of the difference between the BLUEs of $\boldsymbol{\mu} := \mathbf{X}\boldsymbol{\beta}$ under the models $\mathscr{A}$ and $\mathscr{A}_t$. We can denote shortly

$$\mathrm{BLUE}(\mathbf{X}\boldsymbol{\beta} \mid \mathscr{A}) = \tilde{\boldsymbol{\mu}}, \quad \text{and} \quad \mathrm{BLUE}(\mathbf{X}\boldsymbol{\beta} \mid \mathscr{A}_t) = \tilde{\boldsymbol{\mu}}_t. \tag{96}$$

The corresponding considerations for $\mathrm{OLSE}(\mathbf{X}\boldsymbol{\beta}) - \mathrm{BLUE}(\mathbf{X}\boldsymbol{\beta})$ have been made by [5, 6] and for the BLUEs under two models by [25]; see also [14, 24, 33–35].

Suppose that $\mathbf{W} \in \mathscr{W}$. Then the BLUE under the original model $\mathscr{A}$ can be expressed as $\mathbf{Gy}$ where

$$\mathbf{G} = \mathbf{X}(\mathbf{X}'\mathbf{W}^-\mathbf{X})^-\mathbf{X}'\mathbf{W}^-. \tag{97}$$

Moreover, assuming that $\mathbf{X}\boldsymbol{\beta}$ is estimable under the transformed model $\mathscr{A}_t$, the estimator $\mathbf{BFy}$ is the BLUE for $\mathbf{X}\boldsymbol{\beta}$ under $\mathscr{A}_t$ if and only if $\mathbf{B}$ satisfies

$$\mathbf{B}[\mathbf{FX} : \mathbf{FVF}'(\mathbf{FX})^{\perp}] = (\mathbf{X} : \mathbf{0}). \tag{98}$$

One choice for $\mathbf{B}$ is $\mathbf{X}[\mathbf{X}'\mathbf{F}'(\mathbf{FWF}')^-\mathbf{FX}]^-\mathbf{X}'\mathbf{F}'(\mathbf{FWF}')^-$ and so the BLUE of $\mathbf{X}\boldsymbol{\beta}$ under $\mathscr{A}_t$ has representation $\mathbf{G}_t\mathbf{y}$, where

$$\mathbf{G}_t = \mathbf{X}[\mathbf{X}'\mathbf{F}'(\mathbf{FWF}')^-\mathbf{FX}]^-\mathbf{X}'\mathbf{F}'(\mathbf{FWF}')^-\mathbf{F}. \tag{99}$$

We observe that $\mathbf{G}_t\mathbf{G} = \mathbf{G}$ and hence for all $\mathbf{y} \in \mathscr{C}(\mathbf{W})$ we have

$$\begin{aligned}(\mathbf{G}_t - \mathbf{G})\mathbf{y} &= (\mathbf{G}_t - \mathbf{G}_t\mathbf{G})\mathbf{y} \\ &= \mathbf{G}_t(\mathbf{I}_n - \mathbf{G})\mathbf{y} \\ &= \mathbf{G}_t\mathbf{VM}(\mathbf{MVM})^-\mathbf{My},\end{aligned} \tag{100}$$

where we have used (12), i.e.,

$$(\mathbf{I}_n - \mathbf{G})\mathbf{y} = \mathbf{VM}(\mathbf{MVM})^-\mathbf{My} \quad \text{for all } \mathbf{y} \in \mathscr{C}(\mathbf{W}). \tag{101}$$

Notice that in view of (13), the expression $\mathbf{VM}(\mathbf{MVM})^-\mathbf{My}$ is invariant for the choice of $(\mathbf{MVM})^-$ for all $\mathbf{y} \in \mathscr{C}(\mathbf{W})$.

The Euclidean norm of vector $\mathbf{a}$ is of course $\|\mathbf{a}\|_2 = \sqrt{\mathbf{a}'\mathbf{a}}$ and the corresponding matrix norm (spectral norm) $\|\mathbf{A}\|_2$ is defined as the square root of the largest eigenvalue of $\mathbf{A}'\mathbf{A}$. Then, for all $\mathbf{y} \in \mathscr{C}(\mathbf{W})$, we have

$$\begin{aligned}\|\mathbf{G}_t\mathbf{y} - \mathbf{Gy}\|_2^2 &= \|\mathbf{G}_t\mathbf{VM}(\mathbf{MVM})^-\mathbf{My}\|_2^2 \\ &\le \|\mathbf{G}_t\mathbf{VM}\|_2^2\ \|(\mathbf{MVM})^+\|_2^2\ \|\mathbf{My}\|_2^2.\end{aligned} \tag{102}$$

The inequality in (102) follows from the consistency and multiplicativity of the matrix norm $\|\mathbf{A}\|_2$; see, e.g., [15, pp. 19–20].

The special situation when $\mathbf{VM} = \mathbf{0}$, i.e., $\mathscr{C}(\mathbf{V}) \subset \mathscr{C}(\mathbf{X})$, deserves some attention. Notice also, as pointed out by [24, p. 554], that $\mathbf{y}'\mathbf{My} = 0$ for all $\mathbf{y} \in \mathscr{C}(\mathbf{X} : \mathbf{V})$ holds if and only if $\mathbf{VM} = \mathbf{0}$. [21, p. 317] calls a model with property $\mathbf{VM} = \mathbf{0}$ a *degenerated* model. If $\mathscr{A}$ is not a degenerated model then the right-hand side of (102) is zero if and only if

$$\mathbf{G}_t\mathbf{VM} = \mathbf{0}. \tag{103}$$

Noticing that obviously $\mathbf{G}_t$ satisfies $\mathbf{G}_t\mathbf{X} = \mathbf{X}$, we can conclude that (103) means that $\mathbf{G}_t\mathbf{y}$ is a BLUE also under the original model $\mathscr{A}$. Thus, in light of Lemma 5, (103) means also that $\mathbf{Fy}$ is linearly sufficient.

Thus we have proved the following:

Theorem 5 *Suppose that $\boldsymbol{\mu} = \mathbf{X}\boldsymbol{\beta}$ is estimable under the transformed model $\mathscr{A}_t$. Then, using the above notation,*

$$\begin{aligned} \|\tilde{\boldsymbol{\mu}}_t - \tilde{\boldsymbol{\mu}}\|_2^2 &\leq \|\mathbf{G}_t\mathbf{VM}\|_2^2 \, \|(\mathbf{MVM})^+\|_2^2 \, \mathbf{y}'\mathbf{My} \\ &= \frac{a}{\alpha^2}\,\mathbf{y}'\mathbf{My}, \end{aligned} \tag{104}$$

where α is the smallest nonzero eigenvalue of $\mathbf{MVM}$, and a is the largest eigenvalue of $\mathbf{G}_t\mathbf{VMVG}_t'$. Moreover, if $\mathscr{A}$ is not a degenerated model then the right-hand side of (104) *is zero if and only if $\mathbf{Fy}$ is linearly sufficient for $\mathbf{X}\boldsymbol{\beta}$.*

7 Conclusions

The origins of the idea of transforming $\mathscr{A} = \{\mathbf{y}, \mathbf{X}\boldsymbol{\beta}, \mathbf{V}\}$ by a matrix $\mathbf{F}$ of order $f \times n$ follow from a desire of reduction of the initial information delivered by an observed value of a random vector variable $\mathbf{y}$ in such a way that it is still possible to obtain the BLUE of $\mathbf{X}\boldsymbol{\beta}$ from the transformed model $\mathscr{A}_t = \{\mathbf{Fy}, \mathbf{FX}\boldsymbol{\beta}, \mathbf{FVF}'\}$. Hence the concept of the linear sufficiency has an essential role when studying the connection between $\mathscr{A}$ and its transformed version $\mathscr{A}_t$.

In the theory of linear models the classes of matrices

$$\mathscr{W} = \{\mathbf{W} \in \mathbb{R}^{n\times n} : \mathbf{W} = \mathbf{V} + \mathbf{XUU}'\mathbf{X}',\ \mathscr{C}(\mathbf{W}) = \mathscr{C}(\mathbf{X} : \mathbf{V})\}, \tag{105a}$$

$$\mathscr{W}_* = \{\mathbf{W} \in \mathbb{R}^{n\times n} : \mathbf{W} = \mathbf{V} + \mathbf{XTX}',\ \mathscr{C}(\mathbf{W}) = \mathscr{C}(\mathbf{X} : \mathbf{V})\}, \tag{105b}$$

have important roles. In our paper we study in details the properties of these $\mathbf{W}$-matrices related to the concept of linear sufficiency. As far as we know, in all linear sufficiency considerations appearing in literature, it is assumed that $\mathbf{W}$ is nonnegative definite, i.e., $\mathbf{W}$ belongs to set $\mathscr{W}$. We have shown that this is not necessary: it is enough if $\mathbf{W}$ belongs to set $\mathscr{W}_*$.

If $\mathbf{Fy}$ is linearly sufficient then the BLUEs of $\mathbf{X}\boldsymbol{\beta}$ under $\mathscr{A}$ and under $\mathscr{A}_t$ are equal (with probability 1). Hence it might be of interest to describe the relative linear sufficiency of $\mathbf{Fy}$ by comparing the BLUEs under $\mathscr{A}$ and under $\mathscr{A}_t$ by some means. Some suggestions on this matter are made in Sect. 5. The applicability of these measures is left for further research.

Acknowledgements Part of this research was done during the meetings of an International Research Group on Multivariate Models in the Mathematical Research and Conference Center, Będlewo, Poland, March 2015 and October 2015, supported by the Stefan Banach International Mathematical Center.

References

1. Aitken, A.C.: On least squares and linear combination of observations. Proc. R. Soc. Edinburgh Sect. A **55**, 42–49 (1935)
2. Arendacká, B., Puntanen, S.: Further remarks on the connection between fixed linear model and mixed linear model. Stat. Pap. **56**, 1235–1247 (2015)
3. Baksalary, J.K.: An elementary development of the equation characterizing best linear unbiased estimators. Linear Algebra Appl. **388**, 3–6 (2004)
4. Baksalary, J.K., Drygas, H.: A note on the concepts of sufficiency in the general Gauss–Markov model: a coordinate-free approach. Forschungsbericht 92/2, Universität Dortmund, Fachbereich Statistik (1992)
5. Baksalary, J.K., Kala, R.: A bound for the Euclidean norm of the difference between the least squares and the best linear unbiased estimators. Ann. Stat. **6**, 1390–1393 (1978)
6. Baksalary, J.K., Kala, R.: A new bound for the Euclidean norm of the difference between the least squares and the best linear unbiased estimators. Ann. Stat. **8**, 679–681 (1980)
7. Baksalary, J.K., Kala, R.: Linear transformations preserving best linear unbiased estimators in a general Gauss-Markoff model. Ann. Stat. **9**, 913–916 (1981)
8. Baksalary, J.K., Kala, R.: Linear sufficiency with respect to a given vector of parametric functions. J. Stat. Plan. Inf. **14**, 331–338 (1986)
9. Baksalary, J.K., Mathew, T.: Linear sufficiency and completeness in an incorrectly specified general Gauss-Markov model. Sankhyā Ser. A **48**, 169–180 (1986)
10. Baksalary, J.K., Mathew, T.: Rank invariance criterion and its application to the unified theory of least squares. Linear Algebra Appl. **127**, 393–401 (1990)
11. Baksalary, J.K., Puntanen, S.: Weighted-least-squares estimation in the general Gauss–Markov model. In: Dodge, Y. (ed.) Statistical Data Analysis and Inference, pp. 355–368. Elsevier Science Publishers B.V., Amsterdam (1989)
12. Baksalary, J.K., Puntanen, S., Styan, G.P.H.: A property of the dispersion matrix of the best linear unbiased estimator in the general Gauss-Markov model. Sankhyā Ser. A **52**, 279–296 (1990)
13. Baksalary, J.K., Rao, C.R., Markiewicz, A.: A study of the influence of the "natural restrictions" on estimation problems in the singular Gauss-Markov model. J. Stat. Plan. Inf. **31**, 335–351 (1992)
14. Baksalary, O.M., Trenkler, G., Liski, E.: Let us do the twist again. Stat. Pap. **54**, 1109–1119 (2013)
15. Ben-Israel, A., Greville, T.N.E.: Generalized Inverses: Theory and Applications, 2nd edn. Springer, New York (2003)
16. Bloomfield, P., Watson, G.S.: The inefficiency of least squares. Biometrika **62**, 121–128 (1975)
17. Christensen, R.: Plane Answers to Complex Questions: The Theory of Linear Models, 4th edn. Springer, New York (2011)

18. Drygas, H.: The Coordinate-Free Approach to Gauss-Markov Estimation. Springer, Berlin (1970)
19. Drygas, H.: Sufficiency and completeness in the general Gauss-Markov model. Sankhyā Ser. A **45**, 88–98 (1983)
20. Groß, J.: A note on the concepts of linear and quadratic sufficiency. J. Stat. Plan. Inf. **70**, 88–98 (1998)
21. Groß, J.: The general Gauss-Markov model with possibly singular dispersion matrix. Stat. Pap. **45**, 311–336 (2004)
22. Groß, J., Puntanen, S.: Estimation under a general partitioned linear model. Linear Algebra Appl. **321**, 131–144 (2000)
23. Groß, J., Puntanen, S.: Extensions of the Frisch-Waugh-Lovell Theorem. Discussiones Mathematicae - Probab. Stat. **25**, 39–49 (2005)
24. Haslett, S.J., Isotalo, J., Liu, Y., Puntanen, S.: Equalities between OLSE, BLUE and BLUP in the linear model. Stat. Pap. **55**, 543–561 (2014)
25. Hauke, J., Markiewicz, A., Puntanen, S.: Comparing the BLUEs under two linear models. Commun. Stat. Theory Methods **41**, 2405–2418 (2012)
26. Isotalo, J., Puntanen, S.: Linear sufficiency and completeness in the partitioned linear model. Acta Comment Univ. Tartu Math. **10**, 53–67 (2006a)
27. Isotalo, J., Puntanen, S.: Linear prediction sufficiency for new observations in the general Gauss-Markov model. Commun. Stat. Theory Methods **35**, 1011–1023 (2006b)
28. Kala, R., Pordzik, P.R.: Estimation in singular partitioned, reduced or transformed linear models. Stat. Pap. **50**, 633–638 (2009)
29. Kala, R., Puntanen, S., Tian, Y.: Some notes on linear sufficiency. Stat. Pap. (2015). doi:10.1007/s00362-015-0682-2
30. Kornacki, A.: Different kinds of sufficiency in the general Gauss-Markov model. Math. Slovaca **57**, 389–392 (2007)
31. Marsaglia, G., Styan, G.P.H.: Equalities and inequalities for ranks of matrices. Linear and Multilinear Algebra **2**, 269–292 (1974)
32. Müller, J.: Sufficiency and completeness in the linear model. J. Multivar. Anal. **21**, 312–323 (1987)
33. Mäkinen, J.: Bounds for the difference between a linear unbiased estimate and the best linear unbiased estimate. Phys. Chem. Earth Pt A. **25**, 693–698 (2000)
34. Mäkinen, J.: A bound for the Euclidean norm of the difference between the best linear unbiased estimator and a linear unbiased estimator. J. Geodesy **76**, 317–322 (2002)
35. Pordzik, P.R.: A bound for the Euclidean distance between restricted and unrestricted estimators of parametric functions in the general linear model. Stat. Pap. **53**, 299–304 (2012)
36. Puntanen, S., Styan, G.P.H.: The equality of the ordinary least squares estimator and the best linear unbiased estimator (with discussion). Am. Stat. **43**, 151–161 (1989). (Commented by O. Kempthorne on pp. 161–162 and by S. R. Searle on pp. 162–163, Reply by the authors on p. 164)
37. Puntanen, S., Styan, G.P.H., Isotalo, J.: Matrix Tricks for Linear Statistical Models: Our Personal Top Twenty. Springer, Heidelberg (2011)
38. Rao, C.R.: Representations of best linear estimators in the Gauss-Markoff model with a singular dispersion matrix. J. Multivar. Anal. **3**, 276–292 (1973)
39. Rao, C.R., Mitra, S.K.: Generalized Inverse of Matrices and Its Applications. Wiley, New York (1971)
40. Tian, Y., Puntanen, S.: On the equivalence of estimations under a general linear model and its transformed models. Linear Algebra Appl. **430**, 2622–2641 (2009)
41. Watson, G.S.: Serial correlation in regression analysis. I. Biometrika **42**, 327–341 (1955)
42. Zyskind, G., Martin, F.B.: On best linear estimation and general Gauss-Markov theorem in linear models with arbitrary nonnegative covariance structure. SIAM J. Appl. Math. **17**, 1190–1202 (1969)

The Exact and Near-Exact Distributions for the Statistic Used to Test the Reality of Covariance Matrix in a Complex Normal Distribution

Luís M. Grilo and Carlos A. Coelho

Abstract The authors start by approximating the exact distribution of the negative logarithm of the likelihood ratio statistic, used to test the reality of the covariance matrix in a certain complex multivariate normal distribution, by an infinite mixture of Generalized Near-Integer Gamma (GNIG) distributions. Based on this representation they develop a family of near-exact distributions for the likelihood ratio statistic, which are finite mixtures of GNIG distributions and match, by construction, some of the first exact moments. Using a proximity measure based on characteristic functions the authors illustrate the excellent properties of the near-exact distributions. They are very close to the exact distribution but far more manageable and have very good asymptotic properties both for increasing sample sizes as well as for increasing number of variables. These near-exact distributions are much more accurate than the asymptotic approximation considered, namely when the sample size is small and the number of variables involved is large. Furthermore, the corresponding cumulative distribution functions allow for an easy computation of very accurate near-exact quantiles.

Keywords Characteristic function · Beta distribution · Gamma distribution · Small samples · Quantiles

L.M. Grilo (✉)
Unidade Departamental de Matemática, Instituto Politécnico de Tomar, Quinta do Contador, Tomar, Portugal
e-mail: lgrilo@ipt.pt

L.M. Grilo · C.A. Coelho
Centro de Matemática e Aplicações (CMA/FCT-UNL), Quinta da Torre, Caparica, Portugal

C.A. Coelho
Departamento de Matemática (DM/FCT-UNL), Faculdade de Ciências e Tecnologia, Universidade Nova de Lisboa, Quinta da Torre, Caparica, Portugal
e-mail: cmac@fct.unl.pt

N. Bebiano (ed.), *Applied and Computational Matrix Analysis*, Springer Proceedings in Mathematics & Statistics 192,
DOI 10.1007/978-3-319-49984-0_20

1 Introduction

Let $\underline{X}$ be a random vector with a p-variate complex normal distribution, with variance-covariance matrix $\Sigma = \Sigma_1 + \mathrm{i}\Sigma_2$, which is a $p \times p$ positive Hermitian matrix, where Σ_1 is a $p \times p$ symmetric positive-definite matrix, Σ_2 is a $p \times p$ skew-symmetric matrix and $\mathrm{i} = (-1)^{1/2}$.

The multivariate complex normal distribution we refer to is the one defined by Wooding [27] and used in [11, 12], [20, Sect. 8], [3, 22, 23], [1, Probl. 2.64] and [18], where

$$\underset{(p\times 1)}{\underline{X}} = \underset{(p\times 1)}{\underline{Y}} + \mathrm{i}\underset{(p\times 1)}{\underline{Z}}$$

with

$$\begin{bmatrix}\underline{Y}\\ \underline{Z}\end{bmatrix} \sim N_{2p}\left(\begin{bmatrix}\underline{\mu}_1\\ \underline{\mu}_2\end{bmatrix}, \frac{1}{2}\begin{bmatrix}\Sigma_1 & -\Sigma_2\\ \Sigma_2 & \Sigma_1\end{bmatrix}\right),$$

where Σ_2 is a skew-symmetric matrix, with $\Sigma_2 = -\Sigma_2'$, so that

$$E(\underline{X}) = \underline{\mu}_X = \underline{\mu}_1 + \mathrm{i}\,\underline{\mu}_2$$

and

$$\begin{aligned}
\Sigma = Var(\underline{X}) &= E\left[\left(\underline{X} - \underline{\mu}_X\right)\left(\overline{\underline{X} - \underline{\mu}_X}\right)'\right]\\
&= E\left[\left((\underline{Y} + \mathrm{i}\underline{Z}) - (\underline{\mu}_1 + \mathrm{i}\,\underline{\mu}_2)\right)\left((\underline{Y} - \mathrm{i}\underline{Z}) - (\underline{\mu}_1 - \mathrm{i}\,\underline{\mu}_2)\right)'\right]\\
&= E\Big[(\underline{Y} - \underline{\mu}_1)(\underline{Y} - \underline{\mu}_1)' - \mathrm{i}(\underline{Y} - \underline{\mu}_1)(\underline{Z} - \underline{\mu}_2)'\\
&\quad + \mathrm{i}(\underline{Z} - \underline{\mu}_2)(\underline{Y} - \underline{\mu}_1)' - \mathrm{i}^2(\underline{Z} - \underline{\mu}_2)(\underline{Z} - \underline{\mu}_2)'\Big]\\
&= Var(\underline{Y}) - \mathrm{i}\,Cov(\underline{Y}, \underline{Z}) + \mathrm{i}\,Cov(\underline{Z}, \underline{Y}) + Var(\underline{Z})\\
&= \frac{1}{2}\Sigma_1 + \mathrm{i}\frac{1}{2}\Sigma_2 + \mathrm{i}\frac{1}{2}\Sigma_2 + \frac{1}{2}\Sigma_1 = \Sigma_1 + \mathrm{i}\,\Sigma_2\,,
\end{aligned}$$

the p.d.f. of $\underline{X}$ being

$$f_{\underline{X}}(\underline{x}) = \frac{e^{-(\overline{\underline{x}-\underline{\mu}_X})'\Sigma^{-1}(\underline{x}-\underline{\mu}_X)}}{\pi^p\,|\Sigma|}\,,$$

where the overbar denotes the complex conjugate.

To test the reality of Σ, that is to test

$$H_0 : \Sigma_2 = 0 \qquad \text{versus} \qquad H_1 : \Sigma_2 \neq 0\,,$$

we may consider, for a sample of size $n+1$, the power $2/(n+1)$ of the likelihood ratio test statistic, obtained in [22],

$$\Lambda = \frac{|S_1 + \mathrm{i}S_2|}{|S_1|}\,, \tag{1}$$

where $S = S_1 + \mathrm{i}S_2$ is the maximum likelihood estimator of Σ.

When $\Sigma_2 = 0$, the statistic Λ in (1), is shown in [22] to be distributed as a product of independent beta random variables (r.v.'s) with specific parameters. More precisely, for $n \geq p$, if p is even,

$$\Lambda \stackrel{st}{\sim} \prod_{j=1}^{p/2} Y_j \quad \text{with} \quad Y_j \sim Beta\left(n - \frac{p}{2} - j + 1, \frac{p}{2} - \frac{1}{2}\right), \tag{2}$$

where '$\stackrel{st}{\sim}$' means 'stochastically equivalent to' and where the Y_j are $p/2$ independent r.v.'s, or, if p is odd,

$$\Lambda \stackrel{st}{\sim} \prod_{j=1}^{(p-1)/2} Y_j \quad \text{with} \quad Y_j \sim Beta\left(n - \frac{p+1}{2} - j + 1, \frac{p}{2}\right), \tag{3}$$

where the Y_j are $(p-1)/2$ independent r.v.'s, or, for any p in (2) or (3), taking $q^* = \lfloor p/2 \rfloor$ and $q = \lceil p/2 \rceil$, where $\lfloor \cdot \rfloor$ denotes the floor of the argument, that is, the largest integer that does not exceed the argument and $\lceil \cdot \rceil$ denotes the ceiling of the argument, that is, the smallest integer not less than the argument, we may write

$$\Lambda \stackrel{st}{\sim} \prod_{j=1}^{q^*} Y_j \quad \text{with} \quad Y_j \sim Beta\left(n - q - j + 1, q - \frac{1}{2}\right), \tag{4}$$

where the Y_j are q^* independent r.v.'s.

Since for a r.v. X with a Beta distribution, with parameters α and β, the h-th moment of X is given by

$$E(X^h) = \frac{B(\alpha+h,\beta)}{B(\alpha,\beta)} = \frac{\Gamma(\alpha+\beta)}{\Gamma(\alpha)}\,\frac{\Gamma(\alpha+h)}{\Gamma(\alpha+\beta+h)}\,, \quad (h > -\alpha)\,, \tag{5}$$

we may write, for the r.v.'s Y_j in (4),

$$E(Y_j^h) = \frac{\Gamma\left(n - j + \frac{1}{2}\right)}{\Gamma\left(n - q - j + 1\right)}\,\frac{\Gamma\left(n - q - j + 1 + h\right)}{\Gamma\left(n - j + \frac{1}{2} + h\right)}\,, \quad \text{with} \quad h > -(n - q - j + 1)\,,$$

so that, given the independence of the q^* r.v.'s in (4), we may easily obtain, for $h > -(n - p + 1)$

$$E(\Lambda^h) = \prod_{j=1}^{q^*} E(Y_j^h) = \prod_{j=1}^{q^*} \frac{\Gamma\left(n-j+\frac{1}{2}\right)}{\Gamma\left(n-q-j+1\right)} \frac{\Gamma\left(n-q-j+1+h\right)}{\Gamma\left(n-j+\frac{1}{2}+h\right)} . \tag{6}$$

In the next section, we will address the exact distribution of Λ. Based on a factorization of the exact characteristic function of $W = -\log \Lambda$, we first express the exact distribution of W as the distribution of the sum of two independent r.v.'s, one with a Generalized Integer Gamma (GIG) distribution and the other with a distribution of a sum of an independent Logbeta distributions.

Then, in Sect. 3, we will first approximate the exact distribution of W by an infinite mixture of Generalized Near-Integer Gamma (GNIG) distributions and then, based on this representation, we develop near-exact distributions for $W = -\log \Lambda$ which are finite mixtures of GNIG distributions and which equate some of the first exact moments of W. From these we obtain near-exact distributions for Λ, with very manageable cumulative distribution functions (c.d.f.'s), much useful in practice to compute quantiles and p-values. The concept of a near-exact distribution and the procedure used to develop these distributions has already been introduced in a number of papers [6, 13–17].

In Sect. 4 we address the asymptotic distribution in [4, 23] and express it in a manner that is adequate for our purpose of using it to be compared with our near-exact distributions.

The fact that the near-exact distributions developed in Sect. 3 have a much better performance than the asymptotic distribution used in [4, 23] is shown in Sect. 5 where numerical studies are carried out using a measure of proximity between distributions, based on characteristic functions. The numerical studies developed, for different sample sizes, numbers of variables and number of moments equated, show the high closeness of these near-exact distributions to the exact distribution and also their excellent performance, namely when the sample size and the difference between the sample size and the number of variables involved are small.

2 The Exact Distribution of Λ

For a r.v. $X \sim Beta(\alpha, \beta)$, the r.v. $Y = -\log X$ has what is called a Logbeta distribution [21], fact that is denoted by $Y \sim Logbeta(\alpha, \beta)$, and since the Gamma functions in (5) are still valid for any strictly complex h, the characteristic function (cf.) of the r.v. Y is given by

$$\Phi_Y(t) = E(e^{\mathrm{i}tY}) = E(e^{-\mathrm{i}t \log X}) = E(X^{-\mathrm{i}t}) = \frac{\Gamma(\alpha+\beta)}{\Gamma(\alpha)} \frac{\Gamma(\alpha - \mathrm{i}t)}{\Gamma(\alpha+\beta-\mathrm{i}t)}, \tag{7}$$

where $t \in \mathbb{R}$. Considering the independence of the q^* r.v.'s Y_j in (4) and considering (6) and (7) we may write the cf. of $W = -\log \Lambda$ as

$$\begin{aligned}\Phi_W(t) = E(e^{-\mathrm{i}tW}) = E(\Lambda^{-\mathrm{i}t}) &= \prod_{j=1}^{q^*} E(Y_j^{-\mathrm{i}t}) \\ &= \prod_{j=1}^{q^*} \frac{\Gamma\left(n-j+\frac{1}{2}\right)}{\Gamma\left(n-q-j+1\right)} \frac{\Gamma\left(n-q-j+1-\mathrm{i}t\right)}{\Gamma\left(n-j+\frac{1}{2}-\mathrm{i}t\right)}\,.\end{aligned} \tag{8}$$

Using the identity

$$\frac{\Gamma(\alpha+\beta)}{\Gamma(\alpha)} = \prod_{k=0}^{\beta-1}(\alpha+k)\,,$$

for $\beta \in \mathbb{N}$ and α real or complex, we may write

$$\begin{aligned}\Phi_W(t) &= \prod_{j=1}^{q^*} \frac{\Gamma\left(n-j+\frac{1}{2}\right)}{\Gamma\left(n-q-j+1\right)} \frac{\Gamma\left(n-q-j+1-\mathrm{i}t\right)}{\Gamma\left(n-j+\frac{1}{2}-\mathrm{i}t\right)} \\ &= \prod_{j=1}^{q^*} \frac{\Gamma\left(n-j+\frac{1}{2}\right)}{\Gamma\left(n-j\right)} \frac{\Gamma\left(n-j-\mathrm{i}t\right)}{\Gamma\left(n-j+\frac{1}{2}-\mathrm{i}t\right)} \frac{\Gamma\left(n-j\right)}{\Gamma\left(n-q-j+1\right)} \frac{\Gamma\left(n-q-j+1-\mathrm{i}t\right)}{\Gamma\left(n-j-\mathrm{i}t\right)} \\ &= \underbrace{\prod_{j=1}^{q^*} \frac{\Gamma\left(n-j+\frac{1}{2}\right)}{\Gamma\left(n-j\right)} \frac{\Gamma\left(n-j-\mathrm{i}t\right)}{\Gamma\left(n-j+\frac{1}{2}-\mathrm{i}t\right)}}_{\Phi_{1,W}(t)} \\ &\qquad \times \underbrace{\prod_{j=1}^{q^*}\prod_{k=0}^{q-2}(n-q-j+1+k)(n-q-j+1+k-\mathrm{i}t)^{-1}}_{\Phi_{2,W}(t)}\end{aligned} \tag{9}$$

where $\Phi_{1,W}(t)$ is the cf. of the sum of q^* independent r.v.'s with *Logbeta*$(n-j,1/2)$ distributions ($j=1,\ldots,q^*$), and $\Phi_{2,W}(t)$ is the cf. of the sum of $q^*(q-1)$ independent r.v.'s with *Exp*$(n-q-j+1+k)$ distributions ($k=0,\ldots,q-2$; $j=1,\ldots,q^*$).

By identifying the different Exponential distributions that occur in $\Phi_{2,W}(t)$ in (9) and using a counting technique similar to the one used by [26], we may write $\Phi_{2,W}(t)$ as

$$\Phi_{2,W}(t) = \prod_{j=2}^{p-1}(n-j)^{r_j}\,(n-j-\mathrm{i}t)^{-r_j}\,,$$

where

$$r_j = \begin{cases} j-1, & j=2,\ldots,q \\ p-j, & j=q+1,\ldots,p-1\,,\end{cases}$$

that is, $\Phi_{2,W}(t)$ is the cf. of a sum of $p-2$ independent Gamma r.v.'s with integer shape parameters r_j and rate parameters $n-j$ ($j=2,\dots,p-1$), which is a GIG distribution of depth $p-2$ (see [5]).

Therefore, we may write the exact cf. of $W=-\log\Lambda$ as

$$\Phi_W(t)=\underbrace{\prod_{j=1}^{q^*}\frac{\Gamma\left(n-j+\frac{1}{2}\right)}{\Gamma\left(n-j\right)}\frac{\Gamma\left(n-j-\mathrm{i}t\right)}{\Gamma\left(n-j+\frac{1}{2}-\mathrm{i}t\right)}}_{\Phi_{1,W}(t)}\underbrace{\prod_{j=2}^{p-1}(n-j)^{r_j}\left(n-j-\mathrm{i}t\right)^{-r_j}}_{\Phi_{2,W}(t)}\quad(10)$$

which is the cf. of the sum of a GIG distribution, of depth $p-2$, with an independent sum of q^* independent Logbeta distributed r.v.'s.

Since from the two first expressions in Sect. 5 of [25] and also expressions (11) and (14) in the same paper, we may write

$$\frac{\Gamma(a-\mathrm{i}t)}{\Gamma(a+b-\mathrm{i}t)}=\sum_{k=0}^{\infty}p_k(b)(a-\mathrm{i}t)^{-b-k}$$

where $p_0(b)=1$ and for $k=1,2,\dots,$

$$p_k(b)=\frac{1}{k}\sum_{m=0}^{k-1}\left(\frac{\Gamma(1-b-m)}{\Gamma(-b-k)(k-m+1)!}+(-1)^{k+m}b^{k-m+1}\right)p_m(b),$$

we may write the cf. of Y in (7) as

$$\Phi_Y(t)=\sum_{k=0}^{\infty}\underbrace{\frac{\Gamma(\alpha+\beta)}{\Gamma(\alpha)}\frac{p_k(\beta)}{\alpha^{\beta+j}}}_{p_k^*(\alpha,\beta)}\alpha^{\beta+k}(\alpha-\mathrm{i}t)^{-(\beta+k)},\quad(11)$$

which is the cf. of an infinite mixture of $Gamma(\beta+k,\alpha)$ distributions with weights $p_k^*(\alpha,\beta)$.

But then $\Phi_{1,W}(t)$ may be written as

$$\Phi_{1,W}(t)=\prod_{j=1}^{q^*}\sum_{k=0}^{\infty}p_k^*(n-j,1/2)\,(n-j)^{1/2+k}\,(n-j-\mathrm{i}t)^{-(1/2+k)}\quad(12)$$

$$=\sum_{k=0}^{\infty}\sum_{\nu=1}^{K^*}p_{k\nu}^{**}\prod_{j=1}^{q^*}(n-j)^{1/2+\eta_{\nu j}}\,(n-j-\mathrm{i}t)^{-(1/2+\eta_{\nu j})},\quad(13)$$

with

$$K^*=\binom{k+q^*-1}{k},\qquad\sum_{j=1}^{q^*}\eta_{\nu j}=k\;\;(0\le\eta_{\nu j}\le k)$$

and

$$p^{**}_{k\nu} = \prod_{j=1}^{q^*} p^*_{\eta_{\nu j}}(n-j,\, 1/2+\eta_{\nu j})\,, \tag{14}$$

where K^* is the number of different partitions $\mathscr{P}_{k,q^*}$ of the integer k into a sequence of q^* non-negative integers not larger than k and where the ν-th of these partitions is, for $\nu \in \{1, \ldots, K^*\}$, a list with components $\eta_{\nu j}$, for $j = 1, \ldots, q^*$. The weights $p^*_{\eta_{\nu j}}(n-j, 1/2+\eta_{\nu j})$ in (14) are the weights $p^*_k(\alpha, \beta)$ in (11) for $\alpha = n-j$, $\beta = 1/2+\eta_{\nu j}$ and $k = \eta_{\nu j}$.

While the cf. in (12) is the cf. of a sum of q^* independent infinite mixtures of $Gamma(1/2+k, n-j)$ distributions ($j = 1, \ldots, q^*$; $k = 0, 1, \ldots$), (13) is the cf. of an infinite mixture of sums of q^* independent $Gamma(1/2+\eta_{\nu j}, n-j)$ distributions, for $j = 1, \ldots, q^*$.

Although this form of the exact distribution of W may seem more complicated than the one obtained from (10), in the next section we will show how we may use it to develop very sharp near-exact distributions for W and Λ.

3 A Family of Near-Exact Distributions for Λ

Given the fact that the rate parameters of the Gamma distributions in (13) are somewhat similar, with a constant step as a function of j and given the fact that the shape parameters in these Gamma distributions are equal to $1/2+\eta_{\nu j}$, with $\sum_{j=1}^{q^*} \eta_{\nu j} = k$ and we are just adding these q^* Gamma distributions, a somewhat heuristic asymptotic approximation for $\Phi_{1,W}(t)$, for increasing n would be a cf. of an infinite mixture of $Gamma(q^*/2+k, \lambda^*)$ distributions, where λ^* is the rate parameter in

$$\Phi^{**}(t) = \theta(\lambda^*)^{s_1}(\lambda^* - \mathrm{i}t)^{-s_1} + (1-\theta)(\lambda^*)^{s_2}(\lambda^* - \mathrm{i}t)^{-s_2}\,, \tag{15}$$

which is determined together with θ, s_1 and s_2, by solving the system of equations

$$\left.\frac{\partial^h}{\partial t^h}\Phi^{**}(t)\right|_{t=0} = \left.\frac{\partial^h}{\partial t^h}\Phi_{1,W}(t)\right|_{t=0}\,, \qquad h = 1, \ldots, 4\,.$$

This would yield an asymptotic distribution for W which is an infinite mixture of GNIG distributions.

In practice, to obtain a family of near-exact distributions for W we will thus leave $\Phi_{2,W}(t)$ in (10) unchanged and replace $\Phi_{1,W}(t)$ by

$$\Phi^*_{1,W}(t) = \sum_{k=0}^{m^*} \pi_k (\lambda^*)^{q^*/2+k}\,(\lambda^* - \mathrm{i}t)^{-(q^*/2+k)}\,,$$

which is the cf. of a finite mixture of m^*+1 $Gamma(q^*/2+k, \lambda^*)$ distributions, with weights π_k $(k=0,\dots,m^*)$, where λ^* will be the rate parameter in (15) above and the weights π_k, for $k=0,\dots,m^*-1$, are determined by solving the system of linear equations

$$\left.\frac{\partial^h}{\partial t^h}\Phi^*_{1,W}(t)\right|_{t=0} = \left.\frac{\partial^h}{\partial t^h}\Phi_{1,W}(t)\right|_{t=0}, \qquad h=1,\dots,m^*,$$

with $\pi_{m^*}=1-\sum_{k=0}^{m^*-1}\pi_k$.

Near-exact distributions built this way will match the first m^* exact moments of W and yield near-exact cf.'s for W, $\Phi^*_W(t)$, of the form

$$\begin{aligned}\Phi^*_W(t) &= \underbrace{\sum_{k=0}^{m^*}\pi_k(\lambda^*)^{q^*/2+k}(\lambda^*-\mathrm{i}t)^{-(q^*/2+k)}}_{\Phi^*_{1,W}(t)}\underbrace{\prod_{j=2}^{p-1}(n-j)^{r_j}(n-j-\mathrm{i}t)^{-r_j}}_{\Phi_{2,W}(t)}\\ &=\sum_{k=0}^{m^*}\left\{\pi_k(\lambda^*)^{q^*/2+k}(\lambda^*-\mathrm{i}t)^{-(q^*/2+k)}\prod_{j=2}^{p-1}(n-j)^{r_j}(n-j-\mathrm{i}t)^{-r_j}\right\}\end{aligned} \tag{16}$$

which is:

- for odd q^*, the cf. of a mixture of length m^*+1 of GNIG distributions of depth $p-1$ with integer shape parameters r_j $(j=2,\dots,p-1)$ and non-integer shape parameter $q^*/2+k$, and corresponding rate parameters $n-j$ and λ^*;
- for even q^*, the cf. of a mixture of length m^*+1 of GIG distributions of depth $p-1$ with integer shape parameters r_j $(j=2,\dots,p-1)$ and $q^*/2+k$, and corresponding rate parameters $n-j$ and λ^*.

Then, considering the cf. in (16) and the notation in Appendix B of [24], for odd q^*, the near-exact p.d.f.'s of W and Λ, are, respectively

$$f_W(w)=\sum_{k=0}^{m^*}\pi_k f^{GNIG}(w\,|\,r_2,\dots,r_{p-1},q^*/2+k;n-2,\dots,n-p+1,\lambda^*;p-1),$$

for $w>0$, and

$$f_\Lambda(\ell)=\sum_{k=0}^{m^*}\pi_k f^{GNIG}(-\log\ell\,|\,r_2,\dots,r_{p-1},q^*/2+k;n-2,\dots,n-p+1,\lambda^*;p-1)\frac{1}{\ell},$$

for $0<\ell<1$, while the near-exact c.d.f.'s for W and Λ are, respectively, given by

$$F_W(w)=\sum_{k=0}^{m^*}\pi_k F^{GNIG}(w\,|\,r_2,\dots,r_{p-1},q^*/2+k;n-2,\dots,n-p+1,\lambda^*;p-1),$$

for $w > 0$, and

$$F_\Lambda(\ell) = \sum_{k=0}^{m^*} \pi_k\Big(1 - F^{GNIG}(-\log\,\ell \,|\, r_2, \ldots, r_{p-1}, q^*/2+k;$$
$$n-2, \ldots, n-p+1, \lambda^*; p-1)\Big),$$

for $0 < \ell < 1$.

For even q^*, since the GIG distributions are a particular case of GNIG distributions, the p.d.f. and c.d.f. expressions for W and Λ are similar to those presented before, but where the shape parameter $q^*/2 + k$ is an integer. For example, the near-exact c.d.f. of Λ is given by,

$$F_\Lambda(\ell) = \sum_{k=0}^{m^*} \pi_k\Big(1 - F^{GIG}(-\log\,\ell \,|\, r_2, \ldots, r_{p-1}, q^*/2+k;$$
$$n-2, \ldots, n-p+1, \lambda^*; p-1)\Big),$$

for $0 < \ell < 1$.

4 Asymptotic Distribution

In order to compare with the near-exact distributions developed in the previous section, we consider here the asymptotic distribution in [4, 23]. For

$$m = 2n - p - 1/2\,, \quad f = p(p-1)/2 \quad \text{and} \quad \gamma_2 = p(p-1)(p^2 + (p-1)^2 - 8)/96\,,$$

it is used in [4, 23] a Box-type asymptotic distribution for mW which is a two-component mixture of a chi-square with f degrees of freedom and another chi-square with $f + 4$ degrees of freedom, with weights $1 - \gamma_2/m^2$ and γ_2/m^2. This yields for mW the asymptotic cf.

$$\Phi^{**}_{mW}(t) = \left(1 - \frac{\gamma_2}{m^2}\right)\left(\frac{1}{2}\right)^{f/2}\left(\frac{1}{2} - \mathrm{i}t\right)^{-f/2} + \frac{\gamma_2}{m^2}\left(\frac{1}{2}\right)^{f/2+2}\left(\frac{1}{2} - \mathrm{i}t\right)^{-(f/2-2)},$$

which, for W, yields the asymptotic cf.

$$\begin{aligned}\Phi^{**}_{W}(t) &= \left(1 - \frac{\gamma_2}{m^2}\right)\left(\frac{1}{2}\right)^{f/2}\left(\frac{1}{2} - \mathrm{i}\frac{t}{m}\right)^{-f/2} + \frac{\gamma_2}{m^2}\left(\frac{1}{2}\right)^{f/2+2}\left(\frac{1}{2} - \mathrm{i}\frac{t}{m}\right)^{-(f/2-2)} \\ &= \left(1 - \frac{\gamma_2}{m^2}\right)\left(\frac{m}{2}\right)^{f/2}\left(\frac{m}{2} - \mathrm{i}t\right)^{-f/2} + \frac{\gamma_2}{m^2}\left(\frac{m}{2}\right)^{f/2+2}\left(\frac{m}{2} - \mathrm{i}t\right)^{-(f/2-2)}.\end{aligned} \tag{17}$$

5 Numerical Studies

In order to assess the closeness/proximity of the exact distribution to an approximate, near-exact or asymptotic, distribution we use the measure [13–18]

$$\Delta = \frac{1}{2\pi}\int_{-\infty}^{+\infty}\left|\frac{\Phi_W(t)-\Phi_W^+(t)}{t}\right| dt \tag{18}$$

where $\Phi_W(t)$ represents the exact cf. of W and $\Phi_W^+(t)$ its approximate, near-exact or asymptotic, cf.

The measure Δ in (18) is an upper bound on the difference between the exact and the corresponding approximate c.d.f. of W, with

$$\Delta \geq \sup_{w>0}\left|F_W(w)-F_W^+(w)\right| = \sup_{0<\ell<1}\left|F_\Lambda(\ell)-F_\Lambda^+(\ell)\right| ,$$

where $F_W(\cdot)$ and $F_\Lambda(\cdot)$ are, respectively, the exact c.d.f.'s of W and Λ, and $F_W^+(\cdot)$ is the c.d.f. that corresponds to $\Phi_W^+(\cdot)$, being $F_\Lambda^+(\ell) = 1 - F_W^+(-\log \ell)$.

The measure Δ in (18) may be directly derived from the Gil–Pelaez inversion formula for the c.d.f. [10] (see Appendix A) and, as noted in [14], it may also be seen as intimately related with the Berry–Esseen upper-bound [2, 9, 19]. This measure gives us a very accurate assessment of the quality of the approximations, as we have seen in several studies [13–18], with smaller values of this measure showing a better agreement with the exact distribution, both in terms of p-values and quantiles.

In Table 1 we have the results of the numerical studies conducted, using the measure Δ in (18) and the cf.'s in (8), (16) and (17), to assess the behavior of the asymptotic distribution in [4, 23] and the near-exact distributions developed in Sect. 3, for different numbers of exact moments equated (m^*), number of variables (p) and sample sizes $n = p + 1$, $n = 2p$ and $n = 3p$.

We have to point out the excellent performance of the family members of the near-exact distributions, with very low values of the proximity measure, thus showing an extreme closeness to the exact distribution, even for the case where only two exact moments are equated. We can confirm that they are particularly adequate and useful for small sample sizes, that is, for small values of n, or rather, when the values of n and p are close. They also have an asymptotic behavior for increasing number of variables, that is, increasing values of p, while the asymptotic distribution goes the other way around.

One other fact to be pointed out about Table 1 is the fact that, for the two larger values of p, that is, for $p = 35$ and $p = 55$ and the smaller sample sizes associated with these two values of p, that is, respectively for $n = 36$ and $n = 56$, the value of the measure Δ exceeds 1, which, from its definition, should never happen. However it does happen because for these cases with quite large values of p and small sample sizes, the asymptotic distribution is indeed not any more a distribution, with its 'p.d.f.' assuming values below zero and above one. This is a commonly overlooked fact which, for other asymptotic distributions of this type, was already called the attention for in [8] and [7], and which is clearly detected by the measure Δ.

Table 1 Values of the measure Δ in (18) for the asymptotic distribution in [4, 23] and the near-exact distributions developed in Sect. 3

p	n	Asymp. distrib.	Near-exact distributions			
			m^*			
			2	4	6	10
5	6	7.41×10^{-3}	2.58×10^{-7}	7.59×10^{-9}	1.10×10^{-10}	1.14×10^{-13}
	10	2.74×10^{-4}	2.81×10^{-8}	3.11×10^{-10}	2.27×10^{-12}	4.02×10^{-16}
	15	3.31×10^{-8}	4.69×10^{-9}	2.26×10^{-11}	7.53×10^{-14}	4.51×10^{-18}
15	16	2.43×10^{-1}	1.02×10^{-6}	1.91×10^{-9}	5.58×10^{-12}	1.20×10^{-16}
	30	3.44×10^{-3}	2.26×10^{-7}	1.55×10^{-10}	1.60×10^{-13}	3.65×10^{-19}
	45	4.26×10^{-4}	4.86×10^{-8}	1.46×10^{-11}	6.50×10^{-15}	2.71×10^{-21}
25	26	6.83×10^{-1}	3.38×10^{-7}	2.61×10^{-10}	2.96×10^{-13}	9.22×10^{-19}
	50	9.13×10^{-3}	1.14×10^{-7}	3.74×10^{-11}	1.78×10^{-14}	8.80×10^{-21}
	75	1.14×10^{-3}	2.62×10^{-8}	3.79×10^{-12}	7.97×10^{-16}	7.47×10^{-23}
35	36	1.13×10^{0}	1.49×10^{-7}	6.51×10^{-11}	3.90×10^{-14}	3.07×10^{-20}
	70	1.75×10^{-2}	7.05×10^{-8}	1.42×10^{-11}	4.01×10^{-15}	6.52×10^{-22}
	105	2.20×10^{-3}	1.70×10^{-8}	1.52×10^{-12}	1.89×10^{-16}	5.95×10^{-24}
55	56	1.92×10^{0}	4.26×10^{-8}	9.31×10^{-12}	2.50×10^{-15}	3.27×10^{-22}
	110	4.22×10^{-2}	3.63×10^{-8}	3.91×10^{-12}	5.61×10^{-16}	2.10×10^{-23}
	165	5.37×10^{-3}	9.37×10^{-9}	4.50×10^{-13}	2.87×10^{-17}	2.11×10^{-25}

With the near-exact distributions displaying so low values of the measure Δ it can only be assured that there is a very good agreement between the exact and the near-exact distributions throughout the whole range of the random variable and as such that also all near-exact quantiles will display a sharp agreement with the corresponding exact quantiles.

In Tables 2 and 3 we show some near-exact quantiles for $p = 6$ and $n = 6, 7, 8, 9, 10$, since in [4] there were some problems in computing the exact 0.05 quantiles for $n = 6, 7, 8$ and the 0.01 quantiles for $n = 6, 7, 8, 9$ not only in order to make them available for practical use but also in order to show how by increasing the number of exact moments matched we may obtain quantiles which indeed converge, with convergence being assuredly towards the corresponding exact quantiles, given the sharp decrease in the value of the measure Δ that may be observed in Table 1 when the number of exact moments matched is increased.

As we may see from Tables 2 and 3, there are no problems in computing the near-exact quantiles for any combination of number of variables and sample sizes, with the values for the near-exact quantiles for $p = 6$ and $n = 6, 7$ showing that for small sample sizes the asymptotic quantiles show quite some deviation from the exact value, while for $n = 9$ and $n = 10$ the near-exact quantiles match the values presented in [4] for the exact quantiles. To show that there is no problem in computing near-exact quantiles even for quite large numbers of variables, either with very small

Table 2 0.05 quantiles for Λ for $p = 6$ ($n = 6, 7, 8, 9, 10$), $p = 35$ ($n = 36, 70, 105$) and $p = 55$ ($n = 56, 110, 165$), for the near-exact distributions that match $m^* = 2, 4, 6$ and 10 exact moments

p	n	m^*			
		2	4	6	10
6	6	0.0027299422	0.0027297970	0.0027297983	0.0027297983
	7	0.0237278948	0.0237274940	0.0237274976	0.0237274976
	8	0.0599860502	0.0599857342	0.0599857361	0.0599857361
	9	0.1031286563	0.1031284719	0.1031284725	0.1031284725
	10	0.1479772284	0.1479771337	0.1479771339	0.1479771339
35	36	$8.7315394235 \times 10^{-11}$	$8.7315318934 \times 10^{-11}$	$8.7315318962 \times 10^{-11}$	$8.7315318962 \times 10^{-11}$
	70	$1.7094444679 \times 10^{-3}$	$1.7094445181 \times 10^{-3}$	$1.7094445181 \times 10^{-3}$	$1.7094445181 \times 10^{-3}$
	105	$2.3117849266 \times 10^{-2}$	$2.3117848959 \times 10^{-2}$	$2.3117848959 \times 10^{-2}$	$2.3117848959 \times 10^{-2}$
55	56	$1.0453180265 \times 10^{-16}$	$1.0453177384 \times 10^{-16}$	$1.0453177385 \times 10^{-16}$	$1.0453177385 \times 10^{-16}$
	110	$5.7253174091 \times 10^{-5}$	$5.7253171208 \times 10^{-5}$	$5.7253171208 \times 10^{-5}$	$5.7253171208 \times 10^{-5}$
	165	$3.0889621685 \times 10^{-3}$	$3.0889621449 \times 10^{-3}$	$3.0889621449 \times 10^{-3}$	$3.0889621449 \times 10^{-3}$

Table 3 0.01 quantiles for Λ for $p = 6$ ($n = 6, 7, 8, 9, 10$), $p = 35$ ($n = 36, 70, 105$) and $p = 55$ ($n = 56, 110, 165$), for the near-exact distributions that match $m^* = 2, 4, 6$ and 10 exact moments

p	n	m^*			
		2	4	6	10
6	6	0.0005176879	0.0005176208	0.0005176209	0.0005176209
	7	0.0094602783	0.0094594034	0.0094594070	0.0094594070
	8	0.0309225909	0.0309210880	0.0309210943	0.0309210942
	9	0.0610102766	0.0610087013	0.0610087068	0.0610087068
	10	0.0955879031	0.0955865257	0.0955865298	0.0955865298
35	36	$2.6515423321 \times 10^{-11}$	$2.6515370198 \times 10^{-11}$	$2.6515370195 \times 10^{-11}$	$2.6515370195 \times 10^{-11}$
	70	$1.3314442566 \times 10^{-3}$	$1.3314438777 \times 10^{-3}$	$1.3314438777 \times 10^{-3}$	$1.3314438777 \times 10^{-3}$
	105	$1.9963933022 \times 10^{-2}$	$1.9963932207 \times 10^{-2}$	$1.9963932207 \times 10^{-2}$	$1.9963932207 \times 10^{-2}$
55	56	$2.9594023552 \times 10^{-17}$	$2.9594004510 \times 10^{-17}$	$2.9594004509 \times 10^{-17}$	$2.9594004509 \times 10^{-17}$
	110	$4.4855042422 \times 10^{-5}$	$4.4855035977 \times 10^{-5}$	$4.4855035977 \times 10^{-5}$	$4.4855035977 \times 10^{-5}$
	165	$2.6763131139 \times 10^{-3}$	$2.6763130550 \times 10^{-3}$	$2.6763130550 \times 10^{-3}$	$2.6763130550 \times 10^{-3}$

or quite large sample sizes, in those tables are also shown the 0.05 and 0.01 quantiles for $p = 35$ and $p = 55$ for the same sample sizes that the measure Δ was computed in Table 1, with the near-exact distributions that match only 4 exact moments exhibiting quantiles that already match 8–10 significant digits.

We may see how for only 2 or 4 exact moments matched the near-exact quantiles already display a quite large number of decimal places that match those of the corresponding exact quantile, which, in all cases may be taken as the quantile displayed for the near-exact distribution that matches 10 exact moments.

We may also see how the near-exact distributions that exhibit lower values of the measure Δ displaying quantiles which have more decimal places that match the corresponding exact quantiles.

The quantiles shown in Tables 2 and 3 are the 0.05 and the 0.01 quantiles, since the quantiles to be used in testing hypothesis with the statistic Λ will be the left tail quantiles. To show that a similar behavior is displayed by all other quantiles, in Tables B.1, B.2 and B.3 in Appendix B may be analyzed the median and the 0.95 and 0.99 quantiles for the same distributions.

6 Conclusions and Final Remarks

The near-exact distributions developed lie very close to the exact distribution, in terms of cf.'s, moments, c.d.f.'s and quantiles, and the general expressions obtained for the c.d.f.'s are, in fact, very manageable and easily allow the calculation of near-exact quantiles and p-values through the use of some symbolic software. Note that even when we have the expressions for the exact p.d.f.'s and c.d.f.'s available from the literature, these are usually only available for specific numbers of variables and the expressions are highly complex, which renders the computation of exact quantiles too hard.

The comparative analysis conducted allows us to confirm and reinforce the importance of near-exact distributions over the asymptotic ones. The near-exact distributions remain very close to the exact distribution even when the difference between the sample size and the total number of variables, that is, the value of $n - p$, is very small, situation in which the usual asymptotic distributions do not work well, mainly if p is quite large. Furthermore, the near-exact distributions developed also display an asymptotic behavior for increasing number of variables.

Acknowledgements This work was partially supported by Fundação para a Ciência e a Tecnologia (Portuguese Foundation for Science and Technology) through project UID/MAT/00297/2013 (Centro de Matemática e Aplicações – CMA).

Appendix A

Derivation of the Measure Δ in (18) from the Gil–Pelaez Inversion Formula

The measure Δ in (18) may be directly derived from the Gil–Pelaez [10] inversion formula for the c.d.f., which may be written in a number of equivalent forms, as for example

$$F_W(w) = \frac{1}{2} - \frac{1}{2\pi}\int_{-\infty}^{+\infty} \frac{e^{-\mathrm{i}tw}\,\Phi_W(t)}{\mathrm{i}t}\,dt\,.$$

Then, if we take $F_W(\cdot)$ and $F_W^+(\cdot)$ as the c.d.f.'s corresponding to the cf.'s $\Phi_W(\cdot)$ and $\Phi_W^+(\cdot)$ respectively, we have

$$\begin{aligned}\left|F_W(w) - F_W^+(w)\right| &= \frac{1}{2\pi}\left|\int_{-\infty}^{+\infty} \frac{e^{-\mathrm{i}tw}}{\mathrm{i}t}\left(\Phi_W(t)^+ - \Phi_W(t)\right)dt\right| \\ &\le \frac{1}{2\pi}\int_{-\infty}^{+\infty}\left|\frac{e^{-\mathrm{i}tw}}{\mathrm{i}t}\left(\Phi_W(t)^+ - \Phi_W(t)\right)\right| dt\end{aligned}$$

where, for any $t \in \mathbb{R}$ and any $w \in \mathbb{R}$,

$$\left|\frac{e^{-\mathrm{i}tw}}{\mathrm{i}}\right| = 1\,,$$

so that we may write

$$\sup_w \left|F_W(w) - F_W^+(w)\right| \le \frac{1}{2\pi}\int_{-\infty}^{+\infty}\left|\frac{\Phi_W(t) - \Phi_W^+(t)}{t}\right| dt\,.$$

The measure Δ gives thus very sharp upper-bounds on the difference between the c.d.f.'s $F_W(\cdot)$ and $F_W^+(\cdot)$, indeed much sharper than any similar measure that would be based on the more common inversion formula for the c.d.f..

The measure Δ clearly verifies the triangular inequality since if we take

$$\Delta_1 = \frac{1}{2\pi}\int_{-\infty}^{+\infty}\left|\frac{\Phi_1(t) - \Phi_2(t)}{t}\right| dt\,, \qquad \Delta_2 = \frac{1}{2\pi}\int_{-\infty}^{+\infty}\left|\frac{\Phi_1(t) - \Phi_3(t)}{t}\right| dt$$

and

$$\Delta_3 = \frac{1}{2\pi}\int_{-\infty}^{+\infty}\left|\frac{\Phi_2(t) - \Phi_3(t)}{t}\right| dt$$

we have $\Delta_1 \leq \Delta_2 + \Delta_3$ since

$$\begin{aligned}|\Phi_1(t) - \Phi_2(t)| &= |\Phi_1(t) - \Phi_3(t) + \Phi_3(t) - \Phi_2(t)| \\ &\leq |\Phi_1(t) - \Phi_3(t)| + |\Phi_3(t) - \Phi_2(t)|\end{aligned}$$

and, in a similar manner, also $\Delta_2 \leq \Delta_1 + \Delta_3$ and $\Delta_3 \leq \Delta_1 + \Delta_2$.

Appendix B

Median, 0.95 and 0.99 Quantiles for the Statistic Λ

Table B.1 Median (0.5 quantile) for Λ for $p = 6$ ($n = 6, 7, 8, 9, 10$), $p = 35$ ($n = 36, 70, 105$) and $p = 55$ ($n = 56, 110, 165$), for the near-exact distributions that match $m^* = 2, 4, 6$ and 10 exact moments

p	n	m^*			
		2	4	6	10
6	6	0.0442675046	0.0442690653	0.0442690588	0.0442690587
	7	0.1253184924	0.1253201218	0.1253201241	0.1253201239
	8	0.2049594954	0.2049606106	0.2049606139	0.2049606138
	9	0.2758199003	0.2758206080	0.2758206103	0.2758206103
	10	0.3372011435	0.3372015913	0.3372015927	0.3372015927
35	36	$1.1508093035 \times 10^{-9}$	$1.1508099174 \times 10^{-9}$	$1.1508099171 \times 10^{-9}$	$1.1508099171 \times 10^{-9}$
	70	$3.0431000138 \times 10^{-3}$	$3.0431001934 \times 10^{-3}$	$3.0431001934 \times 10^{-3}$	$3.0431001934 \times 10^{-3}$
	105	$3.2440710998 \times 10^{-2}$	$3.2440711269 \times 10^{-2}$	$3.2440711269 \times 10^{-2}$	$3.2440711269 \times 10^{-2}$
55	56	$1.6957648495 \times 10^{-15}$	$1.6957651344 \times 10^{-15}$	$1.6957651343 \times 10^{-15}$	$1.6957651343 \times 10^{-15}$
	110	$1.0147689185 \times 10^{-4}$	$1.0147689497 \times 10^{-4}$	$1.0147689497 \times 10^{-4}$	$1.0147689497 \times 10^{-4}$
	165	$4.3246418891 \times 10^{-3}$	$4.3246419093 \times 10^{-3}$	$4.3246419093 \times 10^{-3}$	$4.3246419093 \times 10^{-3}$

Table B.2 0.95 quantiles for Λ for $p = 6$ ($n = 6, 7, 8, 9, 10$), $p = 35$ ($n = 36, 70, 105$) and $p = 55$ ($n = 56, 110, 165$), for the near-exact distributions that match $m^* = 2, 4, 6$ and 10 exact moments

p	n	m^*			
		2	4	6	10
6	6	0.2218883749	0.2218730281	0.2218729933	0.2218729961
	7	0.3567674618	0.3567569763	0.3567569398	0.3567569405
	8	0.4523223836	0.4523162838	0.4523162649	0.4523162651
	9	0.5234538071	0.5234502656	0.5234502565	0.5234502565
	10	0.5783756332	0.5783735154	0.5783735109	0.5783735109
35	36	$1.0986112877 \times 10^{-8}$	$1.0986102245 \times 10^{-8}$	$1.0986102251 \times 10^{-8}$	$1.0986102251 \times 10^{-8}$
	70	$5.2227076255 \times 10^{-3}$	$5.2227070673 \times 10^{-3}$	$5.2227070674 \times 10^{-3}$	$5.2227070674 \times 10^{-3}$
	105	$4.4573504331 \times 10^{-2}$	$4.4573503657 \times 10^{-2}$	$4.4573503657 \times 10^{-2}$	$4.4573503657 \times 10^{-2}$
55	56	$2.0797150437 \times 10^{-14}$	$2.0797144198 \times 10^{-14}$	$2.0797144200 \times 10^{-14}$	$2.0797144200 \times 10^{-14}$
	110	$1.7572667193 \times 10^{-4}$	$1.7572666234 \times 10^{-4}$	$1.7572666234 \times 10^{-4}$	$1.7572666234 \times 10^{-4}$
	165	$5.9739940469 \times 10^{-3}$	$5.9739939974 \times 10^{-3}$	$5.9739939974 \times 10^{-3}$	$5.9739939974 \times 10^{-3}$

Table B.3 0.99 quantiles for Λ for $p = 6$ ($n = 6, 7, 8, 9, 10$), $p = 35$ ($n = 36, 70, 105$) and $p = 55$ ($n = 56, 110, 165$), for the near-exact distributions that match $m^* = 2, 4, 6$ and 10 exact moments

p	n	m^*			
		2	4	6	10
6	6	0.3427103413	0.3426810423	0.3426805707	0.3426805677
	7	0.4779913382	0.4779750313	0.4779748328	0.4779748314
	8	0.5657922174	0.5657835776	0.5657834997	0.5657834993
	9	0.6280128232	0.6280080804	0.6280080479	0.6280080477
	10	0.6745339034	0.6745311729	0.6745311583	0.6745311582
35	36	$2.5926832547 \times 10^{-8}$	$2.5926767139 \times 10^{-8}$	$2.5926767140 \times 10^{-8}$	$2.5926767140 \times 10^{-8}$
	70	$6.4646154910 \times 10^{-3}$	$6.4646139426 \times 10^{-3}$	$6.4646139425 \times 10^{-3}$	$6.4646139425 \times 10^{-3}$
	105	$5.0538271373 \times 10^{-2}$	$5.0538269683 \times 10^{-2}$	$5.0538269683 \times 10^{-2}$	$5.0538269683 \times 10^{-2}$
55	56	$5.5013242831 \times 10^{-14}$	$5.5013199029 \times 10^{-14}$	$5.5013199029 \times 10^{-14}$	$5.5013199029 \times 10^{-14}$
	110	$2.1914231332 \times 10^{-4}$	$2.1914228517 \times 10^{-4}$	$2.1914228517 \times 10^{-4}$	$2.1914228517 \times 10^{-4}$
	165	$6.8032460452 \times 10^{-3}$	$6.8032459133 \times 10^{-3}$	$6.8032459133 \times 10^{-3}$	$6.8032459133 \times 10^{-3}$

References

1. Anderson, T.W.: An Introduction to Multivariate Statistical Analysis, 3rd edn. Wiley, New York (2003)
2. Berry, A.: The accuracy of the Gaussian approximation to the sum of independent variates. Trans. Am. Math. Soc. **49**, 122–136 (1941)
3. Brillinger, D.R.: Time Series: Data Analysis and Theory. SIAM, Philadelphia (2001)
4. Carter, E.M., Khatri, C.G., Srivastava, M.S.: Nonnull distribution of likelihood ratio criterion for reality of covariance matrix. J. Multivar. Anal. **6**, 176–184 (1976)
5. Coelho, C.A.: The generalized integer gamma distribution - a basis for distributions in multivariate statistics. J. Multivar. Anal. **64**, 86–102 (1998)
6. Coelho, C.A.: The generalized near-integer gamma distribution: a basis for "near-exact" approximations to the distribution of statistics which are the product of an odd number of independent beta random variables. J. Multivar. Anal. **89**, 191–218 (2004)
7. Coelho, C.A.: Near-exact distributions: What are they and why do we need them? In: Proceedings 59th ISI World Statistics Congress, 25–30 August 2013, Hong Kong (Session STS084), pp. 2879–2884 (2013)
8. Coelho, C.A., Marques, F.J.: Near-exact distributions for the likelihood ratio test statistic to test equality of several variance-covariance matrices in elliptically contoured distributions. Commun. Stat. Theory Methods **27**, 627–659 (2012)
9. Esseen, C.-G.: Fourier analysis of distribution functions. A mathematical study of the Laplace-Gaussian Law. Acta Mathematica **77**, 1–125 (1945)
10. Gil-Pelaez, J.: Note on the inversion theorem. Biometrika **38**, 481–482 (1951)
11. Goodman, N.R.: On the joint estimation of the spectra, cospectrum and quadrature spectrum of a two-dimensional stationary Gaussian process. Scientific Paper No. 10, Engineering Statistics Laboratory, New York University/Ph.D. Dissertation, Princeton University (1957)
12. Goodman, N.R.: Statistical analysis based on a certain multivariate complex Gaussian distribution (an introduction). Ann. Math. Stat. **34**, 152–177 (1963)
13. Grilo, L.M.: Development of near-exact distributions for different scenarios of application of the Wilks Lambda statistic (in Portuguese). Ph.D. thesis, Lisbon University of Technology, Lisbon (2005)
14. Grilo, L.M., Coelho, C.A.: Development and comparative study of two near exact approximations to the distribution of the product of an odd number of independent beta random variables. J. Stat. Plan. Inference **137**, 1560–1575 (2007)
15. Grilo, L.M., Coelho, C.A.: Near-exact distributions for the generalized Wilks Lambda statistic. Discuss. Math. Probab. Stat. **30**, 53–86 (2010)
16. Grilo, L.M., Coelho, C.A.: The exact and near-exact distribution for the Wilks Lambda statistic used in the test of independence of two sets of variables. Am. J. Math. Manag. Sci. **30**, 111–140 (2010)
17. Grilo, L.M., Coelho, C.A.: A family of near-exact distributions based on truncations of the exact distribution for the generalized Wilks Lambda statistic. Commun. Stat. Theory Methods **41**, 2321–2341 (2012)
18. Grilo, L.M., Coelho, C.A.: Near-exact distributions for the likelihood ratio statistic used to test the reality of a covariance matrix. AIP Conf. Proc. **1558**, 797–800 (2013)
19. Hwang, H.-K.: On convergence rates in the central limit theorems for combinatorial structures. Eur. J. Comb. **19**, 329–343 (1998)
20. James, A.T.: Distributions of matrix variates and latent roots derived from normal samples. Ann. Math. Stat. **35**, 475–501 (1964)
21. Johnson, N.L., Kotz, S., Balakrishnan, N.: Continuous Univariate Distributions, vol. 2. Wiley, New York (1995)
22. Khatri, C.G.: Classical statistical analysis based on a certain multivariate complex Gaussian distribution. Ann. Math. Stat. **36**, 98–114 (1965)
23. Khatri, C.G.: A test for reality of a covariance matrix in a certain complex Gaussian distribution. Ann. Math. Stat. **36**, 115–119 (1965)

24. Marques, F.J., Coelho, C.A., Arnold, B.C.: A general near-exact distribution theory for the most common likelihood ratio test statistics used in multivariate analysis. TEST **20**, 180–203 (2011)
25. Tricomi, F.G., Erdélyi, A.: The asymptotic expansion of a ratio of gamma functions. Pac. J. Math. **1**, 133–142 (1951)
26. Wald, A., Brookner, R.J.: On the distribution of Wilk's statistic for testing the independence of several groups of variates. Ann. Math. Stat. **12**, 137–152 (1941)
27. Wooding, R.A.: The multivariate distribution of complex normal variables. Biometrika **43**, 212–215 (1956)

Variance Components Estimation in Mixed Linear Model—The Sub-diagonalization Method

A. Silva, M. Fonseca and J. Mexia

Abstract This work aims to introduce a new method of estimating the variance components in mixed linear models. The approach will be done firstly for models with 3 variances components and secondly attention will be devoted to general case of models with an arbitrary number of variance components. In our approach, we construct and apply a finite sequence of orthogonal matrices to the mixed linear model variance-covariance structure in order to produce a set of Gauss–Markov sub-models which will be used to create pooled estimators for the variance components. Numerical results will be given, comparing the performance of our proposed estimator to the one based on likelihood procedure.

Keywords Mixed linear model · Variance components · Orthogonal matrices · Simultaneous diagonalization

1 Introduction

Mixed linear models (*MLM*) arise due to the necessity of assessing the amount of variation caused by certain sources in a statistical designs with fixed effects (see Khuri [7]), for example, the amount of variations that are not controlled by the experimenters and those whose levels are selected at random. The variances of such sources of variation, currently refereed to as variance components, has been widely investigated in the last fifty years of the last century (see Khuri and Sahai [8], Searle [13, 14], among others) and during the period ranging somewhat from early 1960

A. Silva (✉)
UniCV, Praia, Cabo Verde
e-mail: adilson.dasilva@docente.unicv.edu.cv; ad.silva@campus.fct.unl.pt

A. Silva · M. Fonseca · J. Mexia
UNL, Lisbon, Portugal
e-mail: fmig@fct.unl.pt

J. Mexia
e-mail: jtm@fct.unl.pt

N. Bebiano (ed.), *Applied and Computational Matrix Analysis*,
Springer Proceedings in Mathematics & Statistics 192,
DOI 10.1007/978-3-319-49984-0_21

to 1990, due to the proliferation of investigation on genetic and animal breeding as well as industrial quality control and improvement (for more details, see Anderson [1–3], Anderson and Crump [4], Searle [13], among others), several techniques of estimation have been proposed. Among those techniques we highlight the ANOVA and the maximum likelihood - based methods (see, for example, Searle et al. [15] and Casella and Berger [5]). Nevertheless, notwithstanding the ANOVA method adapt readily to mixed models with balanced data and save the unbiasedness, it does not adapt in situation with unbalanced data (mostly because it use computations derived from fixed effect models rather than mixed models). On its turn, the maximum likelihood - based methods, highlighting the ML and the restricted ML (REML) methods, provide estimators with several statistical optimal properties such as consistency and asymptotic normality either for models with balanced data, or for those with unbalanced data. For these optimal properties we recommend Miller [9], and for some details on applications of such methods we recommend, for example, Anderson [2] and Hartley and Rao [6].

This paper is organized as follows. In Sect. 2 (notation and basic concepts on matrix theory) we review some needed notions and results on matrix theory, mainly on matrix diagonalization. A new method to estimate the variance components in the *MLM* is summarized in Sect. 3, and numerical results ensuring their optimality will be available in Sect. 4.

2 Notation and Basic Concepts on Matrix Theory

In this section we summarize a few needed notions and results on matrix diagonalization. The proofs for the results can be found in Schott [12].

Let $\mathcal{M}^{n\times m}$ and $\mathcal{S}^n = \{A : A \in \mathcal{M}^{n\times n}, A = A^\top\}$ stands for the set of the matrices with n rows and m columns and the set of the $n \times n$ symmetric matrices, respectively. The *range* and the *rank* of a matrix A will be respectively denoted by $R(A)$ and $r(A)$, and the *projection matrix* onto the range space of A denoted by $P_{R(A)}$ (see Schott [12, Chap. 2, Sect. 7] for *projection matrix* notion). We will denote by $tr(A)$ the *trace* of A.

If the eigenvalues $\lambda_1, \ldots, \lambda_r$ of the matrix $M \in \mathcal{M}^{r\times r}$ are all distinct, it follows from the Theorem 3.6 of Schott [12] that the matrix X, whose columns are the eigenvectors associated to those eigenvalues, is non-singular. Thus, by the eigenvalue - eigenvector equation $MX = XD$ or, equivalently, $X^{-1}MX = D$, with $D = diag(\lambda_1 \ldots \lambda_r)$, and the Theorem 3.2.($d$) of Schott [12], the eigenvalues of D are the same as those of M. Meanwhile, since M can be transformed into a diagonal matrix by postmultiplication by the non-singular matrix X and premultiplication by its inverse X^{-1} it is said to be diagonalizable.

If the matrix M is symmetric we will have that the eigenvectors associated to its different eigenvalues will be orthogonal (see Schott [12]). Indeed, if we consider two different eigenvalues λ_i and λ_j whose associated eigenvectors are $\mathbf{x}_i$ and $\mathbf{x}_j$, respectively, we see that, since M is symmetric,

$$\lambda_i \mathbf{x}_i^\top \mathbf{x}_j = (M\mathbf{x}_i)^\top \mathbf{x}_j = \mathbf{x}_i^\top (M\mathbf{x}_j) = \lambda_j \mathbf{x}_i^\top \mathbf{x}_j.$$

So, since $\lambda_i \neq \lambda_j$, we must have $\mathbf{x}_i^\top \mathbf{x}_j = 0$.

According with Theorem 3.10 of Schott [12], without lost in generality, the columns of the matrix X can be taken to be orthonormal so that X is an orthogonal matrix. Thus, the eigenvalue - eigenvector equation can now be written as

$$X^\top M X = D \text{ or, equivalently, } M = XDX^\top,$$

which is known as spectral decomposition of M.

Definition 1 Let

$$A = \begin{bmatrix} A_{11} & \dots & A_{1n} \\ \vdots & \ddots & \vdots \\ A_{n1} & \dots & A_{nn} \end{bmatrix}$$

be a diagonal blockwise matrix. We say that a matrix T sub-diagonalizes A if the TA produces a blockwise matrix whose matrices in the diagonal are all diagonal matrices, that is T diagonalizes the matrices $A_{11}, \dots, A_{nn}$ in the diagonal of A.

3 Inference

Variance components estimation in linear models (with mixed and/or fixed effects) have been widely investigated and consequently several methods for estimation with important properties have been derived. Some of this methods are summarized in Searle et al. [15].

In this section we will sub-diagonalize the variance-covariance matrix

$$V = \sum_{d=1}^{r+1} \gamma_d N_d$$

in the Normal *MLM*

$$z \sim \mathcal{N}_m (X\beta,\ V), \tag{1}$$

with $\gamma_d > 0, d = 1, \dots, r$, unknown parameters, $N_d = X_d X_d^\top \in \mathcal{S}^m$, $X_d \in \mathcal{M}^{m\times s}$ known matrices, and $N_{r+1} = I_m$, and develop optimal estimators for the variance components $\gamma_1, \dots, \gamma_{r+1}$.

Since the components we want to estimate depends only on the random effect part, it is of our interest to remove the dependence of the distribution of z on the fixed effect part. With $P_o = P_{R(X)}$ denoting the projection matrix onto the column space of the matrix X, so that $I_m - P_o$ will be the projection matrix onto its orthogonal complement, there is a matrix B_o whose columns are the eigenvectors associated to the null eigenvalues of P_o such that

$$B_o^\top B_o = I_{m-r(P_o)} \text{ and } B_o B_o^\top = I_m - P_o.$$

Thus, instead of the model (1) we will approach the restricted model:

$$y = B_o^\top z \sim \mathcal{N}_n\left(\mathbf{0}_n, \sum_{d=1}^{r+1} \gamma_d M_d\right), \tag{2}$$

with $M_d = B_o^\top N_d B_o$, $n = m - r(P_o)$, and $\mathbf{0}_n$ denotes an $n \times 1$ vector of zeros; that is, we will diagonalize the variance-covariance matrix

$$V^* = \sum_{d=1}^{r+1} \gamma_d M_d$$

instead of V.

3.1 The Case $r = 2$

In this subsection we will sub-diagonalize the variance-covariance matrix in the *MLM* for $r = 2$ (recall the general model in (2)), that is

$$y \sim \mathcal{N}_n\left(\mathbf{0}_n,\ \gamma_1 M_1 + \gamma_2 M_2 + \gamma_3 I_n\right). \tag{3}$$

There exists (see Schott [12, Chap. 4, Sects. 3 and 4]) an orthogonal matrix $P_1 = \begin{bmatrix} A_{11} \\ \vdots \\ A_{1h_1} \end{bmatrix} \in \mathcal{M}^{\left(\sum_{i=1}^{h_1} g_i\right)\times n}$, with $A_{1i} \in \mathcal{M}^{g_i \times n}$ ($\sum_{i=1}^{h_1} g_i = n$), such that $M_1 = P_1^\top D_1 P_1$, or equivalently $P_1 M_1 P_1^\top = D_1$, where

$$D_1 = \begin{bmatrix} \theta_{11} I_{g_1} & 0 & \dots & 0 \\ 0 & \theta_{12} I_{g_2} & \dots & 0 \\ \vdots & \vdots & \ddots & \vdots \\ 0 & 0 & \dots & \theta_{1h_1} I_{g_{h_1}} \end{bmatrix} \tag{4}$$

is a diagonal matrix whose diagonal entries θ_{1i}, $i = 1, \dots, h_1$, are the eigenvalues of the matrix M_1 with corresponding roots $g_i = r(A_{1i}^\top)$, $i = 1, \dots, h_1$. It must be noted that the set of columns of each matrix $A_{1i}^\top$ forms a set of g_i orthonormal vectors associated to the eigenvalue θ_{1i} of the matrix M_1 (Theorem 3.10. of Schott [12] guarantees the existence of such matrix $A_{1i}^\top$), so that $A_{1i} A_{1i}^\top = I_{g_i}$ and $A_{1i}^\top A_{1i} =$

$P_{R(A_{1i}^\top)}$. Hence $P_1 P_1^\top = I_n$, and

$$\begin{aligned} P_1^\top P_1 &= A_{11}^\top A_{11} + \cdots + A_{1h_1}^\top A_{1h_1} \\ &= P_{R(A_{11}^\top)} + \cdots + P_{R(A_{1h_1}^\top)} \\ &= I_n. \end{aligned} \tag{5}$$

With

$$A_{1i} M_2 A_{1s}^\top = \begin{cases} M_{ii}^2 & i = s \\ W_{is}^2 & i \neq s \end{cases} \tag{6}$$

and $cov(v)$ denoting the variance-covariance matrix of a random vector v, we will have that

$$\begin{aligned} cov(P_1 y) &= \gamma_1 P_1 M_1 P_1^\top + \gamma_2 P_1 M_2 P_1^\top + \gamma_3 P_1 P_1^\top \\ &= \gamma_1 \begin{bmatrix} \theta_{11} I_{g_1} & 0 & \dots & 0 \\ 0 & \theta_{12} I_{g_2} & \dots & 0 \\ \vdots & \vdots & \ddots & \vdots \\ 0 & 0 & \dots & \theta_{1h_1} I_{g_{h_1}} \end{bmatrix} + \gamma_2 \begin{bmatrix} M_{11}^2 & W_{12}^2 & \dots & W_{1h_1}^2 \\ W_{21}^2 & M_{22}^2 & \dots & W_{2h_1}^2 \\ \vdots & \vdots & \ddots & \vdots \\ W_{h_1 1}^2 & W_{h_1 2}^2 & \dots & M_{h_1 h_1}^2 \end{bmatrix} \\ &\quad + \gamma_3 \begin{bmatrix} I_{g_1} & 0 & \dots & 0 \\ 0 & I_{g_2} & \dots & 0 \\ \vdots & \vdots & \ddots & \vdots \\ 0 & 0 & \dots & I_{g_{h_1}} \end{bmatrix} \\ &= \gamma_1 D(\theta_1 I_{g_1} \dots \theta_{h1} I_{g_{h_1}}) + \gamma_2 \Gamma + \gamma_3 D(I_{g_1} \dots I_{g_{h_1}}), \end{aligned} \tag{7}$$

where

$$\Gamma = \begin{bmatrix} M_{11}^2 & W_{12}^2 & \dots & W_{1h_1}^2 \\ W_{21}^2 & M_{22}^2 & \dots & W_{2h_1}^2 \\ \vdots & \vdots & \ddots & \vdots \\ W_{h_1 1}^2 & W_{h_1 2}^2 & \dots & M_{h_1 h_1}^2 \end{bmatrix}.$$

It is clear that for the three matrices $D(\theta_1 I_{g_1} \dots \theta_{h1} I_{g_{h_1}})$, $D(I_{g_1} \dots I_{g_{h_1}})$ and Γ appearing in (7), the blockwise matrix Γ is the only one which is not a diagonal matrix.

Next we diagonalize the symmetric matrices M_{ii}^2, $i = 1, \dots, h_1$, that appear in the diagonal of the matrix Γ, i.e, we sub-diagonalize the matrix Γ.

Since M_{ii}^2 is symmetric there exists (see Schott [12, Chap. 4, Sects. 3 and 4]) an orthogonal matrix $P_{2i} = \begin{bmatrix} A_{2i1} \\ \vdots \\ A_{2ih_{2i}} \end{bmatrix} \in \mathcal{M}^{\left(\sum_{j=1}^{h_{2i}} g_{ij}\right) \times g_i}$, where $A_{2ij} \in \mathcal{M}^{g_{ij} \times g_i}$ $(\sum_{j=1}^{h_{2i}} g_{ij} = g_i)$, such that

$$D_{ii}^2 = P_{2i} M_{ii}^2 P_{2i}^\top = \begin{bmatrix} \theta_{2i1} I_{g_{i1}} & 0 & \dots & 0 \\ 0 & \theta_{2i2} I_{g_{i2}} & \dots & 0 \\ \vdots & \vdots & \ddots & \vdots \\ 0 & 0 & \dots & \theta_{2ih_{2i}} I_{g_{ih_{2i}}} \end{bmatrix}, \; i = 1, \dots, h_1. \quad (8)$$

It must be noted that the matrix $A_{2ij}^\top, i = 1, \dots, h_1, j = 1, \dots, h_{2i}$, is an orthogonal matrix whose columns form a set of $g_{ij} = r(A_{2ij}^\top)$ orthonormal eigenvectors associated to the eigenvalue θ_{2ij} of the matrix M_{ii}^2; that is, g_{ij} is the multiplicity of the eigenvalues θ_{2ij}, and $A_{2ij}^T A_{2ij} = P_{R\left(A_{2ij}^T\right)}$ and $A_{2ij} A_{2ij}^T = I_{g_{ij}}$.

Thus, with

$$P_2 = \begin{bmatrix} P_{21} & 0 & \dots & 0 \\ 0 & P_{22} & \dots & 0 \\ \vdots & \vdots & \ddots & \vdots \\ 0 & 0 & \dots & P_{2h_1} \end{bmatrix} \in \mathscr{M}^{\left(\sum_{i=1}^{h_1} \sum_{j=1}^{h_{2i}} g_{ij}\right) \times \left(\sum_{i=1}^{h_1} g_i\right)},$$

the new model $w_2 = P_2 P_1 y$ will have variance-covariance matrix

$$\begin{aligned} cov(w_2) = \Sigma(P_2 P_1 y) &= \gamma_1 P_2 D(\theta_{11} I_{g_1} \dots \theta_{1h1} I_{g_{h_1}}) P_2^\top + \gamma_2 P_2 \Gamma P_2^\top + \gamma_3 P_2 D(I_{g_1} \dots I_{g_{h_1}}) P_2^\top \\ &= \gamma_1 \begin{bmatrix} \theta_{11} P_{21} P_{21}^\top & 0 & \dots & 0 \\ 0 & \theta_{12} P_{22} P_{22}^\top & \dots & 0 \\ \vdots & \vdots & \ddots & \vdots \\ 0 & 0 & \dots & \theta_{1h_1} P_{2h_1} P_{2h_1}^\top \end{bmatrix} \\ &+ \gamma_2 \begin{bmatrix} D_{11}^2 & P_{21} W_{12}^2 P_{22}^\top & \dots & P_{21} W_{1h_1}^2 P_{2h_1}^\top \\ P_{22} W_{21}^2 P_{21}^\top & D_{22}^2 & \dots & P_{22} W_{2h_1}^2 P_{2h_1}^\top \\ \vdots & \vdots & \ddots & \vdots \\ P_{2h_1} W_{h_1 1}^2 P_{21}^\top & P_{2h_1} W_{h_1 2}^2 P_{22}^\top & \dots & D_{h_1 h_1}^2 \end{bmatrix} \\ &+ \gamma_3 \begin{bmatrix} P_{21} P_{21}^\top & 0 & \dots & 0 \\ 0 & P_{22} P_{22}^\top & \dots & 0 \\ \vdots & \vdots & \ddots & \vdots \\ 0 & 0 & \dots & P_{2h_1} P_{2h_1}^\top \end{bmatrix}, \end{aligned} \quad (9)$$

where

$$P_{2i} P_{2i}^\top = \begin{bmatrix} A_{2i1} A_{2i1}^\top & 0 & \dots & 0 \\ 0 & A_{2i2} A_{2i2}^\top & \dots & 0 \\ \vdots & \vdots & \ddots & \vdots \\ 0 & 0 & \dots & A_{2ih_{2i}} A_{2ih_{2i}}^\top \end{bmatrix} = \begin{bmatrix} I_{g_{i1}} & 0 & \dots & 0 \\ 0 & I_{g_{i2}} & \dots & 0 \\ \vdots & \vdots & \ddots & \vdots \\ 0 & 0 & \dots & I_{g_{ih_{2i}}} \end{bmatrix},$$

and, with $i \neq s$,

$$P_{2i}W_{is}^2P_{2s}^\top = \begin{bmatrix} A_{2i1}W_{is}^2A_{2s1}^\top & A_{2i1}W_{is}^2A_{2s2}^\top & \cdots & A_{2i1}W_{is}^2A_{2sh_{2s}}^\top \\ A_{2i2}W_{is}^2A_{2s1}^\top & A_{2i2}W_{is}^2A_{2s2}^\top & \cdots & A_{2i2}W_{is}^2A_{2sh_{2s}}^\top \\ \vdots & \vdots & \ddots & \vdots \\ A_{2ih_{2i}}W_{is}^2A_{2s1}^\top & A_{2ih_{2i}}W_{is}^2A_{2s2}^\top & \cdots & A_{2ih_{2i}}W_{is}^2A_{2sh_{2s}}^\top \end{bmatrix}.$$

The matrix $D_{ii}^2 = P_{2i}M_{ii}^2P_{2i}^\top, i = 1, \ldots, h_1$, appearing in the diagonal at the right side of (9) is defined in (8).

Note that

$$w_2 = P_2P_1y = \begin{bmatrix} A_{211}A_{11}y \\ \vdots \\ A_{21h_{21}}A_{11}y \\ A_{221}A_{12}y \\ \vdots \\ A_{22h_{22}}A_{12}y \\ \vdots \\ \vdots \\ A_{2h_11}A_{1h_1}y \\ \vdots \\ A_{2h_1h_{2h_1}}A_{1h_1}y \end{bmatrix}.$$

The distribution of the sub-models

$$y_{ij} = A_{2ij}A_{1i}y, \ i = 1, \ldots, h_1, \ j = 1, \ldots, h_{2i}$$

is summarized in the following result.

Proposition 1

$$y_{ij} \sim \mathcal{N}_{g_{ij}}\left(\boldsymbol{0}_{g_{ij}}, \ \lambda_{ij}I_{g_{ij}}\right), \ i = 1, \ldots, h_1; \ j = 1, \ldots, h_{2i},$$

where $\lambda_{ij} = \gamma_1\theta_{1i} + \gamma_2\theta_{2ij} + \gamma_3$.

Proof Recalling that $A_{2ij}A_{1i} \in \mathcal{M}^{g_{ij}\times n}$ and $g_{ij} \leq n$, according with Moser [10, Theorem 2.1.2] we will have that

$$y_{ij} \sim \mathcal{N}_{g_{ij}}\left(\boldsymbol{0}_{g_{ij}}, \sum_{d=1}^{2}\gamma_d A_{2ij}A_{1i}M_dA_{1i}^\top A_{2ij}^\top + \gamma_3A_{2ij}A_{1i}A_{1i}^\top A_{2ij}^\top\right).$$

The portions $\sum_{d=1}^{2}\gamma_d A_{2ij}A_{1i}M_dA_{1i}^\top A_{2ij}^\top$ and $\gamma_3A_{2ij}A_{1i}A_{1i}^\top A_{2ij}^\top$ in the variance-covariance matrix yield:

$$\sum_{d=1}^{2} \gamma_d A_{2ij} A_{1i} M_d A_{1i}^\top A_{2ij}^\top = \gamma_1 A_{2ij} \left(\theta_{1i} I_{g_i}\right) A_{2ij}^\top + \gamma_2 A_{2ij} M_{ii}^2 A_{2ij}^\top$$
$$= \gamma_1 \theta_{1i} I_{g_{ij}} + \gamma_2 \theta_{2ij} I_{g_{ij}};$$

and

$$\gamma_3 A_{2ij} A_{1i} A_{1i}^\top A_{2ij}^\top = \gamma_3 A_{2ij} I_{g_i} A_{2ij}^\top = \gamma_3 I_{g_{ij}}$$

which, clearly, completes the proof. □

With $\mathbf{0}$ denoting an adequate null matrix and $cov(\nu, \upsilon)$ denoting the cross-covariance between the random vectors ν and υ, from (9) one might note that the cross-covariance matrix between the sub-models $y_{ij} = A_{2ij} A_i y$ and $y_{sk} = A_{2sk} A_s y$, $i, s = 1, \ldots, h_1$, $j, k = 1, \ldots, h_{2i}$ is given by

$$cov(y_{ij},\ y_{sk}) = \gamma_2 A_{2ij} A_{1i} M_2 A_{1s}^\top A_{2sk}^\top = \begin{cases} \mathbf{0} & i = s;\ j \neq k \\ \lambda_{ij} & i = s;\ j = k \\ \gamma_2 A_{2ij} W_{is}^2 A_{2sk}^\top & i \neq s \end{cases} \quad (10)$$

with $i \leq s$, $j \leq k$ (symmetry applies), so that, for $i \neq s$, the sub-models y_{ij} and y_{sk} are correlated and for $i = s$ they are not.

3.2 *Estimation for r* = 2

From the Sect. 3.1 we see that (with i and j respectively replaced by i_1 and i_2, for convenience) $w_2 = P_2 P_1 y$ produces the following sub-models

$$y_{i_1 i_2} \sim \mathcal{N}_{g_{i_1 i_2}}(\mathbf{0}_{g_{i_1 i_2}},\ \lambda_{i_1 i_2} I_{g_{i_1 i_2}}),\ i_1 = 1, \ldots, h_1,\ i_2 = 1, \ldots, h_{2i_1}, \quad (11)$$

of the model $y \sim \mathcal{N}_n(\mathbf{0}_n,\ \gamma_1 M_1 + \gamma_2 M_2 + \gamma_3 I_n)$, where

$$\lambda_{i_1 i_2} = \gamma_1 \theta_{1i_1} + \gamma_2 \theta_{2i_1 i_2} + \gamma_3.$$

An unbiased estimator of $\lambda_{i_1 i_2}$ for model (11) is (one based on its maximum likelihood estimator $\hat{\lambda}_{i_1 i_2}$)

$$S_{i_1 i_2}^2 = \frac{y_{i_1 i_2}^\top y_{i_1 i_2}}{g_{i_1 i_2}},$$
$$i_1 = 1, \ldots, h_1,\ i_2 = 1, \ldots, h_{2i_1}.$$

Indeed (see Rencher and Schaalje [11, Theorem 5.2a]),

$$E(S^2_{i_1i_2}) = \frac{1}{g_{i_1i_2}} tr\left\{\lambda_{i_1i_2} I_{g_{i_1i_2}}\right\}$$
$$= \lambda_{i_1i_2}. \tag{12}$$

Thus

$$E(S^2_{i_1i_2}) = \lambda_{i_1i_2} = \gamma_1\theta_{1i_1} + \gamma_2\theta_{2i_1i_2} + \gamma_3, \quad i_1 = 1, \ldots, h_1, \; i_2 = 1, \ldots, h_{2i_1}$$

so that, with $S = \begin{bmatrix} S^2_{11} \\ \cdots \\ S^2_{1h_{21}} \\ S^2_{21} \\ \cdots \\ S^2_{2h_{22}} \\ \cdots \\ \cdots \\ S^2_{h_11} \\ \cdots \\ S^2_{h_1h_{2h_1}} \end{bmatrix}$, $\Theta = \begin{bmatrix} \theta_{11} & \theta_{211} & 1 \\ \cdots & \cdots & \cdots \\ \theta_{11} & \theta_{21h_{21}} & 1 \\ \theta_{12} & \theta_{221} & 1 \\ \cdots & \cdots & \cdots \\ \theta_{12} & \theta_{22h_{22}} & 1 \\ \cdots & \cdots & \cdots \\ \cdots & \cdots & \cdots \\ \theta_{1h_1} & \theta_{2h_11} & 1 \\ \cdots & \cdots & \cdots \\ \theta_{1h_1} & \theta_{2h_1h_{2h_1}} & 1 \end{bmatrix}$, and $\gamma = \begin{bmatrix} \gamma_1 \\ \gamma_2 \\ \gamma_3 \end{bmatrix}$, we will have

$$E(S) = \Theta\gamma. \tag{13}$$

Thus, for $i_1 = 1, \ldots, h_1, \; i_2 = 1, \ldots, h_{2i_1}$, equalizing the variances $\lambda_{i_1i_2}$ to the correspondent estimators $S^2_{i_1i_2}$ it yields the following system of equations:

$$\begin{aligned}
S^2_{11} &= \gamma_1\theta_{11} + \gamma_2\theta_{211} + \gamma_3; \\
\cdots & \quad \cdots\cdots\cdots\cdots\cdots; \\
S^2_{1h_{21}} &= \gamma_1\theta_{11} + \gamma_2\theta_{21h_{21}} + \gamma_3; \\
S^2_{21} &= \gamma_1\theta_{12} + \gamma_2\theta_{221} + \gamma_3; \\
\cdots & \quad \cdots\cdots\cdots\cdots\cdots \\
S^2_{2h_{22}} &= \gamma_1\theta_{12} + \gamma_2\theta_{22h_{22}} + \gamma_3; \\
\cdots & \quad \cdots\cdots\cdots\cdots\cdots; \\
\cdots & \quad \cdots\cdots\cdots\cdots\cdots; \\
S^2_{h_11} &= \gamma_1\theta_{1h_1} + \gamma_2\theta_{2h_11} + \gamma_3; \\
\cdots & \quad \cdots\cdots\cdots\cdots\cdots; \\
S^2_{h_1h_{2h_1}} &= \gamma_1\theta_{1h_1} + \gamma_2\theta_{2h_1h_{2h_1}} + \gamma_3;
\end{aligned}$$

which in matrix notation becomes

$$S = \Theta\gamma. \tag{14}$$

Since by construction $\theta_{1i_1} \neq \theta_{1i_1'}$, $i_1 \neq i_1' = 1, \ldots, h_1$ (they are the different eigenvalues of M_1) and $\theta_{2i_1i_2} \neq \theta_{2i_1i_2'}$, $i_2 \neq i_2' = 1, \ldots, h_{2i_1}$ (they are the distinct eigenvalues of $M_{ii}^2 = A_{1i_1} M_2 A_{1i_1}^\top$), it is easily seen that the matrix Θ is a full rank one; that is $r(\Theta) = 3$.

By Rencher and Schaalje [11, Theorem 2.6d] the matrix

$$\Theta^\top \Theta = \begin{bmatrix} \sum_{i_1}^{h_1} \sum_{i_2}^{h_{2i_1}} \theta_{1i_1}^2 & \sum_{i_1}^{h_1} \sum_{i_2}^{h_{2i_1}} \theta_{1i_1} \theta_{2i_1i_2} & \sum_{i_1}^{h_1} \sum_{i_2}^{h_{2i_1}} \theta_{1i_1} \\ \sum_{i_1}^{h_1} \sum_{i_2}^{h_{2i_1}} \theta_{1i_1} \theta_{2i_1i_2} & \sum_{i_1}^{h_1} \sum_{i_2}^{h_{2i_1}} \theta_{2i_1i_2}^2 & \sum_{i_1}^{h_1} \sum_{i_2}^{h_{2i_1}} \theta_{2i_1i_2} \\ \sum_{i_1}^{h_1} \sum_{i_2}^{h_{2i_1}} \theta_{1i_1} & \sum_{i_1}^{h_1} \sum_{i_2}^{h_{2i_1}} \theta_{2i_1i_2} & \sum_{i_1}^{h_1} \sum_{i_2}^{h_{2i_1}} \end{bmatrix}$$

is positive-definite, and by Rencher and Schaalje [11, Corollary 1], $\Theta^\top \Theta$ is non-singular; we, thus, take its inverse to be $(\Theta^\top \Theta)^{-1}$.

Now, premultiplying the system (14) in both side by $\Theta^\top$ the resulting system of equations will be

$$\Theta^\top S = \Theta^\top \Theta \gamma, \tag{15}$$

whose unique solution (and therefore an estimator of γ) is

$$\hat{\gamma} = (\Theta^\top \Theta)^{-1} \Theta^\top S. \tag{16}$$

$\hat{\gamma} = \begin{bmatrix} \hat{\gamma}_1 \\ \hat{\gamma}_2 \\ \hat{\gamma}_3 \end{bmatrix}$ will be referred to as *Sub-D estimator* and the underlying method referred to as *Sub-D method.*

Proposition 2 *$\hat{\gamma}$ is an unbiased estimator of γ, with $\gamma = \begin{bmatrix} \gamma_1 \\ \gamma_2 \\ \gamma_3 \end{bmatrix}$.*

Proof Indeed, $E(\hat{\gamma}) = E\left((\Theta^\top \Theta)^{-1} \Theta^\top S\right) = (\Theta^\top \Theta)^{-1} \Theta^\top E(S) = (\Theta^\top \Theta)^{-1}$ $\Theta^\top \Theta \gamma = \gamma$. □

Proposition 3 *With $i \leq i^*$, $j \leq j^*$ (symmetry applies),*

$$cov\left(S_{ij}^2,\ S_{i^*j^*}^2\right) = \begin{cases} (\boldsymbol{a})\ i = i^*;\ j \neq j^*: & 0, \\ (\boldsymbol{b})\ i = i^*;\ j = j^*: & 2\frac{\lambda_{ij}^2}{g_{ij}}, \\ (\boldsymbol{c})\ i \neq i^*: & 2\gamma_2^2 tr(\Omega M_2), \end{cases}$$

where $\Omega = \nabla_{ij} M_2 \nabla_{i^*j^*}$, with $\nabla_{ij} = \frac{A_{1i}^\top A_{2ij}^\top A_{2ij} A_{1i}}{g_{ij}}$.

Proof We have that

$$
\begin{aligned}
cov\left(S_{ij}^2, S_{i^*j^*}^2\right) &= cov\left(\frac{y_{ij}^\top y_{ij}}{g_{ij}}, \frac{y_{i^*j^*}^\top y_{i^*j^*}}{g_{i^*j^*}}\right) \\
&= cov\left(y^\top\left(\frac{A_{1i}^\top A_{2ij}^\top A_{2ij}A_{1i}}{g_{ij}}\right)y,\ y^\top\left(\frac{A_{1i^*}^\top A_{2i^*j^*}^\top A_{2i^*j^*}A_{1i^*}}{g_{i^*j^*}}\right)y\right) \\
&= cov\left(y^\top \nabla_{ij} y,\ y^\top \nabla_{i^*j^*} y\right) \\
&= 2tr\left(\nabla_{ij} V \nabla_{i^*j^*} V\right) \\
&= 2\gamma_1^2 tr(\nabla_{ij}M_1\nabla_{i^*j^*}M_1) + 2\gamma_1\gamma_2 tr(\nabla_{ij}M_1\nabla_{i^*j^*}M_2) + 2\gamma_1\gamma_3 tr(\nabla_{ij}M_1\nabla_{i^*j^*}) \\
&+ 2\gamma_2\gamma_1 tr(\nabla_{ij}M_2\nabla_{i^*j^*}M_1) + 2\gamma_2^2 tr(\nabla_{ij}M_2\nabla_{i^*j^*}M_2) + 2\gamma_2\gamma_3 tr(\nabla_{ij}M_2\nabla_{i^*j^*}) \\
&+ 2\gamma_3\gamma_1 tr(\nabla_{ij}\nabla_{i^*j^*}M_1) + 2\gamma_3\gamma_2 tr(\nabla_{ij}\nabla_{i^*j^*}M_2) + 2\gamma_3^2 tr(\nabla_{ij}\nabla_{i^*j^*}) \\
&= \begin{cases} i = i^*;\ j \neq j^*: & 0, \\ i = i^*;\ j = j^*: & 2\frac{\lambda_{ij}^2}{g_{ij}}, \\ i \neq i^*: & 2\gamma_2^2 tr(\nabla_{ij}M_2\nabla_{i^*j^*}M_2). \end{cases}
\end{aligned}
$$

For the case (**a**), that is $i = i^*;\ j \neq j^*$, we have that

$$
\begin{aligned}
\nabla_{ij}M_1\nabla_{ij^*} &= \frac{1}{g_{ij}g_{ij^*}} A_{1i}^\top A_{2ij}^\top A_{2ij}A_{1i}M_1A_{1i}^\top A_{2ij^*}^\top A_{2ij^*}A_{1i} \\
&= \frac{1}{g_{ij}g_{ij^*}} A_{1i}^\top A_{2ij}^\top A_{2ij}\left(\theta_{1i}I_{g_i}\right)A_{2ij^*}^\top A_{2ij^*}A_{1i} \\
&= \mathbf{0}_{g_i\times g_i} \text{ (see (4) for the explanation);} \qquad (17)
\end{aligned}
$$

$$
\begin{aligned}
\nabla_{ij}M_2\nabla_{ij^*} &= \frac{1}{g_{ij}g_{ij^*}} A_{1i}^\top A_{2ij}^\top A_{2ij}A_{1i}M_2A_{1i}^\top A_{2ij^*}^\top A_{2ij^*}A_{1i} \\
&= \frac{1}{g_{ij}g_{ij^*}} A_{1i}^\top A_{2ij}^\top A_{2ij}\left(M_{ii}^2\right)A_{2ij^*}^\top A_{2ij^*}A_{1i} \\
&= \mathbf{0}_{g_i\times g_i} \text{ (see (8) for the explanation);} \qquad (18)
\end{aligned}
$$

$$
\begin{aligned}
\nabla_{ij}\nabla_{ij^*} &= \frac{1}{g_{ij}g_{ij^*}} A_{1i}^\top A_{2ij}^\top \left(\mathbf{0}_{g_{ii}\times g_{ij^*}}\right) A_{2ij^*}A_{1i} \\
&= \mathbf{0}_{g_i\times g_i}. \qquad (19)
\end{aligned}
$$

Therefore, (17)–(19) together with Schott [12, Theorem 1.3.(d)] proves the case (**a**).

For the case (**c**), that is $i \neq i^*$, the desired result becomes clear if use the Theorem 1.3.(d) of Schott [12] and note that

$$A_{1i}M_1A_{1i^*} = A_{1i}A_{1i^*} = \mathbf{0}_{g_i\times g_{i^*}}.$$

Finally, for the case (**b**), that is $i = i^*$; $j = j^*$, recalling $y_{ij} \sim \mathcal{N}_n\left(\mathbf{0}_{g_{ij}},\ \lambda_{ij} I_{g_{ij}}\right)$, it holds

$$cov\left(S^2_{ij}\right) = \Sigma\left(\frac{y_{ij}^\top y_{ij}}{g_{ij}}, \frac{y_{ij}^\top y_{ij}}{g_{ij}}\right) = 2tr\left\{\frac{\lambda_{ij}}{g_{ij}} I_{g_{ij}} \frac{\lambda_{ij}}{g_{ij}} I_{g_{ij}}\right\} = 2\frac{\lambda^2_{ij}}{g^2_{ij}} tr\left\{I_{g_{ij}}\right\}$$
$$= 2\frac{\lambda^2_{ij}}{g_{ij}}, \tag{20}$$

and therefore the proof is complete. □

The next result introduce the variance-covariance matrix of the sub-diagonalization estimator:

$$\hat{\gamma} = (\Theta^\top \Theta)^{-1} \Theta^\top S.$$

Proposition 4 *In order to simplify the notation, let* $\Sigma_{S_{ij} S_{kl}}$ *denote* $cov(S^2_{ij},\ S^2_{kl})$. *Then,*

$$cov(\hat{\gamma}) = (\Theta^\top \Theta)^{-1} \Theta^\top cov(S) \Theta (\Theta^\top \Theta)^{-1}, \tag{21}$$

where $cov(S) = \begin{bmatrix} D_1 & \Lambda_{12} & \Lambda_{13} & \dots & \Lambda_{1h_1} \\ \Lambda_{21} & D_2 & \Lambda_{23} & \dots & \Lambda_{2h_1} \\ \Lambda_{31} & \Lambda_{32} & D_3 & \dots & \Lambda_{3h_1} \\ \vdots & \vdots & \vdots & \ddots & \vdots \\ \Lambda_{h_1 1} & \Lambda_{h_1 2} & \Lambda_{h_1 3} & \dots & D_{h_1} \end{bmatrix}$, *with* $D_i = 2 \begin{bmatrix} \frac{\lambda^2_{i1}}{g_{i1}} & 0 & \dots & 0 \\ 0 & \frac{\lambda^2_{i2}}{g_{i2}} & \dots & 0 \\ \vdots & \vdots & \ddots & \vdots \\ 0 & 0 & \dots & \frac{\lambda^2_{ih_{2i}}}{g_{ih_{2i}}} \end{bmatrix}$ *and*

$$\Lambda_{ks} = \begin{bmatrix} \Sigma_{S_{k1} S_{s1}} & \Sigma_{S_{k1} S_{s2}} & \dots & \Sigma_{S_{k1} S_{sh_{2s}}} \\ \Sigma_{S_{k2} S_{s1}} & \Sigma_{S_{k2} S_{s2}} & \dots & \Sigma_{S_{k2} S_{sh_{2s}}} \\ \vdots & \vdots & \ddots & \vdots \\ \Sigma_{S_{kh_{2k}} S_{s1}} & \Sigma_{S_{kh_{2k}} S_{s2}} & \dots & \Sigma_{S_{kh_{2k}} S_{sh_{2s}}} \end{bmatrix}.$$

Proof The proof is a consequence of the Proposition 3. □

3.3 *The General Case: r ≥ 1*

Now, without lost in generality, lets consider the general *MLM* in (2):

$$y \sim \mathcal{N}_n\left(\mathbf{0}_n, \sum_{d=1}^{r+1} \gamma_d M_d\right), \text{ with } M_d = X_d X_d^\top \in \mathcal{S}^n \text{ and } M_{r+1} = I_n.$$

One may note that $y = \sum_{d=1}^{r+1} B_o^\top X_d \beta_d$, where $\beta_d \sim \mathcal{N}(0,\ \gamma_d I)$, $d = 1, \ldots, r$, $\beta_{r+1} \sim \mathcal{N}(0,\ \gamma_d I_n)$, and $\beta_1, \ldots, \beta_{r+1}$ are not correlated.

With $i_1 = 1, \ldots, h_1$, $i_j = 1, \ldots, h_{j,i_1,\ldots,i_{j-1}}$, consider the finite sequence of r matrices P_1, P_2, …, P_r defined as follow:

$$P_1 = \begin{bmatrix} A_{11} \\ A_{12} \\ \vdots \\ A_{1h_1} \end{bmatrix} \in \mathcal{M}^{\left(\sum_{i_1}^{h_1} g_{i_1}\right)\times n}, \text{ with } A_{1i_1} \in \mathcal{M}^{(g_{i_1})\times n} \left(\text{note: } \sum_{i_1}^{h_1} g_{i_1} = n\right); \quad (22)$$

$$P_2 = \begin{bmatrix} P_{21} & 0 & \ldots & 0 \\ 0 & P_{22} & \ldots & 0 \\ \vdots & \vdots & \ddots & \vdots \\ 0 & 0 & \ldots & P_{2h_1} \end{bmatrix} \in \mathcal{M}^{\left(\sum_{i_1}^{h_1} \sum_{i_2}^{h_{2,i_1}} g_{i_1 i_2}\right)\times\left(\sum_{i_1}^{h_1} g_{i_1}\right)}, \text{ where}$$

$$P_{2i_1} = \begin{bmatrix} A_{2i_1 1} \\ A_{2i_1 2} \\ \vdots \\ A_{2i_1 h_{2i_1}} \end{bmatrix} \in \mathcal{M}^{\left(\sum_{i_2}^{h_{2,i_1}} g_{i_1 i_2}\right)\times g_{i_1}}, \text{ with } \sum_{i_2}^{h_{2,i_1}} g_{i_1 i_2} = g_{i_1} \text{ and } A_{2i_1 i_2} \in \mathcal{M}^{g_{i_1 i_2}\times g_{i_1}};$$

$$P_3 = \begin{bmatrix} P_{31} & 0 & \ldots & 0 \\ 0 & P_{32} & \ldots & 0 \\ \vdots & \vdots & \ddots & \vdots \\ 0 & 0 & \ldots & P_{3h_1} \end{bmatrix} \in \mathcal{M}^{\left(\sum_{i_1}^{h_1} \sum_{i_2}^{h_{2,i_1}} \sum_{i_3}^{h_{3,i_1,i_2}} g_{i_1 i_2 i_3}\right)\times\left(\sum_{i_1}^{h_1} \sum_{i_2}^{h_{2,i_1}} g_{i_1 i_2}\right)},$$

$$\text{where } P_{3i_1} = \begin{bmatrix} P_{3i_1 1} & 0 & \ldots & 0 \\ 0 & P_{3i_1 2} & \ldots & 0 \\ \vdots & \vdots & \ddots & \vdots \\ 0 & 0 & \ldots & P_{3i_1 h_{2,i_1}} \end{bmatrix} \in \mathcal{M}^{\left(\sum_{i_2}^{h_{2,i_1}} \sum_{i_3}^{h_{3,i_1,i_2}} g_{i_1 i_2 i_3}\right)\times\left(\sum_{i_2}^{h_{2,i_1}} g_{i_1 i_2}\right)} \text{ and}$$

$$P_{3i_1 i_2} = \begin{bmatrix} A_{3i_1 i_2 1} \\ A_{3i_1 i_2 2} \\ \vdots \\ A_{3i_1 i_2 h_{3,i_1,i_2}} \end{bmatrix} \in \mathcal{M}^{\left(\sum_{i_3}^{h_{3,i_1,i_2}} g_{i_1 i_2 i_3}\right)\times g_{i_1 i_2}}, \text{ with } \sum_{i_3}^{h_{3,i_1,i_2}} g_{i_1 i_2 i_3} = g_{i_1 i_2} \text{ and}$$

$$A_{3i_1 i_2 i_3} \in \mathcal{M}^{g_{i_1 i_2 i_3}\times g_{i_1 i_2}};$$

Thus, for $r \geq 2$, each matrix P_r will be given by (P_1 is given in (22)):

$$P_r = \begin{bmatrix} P_{r1} & 0 & \dots & 0 \\ 0 & P_{r2} & \dots & 0 \\ \vdots & \vdots & \ddots & \vdots \\ 0 & 0 & \dots & P_{rh_1} \end{bmatrix} \in \mathscr{M}^{\left(\sum_{i_1}^{h_1} \cdots \sum_{i_r}^{h_{r,i_1,\dots,i_{r-1}}} g_{i_1 \dots i_r}\right) \times \left(\sum_{i_1}^{h_1} \cdots \sum_{i_{(r-1)}}^{h_{(r-1),i_1,\dots,i_{r-2}}} g_{i_1 \dots i_{(r-1)}}\right)}, \tag{23}$$

where

$$P_{ri_1} = \begin{bmatrix} P_{ri_11} & 0 & \dots & 0 \\ 0 & P_{ri_12} & \dots & 0 \\ \vdots & \vdots & \ddots & \vdots \\ 0 & 0 & \dots & P_{ri_1h_{2,i_1}} \end{bmatrix} \in \mathscr{M}^{\left(\sum_{i_2}^{h_{2,i_1}} \cdots \sum_{i_r}^{h_{r,i_1,\dots,i_{r-1}}} g_{i_1 \dots i_r}\right) \times \left(\sum_{i_2}^{h_{2,i_1}} \cdots \sum_{i_{(r-1)}}^{h_{(r-1),i_1,\dots,i_{r-2}}} g_{i_1 \dots i_{(r-1)}}\right)},$$

$$\dots \dots \dots \dots \dots$$

$$P_{ri_1 \dots i_{(r-2)}} = \begin{bmatrix} P_{ri_1 \dots i_{(r-2)}1} & 0 & \dots & 0 \\ 0 & P_{ri_1 \dots i_{(r-2)}2} & \dots & 0 \\ \vdots & \vdots & \ddots & \vdots \\ 0 & 0 & \dots & P_{ri_1 \dots i_{(r-2)}h_{r-1,i_1,\dots,i_{r-2}}} \end{bmatrix} \in \mathscr{M}^{\left(\sum_{i_{(r-1)}}^{h_{(r-1),i_1,\dots,i_{r-2}}} \sum_{i_r}^{h_{r,i_1,\dots,i_{r-1}}} g_{i_1 \dots i_r}\right) \times \left(\sum_{i_{(r-1)}}^{h_{(r-1),i_1,\dots,i_{r-2}}} g_{i_1 \dots i_{(r-1)}}\right)},$$

$$\text{and } P_{ri_1 \dots i_{(r-1)}} = \begin{bmatrix} A_{ri_1 \dots i_{(r-1)}1} \\ A_{ri_1 \dots i_{(r-1)}2} \\ \vdots \\ A_{ri_1 \dots i_{(r-1)}h_{r,i_1,\dots,i_{r-1}}} \end{bmatrix} \in \mathscr{M}^{\left(\sum_{i_r}^{h_{r,i_1,\dots,i_{r-1}}} g_{i_1 \dots i_r}\right) \times g_{i_1 \dots i_{(r-1)}}},$$

$$\text{with } \sum_{i_r}^{h_{r,i_1,\dots,i_{r-1}}} g_{i_1 \dots i_r} = g_{i_1 \dots i_{(r-1)}}, \sum_{i_1}^{h_1} g_{i_1} = n, \; A_{ri_1 \dots i_r} \in \mathscr{M}^{g_{i_1 \dots i_r} \times g_{i_1 \dots i_{(r-1)}}};$$

Theorem 1 *Let the matrices $P_1, P_2, \dots, P_r$ defined above be such that:*

(c_1) *The columns of $A_{1i_1}^\top$, $i_1 = 1, \dots, h_1$, form a set of $g_{i_1} = r(A_{1i_1}^\top)$ orthonormal eigenvectors associated to the eigenvalues θ_{1i_1} of the matrix M_1 (θ_{1i_1} has multiplicity g_{i_1});*

(c_2) *The columns of* $A^{\top}_{2i_1i_2}$, $i_2 = 1, \ldots, h_{2,i_1}$, *form a set of* $g_{i_1i_2} = r(A^{\top}_{2i_1i_2})$ *orthonormal eigenvectors associated to the eigenvalues* $\theta_{2i_1i_2}$ *of the matrix* $M^2_{i_1i_1} = A_{1i_1}M_2A^{\top}_{1i_1}$ ($\theta_{2i_1i_2}$ *has multiplicity* $g_{i_1i_2}$);

(c_3) *The columns of* $A^{\top}_{3i_1i_2i_3}$, $i_3 = 1, \ldots, h_{3,i_1,i_2}$, *form a set of* $g_{i_1i_2i_3} = r(A^{\top}_{3i_1i_2i_3})$ *orthonormal eigenvectors associated to the eigenvalues* $\theta_{3i_1i_2i_3}$ *of the matrix*

$$A_{2i_1i_2}M^3_{i_1i_1}A^{\top}_{2i_1i_2} = A_{2i_1i_2}A_{1i_1}M_3A^{\top}_{1i_1}A_{2i_1i_2}$$

($\theta_{3i_1i_2i_3}$ *has multiplicity* $g_{i_1i_2i_3}$);

....

(c_r) *The columns of* $A^{\top}_{ri_1\ldots i_r}$, $i_r = 1, \ldots, h_{r,i_1,\ldots,i_{r-1}}$, *form a set of* $g_{i_1\ldots i_r} = r(A^{\top}_{ri_1\ldots i_r})$ *orthonormal eigenvectors associated to the eigenvalues* $\theta_{ri_1\ldots i_r}$ *of the matrix*

$$A_{(r-1)i_1\ldots i_{(r-1)}} \cdots A_{1i_1}M_rA^{\top}_{1i_1} \cdots A^{\top}_{(r-1)i_1\ldots i_{(r-1)}}$$

($\theta_{ri_1\ldots i_r}$ *has multiplicity* $g_{i_1\ldots i_r}$).

Then each matrix P_d, $d = 1, \ldots, r$, in the finite sequence of matrices $P_1, P_2, \ldots, Pr$ will be an orthogonal matrix.

Proof By the way P_d is defined (see (23)), since

$$P_{di_1\ldots i_{(d-1)}} = \begin{bmatrix} A_{di_1\ldots i_{(d-1)}1} \\ A_{di_1\ldots i_{(d-1)}2} \\ \vdots \\ A_{di_1\ldots i_{(d-1)}h_{d,i_1,\ldots,i_{d-1}}} \end{bmatrix}, \ i_{(d-1)} = 1, \ldots, h_{(d-1),i_1,\ldots,i_{d-2}},$$

and according with condition c_d we see that the matrices $P_{di_1\ldots i_{(d-1)}}$ are orthogonal. Thus, the desired result comes if we see that $P_d^{\top}P_d$ will be a diagonal blockwise matrix whose diagonal entries are $P^{\top}_{di_1}P_{di_1}$, $i_1 = 1, \ldots, h_1$. The diagonal entries $P^{\top}_{di_1}P_{di_1}$ will be diagonal blockwise matrices whose diagonal entries will be $P^{\top}_{di_1i_2}P_{di_1i_2}$, $i_2 = 1, \ldots, h_{2,i_1}$. Proceeding this way $d-2$ times, we will find that the diagonal entries of the blockwise matrices $P^{\top}_{di_1\ldots i_{(d-2)}}P_{di_1\ldots i_{(d-2)}}$, $i_{(d-2)} = 1, \ldots, h_{(d-2),i_1,\ldots,i_{d-3}}$, will be

$$\begin{aligned} P^{\top}_{di_1\ldots i_{(d-1)}}P_{di_1\ldots i_{(d-1)}} &= A^{\top}_{di_1\ldots i_{(d-1)}1}A_{di_1\ldots i_{(d-1)}1} \\ &\quad + \cdots + A^{\top}_{di_1\ldots i_{(d-1)}h_{d,i_1,\ldots,i_{d-1}}}A_{di_1\ldots i_{(d-1)}h_{d,i_1,\ldots,i_{d-1}}} \\ &= I_{g_{i_1\ldots i_{(d-1)}}}, \end{aligned}$$

reaching, therefore, the desired result. Proceeding in same way we would also see that $P_{di_1\ldots i_{(d-1)}}P^{\top}_{di_1\ldots i_{(d-1)}}$ is a Blockwise diagonal matrix whose diagonal entries are $A_{di_1\ldots i_{(d-1)}1}A^{\top}_{di_1\ldots i_{(d-1)}j}$, $j = 1, \ldots, h_{d,i_1,\ldots,i_{d-1}}$, so that $P_dP_d^{\top}$ is an identity matrix. □

The model $w_r = P_r \ldots P_2 P_1 y$ will produces the following sub - models:

$$y_{i_1 \ldots i_r} = A_{r i_1 \ldots i_r} A_{(r-1) i_1 \ldots i_{(r-1)}} \cdots A_{2 i_1 i_2} A_{1 i_1} y,$$
$$i_1 = 1, \ldots, h_1, i_j = 1, \ldots, h_{j, i_1, \ldots, i_{j-1}}.$$

We summarize the distribution of each of the sub-model $y_{i_1 \ldots i_r}$ in the following result.

Proposition 5

$$y_{i_1 \ldots i_r} \sim \mathcal{N}_{g_{i_1 \ldots i_r}} \left(0_{g_{i_1 \ldots i_r}}, \ \lambda_{i_1 \ldots i_r} I_{g_{i_1 \ldots i_r}} \right),$$

where $\lambda_{i_1 \ldots i_r} = \sum_{d=1}^{r} \gamma_d \theta_{d i_1 \ldots i_d} + \gamma_{r+1}$.

Proof The proof becomes obvious after looking to the proofs of the Proposition 1. □

From the results about cross-covariance on the preceding sections we easily conclude that the cross-covariance matrix between the sub-models $y_{i_1 \ldots i_r}$ and $y_{i_1^* \ldots i_r^*}$, with $i_1, i_1^* = 1, \ldots, h_1; i_j, i_j^* = 1, \ldots, h_{j, i_1, \ldots, i_{j-1}}$, is given by

$$cov(y_{i_1 \ldots i_r}, \ y_{i_1^* \ldots i_r^*}) = \begin{cases} 0 & i_1 = i_1^*, \\ \lambda_{i_1 \ldots i_r} & i_j = i^* j \\ & j = 1, \ldots, h_{j, i_1, \ldots, i_{j-1}} \\ \sum_{d=2}^{r} \gamma_d A_{r i_1 \ldots i_r} \ldots A_{1i} M_d A_{1 i_1^*}^{\top} \ldots A_{r i_1^* \ldots i_r^*} & i_1 \neq i_1^* \end{cases}$$

so that, for $i_1 \neq i_1^*$, the sub-models $y_{i_1 \ldots i_r}$ and $y_{i_1^* \ldots i_r^*}$ are correlated and for $i_1 = i_1^*$ they are not.

3.4 Estimation for the General Case: $r \geq 1$

Recalling that for the *MLM* in (1), $P_r \ldots P_2 P_1 y$ produces the following sub-models

$$y_{i_1 i_2 \ldots i_r} \sim \mathcal{N}_{g_{i_1 i_2 \ldots i_r}} (0_{g_{i_1 \ldots i_r}}, \ \lambda_{i_1 i_2 \ldots i_r} I_{g_{i_1 i_2 \ldots i_r}}),$$
$$i_1 = 1, \ldots, h_1, i_j = 1, \ldots, h_{j, i_1, \ldots, i_{j-1}} \quad (24)$$

where

$$\lambda_{i_1 i_2 \ldots i_r} = \sum_{d=1}^{r} \gamma_d \theta_{d i_1 \ldots i_d} + \gamma_{r+1}.$$

The matrices P_d, $d = 1, \ldots, r$, are defined in the Sect. 3.3.

An unbiased estimator of $\lambda_{i_1 i_2 \ldots i_r}$ in the sub-model (24) is (the one based on its maximum likelihood estimator $\hat{\lambda}_{i_1 i_2 \ldots i_r}$)

$$S^2_{i_1 i_2 \ldots i_r} = \frac{1}{g_{i_1 i_2 \ldots i_r}} y^\top_{i_1 i_2 \ldots i_r} y_{i_1 i_2 \ldots i_r}$$

Indeed (see Rencher and Schaalje [11], Theorem 5.2(a), and the explanation for (12)),

$$\begin{aligned} E\left(S^2_{i_1 i_2 \ldots i_r}\right) &= \frac{\lambda_{i_1 i_2 \ldots i_r}}{g_{i_1 i_2 \ldots i_r}} tr\left[I_{g_{i_1 i_2 \ldots i_r}}\right] \\ &= \lambda_{i_1 i_2 \ldots i_r}. \end{aligned} \tag{25}$$

For convenience, in what follows, instead of $S^2_{i_1 i_2 \ldots i_r}$, we may sometimes use the notation $S^2_{i_1 i_2 \ldots i_{(r-1)} i_r}$.

Thus

$$\begin{aligned} E(S^2_{i_1 i_2 \ldots i_{(r-1)} i_r}) &= \sum_{d=1}^{r} \gamma_d \theta_{d i_1 \ldots i_d} + \gamma_{r+1} \\ &= \gamma_1 \theta_{1 i_1} + \gamma_2 \theta_{2 i_1 i_2} + \cdots + \gamma_r \theta_{r i_1 i_2 \ldots i_{(r-1)} i_r} + \gamma_{r+1}, \end{aligned}$$

$$i_1 = 1, \ldots, h_1; i_j = 1, \ldots, h_{j, i_1, \ldots, i_{j-1}}$$

so that, with $S = \begin{bmatrix} S^2_{11\ldots11} \\ S^2_{11\ldots12} \\ \ldots \\ S^2_{11\ldots1h_{r,1,\ldots,1}} \\ S^2_{11\ldots21} \\ \ldots \\ S^2_{11\ldots2h_{r,1,\ldots,2}} \\ \ldots \\ \ldots \\ \ldots \\ S^2_{h_1 1\ldots11} \\ \ldots \\ \ldots \\ \ldots \\ S^2_{h_1 h_{2,h_1} \ldots h_{r,h_1,\ldots,h_{r-1}}} \end{bmatrix}$,

$$\Theta = \begin{bmatrix} \theta_{11} & \theta_{211} & \theta_{3111} & \dots & \theta_{r11\dots11} & 1 \\ \theta_{11} & \theta_{211} & \theta_{3111} & \dots & \theta_{r11\dots12} & 1 \\ \dots & \dots & \dots & \dots & \dots & \dots \\ \theta_{11} & \theta_{211} & \theta_{3111} & \dots & \theta_{r11\dots1h_{r,1,\dots,1,h_{r-1}}} & 1 \\ \theta_{11} & \theta_{211} & \theta_{3111} & \dots & \theta_{r11\dots21} & 1 \\ \dots & \dots & \dots & \dots & \dots & \dots \\ \theta_{11} & \theta_{211} & \theta_{3111} & \dots & \theta_{r11\dots2h_{r,1,\dots,2,h_{r-1}}} & 1 \\ \dots & \dots & \dots & \dots & \dots & \dots \\ \dots & \dots & \dots & \dots & \dots & \dots \\ \dots & \dots & \dots & \dots & \dots & \dots \\ \theta_{1h_1} & \theta_{2h_11} & \theta_{3h_111} & \dots & \theta_{rh_11\dots11} & 1 \\ \dots & \dots & \dots & \dots & \dots & \dots \\ \dots & \dots & \dots & \dots & \dots & \dots \\ \dots & \dots & \dots & \dots & \dots & \dots \\ \theta_{1h_1} & \theta_{2h_1h_{2,h_1}} & \theta_{3h_1h_{2,h_1}h_{3,h_1,h_2}} & \dots & \theta_{rh_1h_{2,h_1}\dots h_{(r-1),h_1,\dots,h_{r-2}}h_{r,h_1,\dots,h_{r-1}}} & 1 \end{bmatrix},$$

and $\gamma = \begin{bmatrix} \gamma_1 \\ \gamma_2 \\ \gamma_3 \\ \dots \\ \dots \\ \gamma_r \\ \gamma_{(r+1)} \end{bmatrix}$, we will have

$$E(S) = \Theta\gamma. \tag{26}$$

Thus, for $i_1 = 1, \dots, h_1, i_j = 1, \dots, h_{j,i_1,\dots,i_{j-1}}, j > 1$, equalizing the variances $\lambda_{i_1i_2\dots i_r}$ to the correspondent estimators $S^2_{i_1i_2\dots i_r}$ it yields the following system of equations (in matrix notation)

$$S = \Theta\gamma. \tag{27}$$

Since by construction $\theta_{1i_1} \neq \theta_{1i_1'}$ (they are the different eigenvalues of M_1), $\theta_{2i_1i_2} \neq \theta_{2i_1i_2'}$ (they are the distinct eigenvalues of $M^2_{ii} = A_{1i_1}M_2A^\top_{1i_1}$), $\theta_{3i_1i_2i_3} \neq \theta_{3i_1i_2i_3'}$ (they are the distinct eigenvalues of $A_{2i_1i_2}A_{1i_1}M_2A^\top_{1i_1}A^\top_{2i_1i_2}$), ..., $\theta_{ri_1i_2\dots i_{(r-1)}i_r} \neq \theta_{ri_1i_2\dots i_{(r-1)}i_r'}$ (they are the distinct eigenvalues of $A_{(r-1)i_1i_2\dots i_{(r-1)}}\dots A_{1i_1}M_rA^\top_{1i_1}\dots A^\top_{(r-1)i_1i_2\dots i_{(r-1)}}$) where $i_j \neq i_j^{'}$, $j = 1, \dots, r$, it is easily seen that the matrix Θ is of full rank; that is $r(\Theta) = r + 1$.

According with Theorem 2.6*d* (Rencher and Schaalje [11]), with $\sum$ denoting $\sum_{i_1}^{h_1}\sum_{i_2}^{h_{2,i_1}}\dots\sum_{i_r}^{h_{r,i_1,\dots,i_{r-1}}}$, the matrix

$$\Theta^\top \Theta = \begin{bmatrix} \sum\theta_{1i_1}^2 & \sum\theta_{1i_1}\theta_{2i_1i_2} & \sum\theta_{1i_1}\theta_{3i_1i_2i_3} & \cdots & \sum\theta_{1i_1}\theta_{ri_1...ir} & \sum\theta_{1i_1} \\ \sum\theta_{1i_1}\theta_{2i_1i_2} & \sum\theta_{2i_1i_2}^2 & \theta_{2i_1i_2}\theta_{3i_1i_2i_3} & \cdots & \sum\theta_{2i_1i_2}\theta_{ri_1...ir} & \sum\theta_{2i_1i_2} \\ \sum\theta_{1i_1}\theta_{3i_1i_2i_3} & \sum\theta_{2i_1i_2}\theta_{3i_1i_2i_3} & \sum\theta_{3i_1i_2i_3}^2 & \cdots & \sum\theta_{3i_1i_2i_3}\theta_{ri_1...ir} & \sum\theta_{3i_1i_2i_3} \\ \vdots & \vdots & \vdots & \ddots & \vdots & \vdots \\ \sum\theta_{1i_1}\theta_{ri_1...ir} & \sum\theta_{2i_1i_2}\theta_{ri_1...ir} & \sum\theta_{3i_1i_2i_3}\theta_{ri_1...ir} & \cdots & \sum\theta_{ri_1...ir}^2 & \sum\theta_{ri_1...ir} \\ \sum\theta_{1i_1} & \sum\theta_{2i_1i_2} & \sum\theta_{3i_1i_2i_3} & \cdots & \sum\theta_{ri_1...ir} & \sum \end{bmatrix}$$

is positive-definite, and according with Corollary 1 of (Rencher and Schaalje [11], p. 27) $\Theta^\top\Theta$ is non-singular; that is, it is invertible. We denote its inverse by $(\Theta^\top\Theta)^{-1}$.

Now, premultiplying the system (27) in both side by $\Theta^\top$ the resulting system of equations will be

$$\Theta^\top S = \Theta^\top\Theta\gamma, \tag{28}$$

whose unique solution (and therefore an estimator of γ) will be the *Sub-D* estimator

$$\hat{\gamma} = (\Theta^\top\Theta)^{-1}\Theta^\top S. \tag{29}$$

Proposition 6 *$\hat{\gamma} = (\Theta^\top\Theta)^{-1}\Theta^\top S$ is an unbiased estimator of*

$$\gamma = \begin{bmatrix} \gamma_1 \\ \gamma_2 \\ \gamma_3 \\ \cdots \\ \cdots \\ \gamma_r \\ \gamma_{(r+1)} \end{bmatrix}, \text{ where } \begin{bmatrix} \hat{\gamma}_1 \\ \hat{\gamma}_2 \\ \hat{\gamma}_3 \\ \cdots \\ \cdots \\ \hat{\gamma}_r \\ \hat{\gamma}_{(r+1)} \end{bmatrix}.$$

Indeed, $E(\hat{\gamma}) = E\left((\Theta^\top\Theta)^{-1}\Theta^\top S\right) = (\Theta^\top\Theta)^{-1}\Theta^\top E(S) = (\Theta^\top\Theta)^{-1}\Theta^\top\Theta\gamma = \gamma$.

4 Numerical Results

In this section we carry numerical tests to the sub-diagonalization method for the case $r = 2$, that is for a model with 3 variances components. For this case we pick the particular model $z \sim \mathcal{N}_{21}\left(X\beta, \gamma_1 N_1 + \gamma_2 N_2 + \gamma_3 I_{21}\right)$, where $N_j = X_j X_j^\top$, $j = 1, 2$, with design matrices

$$X_1 = \begin{bmatrix} 1_5 & 0_5 & 0_5 \\ 0_9 & 1_9 & 0_9 \\ 0_7 & 0_7 & 1_7 \end{bmatrix}, X_2 = \begin{bmatrix} 1_2 & 0_2 & 0_2 \\ 0_4 & 1_4 & 0_4 \\ 0_8 & 0_8 & 1_8 \\ 1_4 & 0_4 & 0_4 \\ 0_3 & 1_3 & 0_3 \end{bmatrix},$$

and $X = 1_{21}$. 1_k and 0_k denote, respectively, $k \times 1$ vectors of 1 and 0.

Let B_o be a matrix whose columns are the eigenvectors associated to the null eigenvalues of $\frac{1}{21}J_{21}$. Then $B_o B_o^\top = I_{21} - \frac{1}{21}J_{21}$ and $B_o^\top B_o = I_{20}$, and so the new model will be

$$y = B_o^\top z \sim \mathcal{N}_{20}\left(\mathbf{0}_{20},\ \gamma_1 M_1 + \gamma_2 M_2 + \gamma_3 I_{20}\right),$$

where $M_d = B_o^\top N_d B_o$.

Since $r(N_1) = 3$ we have that (see Schott [12, Theorem 2.10c]) $r(M_1) = r(B_0^\top N_1 B_0) = 3$. The eigenvalues of M_1 are $\theta_{11} = 7.979829$, $\theta_{12} = 5.639219$, and $\theta_{13} = 0$ (θ_{13} with multiplicity (root) equal to 18). Thus we have that $M_{11}^2 = A_{11}M_2A_{11}^\top = 5.673759$ and $M_{22}^2 = A_{12}M_2A_{12}^\top = 0.6246537$ will be 1×1 matrices, and $M_{33}^2 = A_{13}M_2A_{13}^\top$ an 18×18 matrix.

We have the following: M_{11}^2 has eigenvalue $\theta_{211} = 5.673759$; M_{22}^2 has eigenvalue $\theta_{221} = 0.6246537$; M_{33}^2 has 3 eigenvalues: $\theta_{231} = 6.390202$; $\theta_{232} = 1.216148$; $\theta_{233} = 0$ (θ_{233} with multiplicity equal to 16).

Finally we found that

$$S^\top = [190.779246\ \ 8.866357\ \ 5.234293\ \ 53.654627\ \ 1.334877]$$

and $$\Theta = \begin{bmatrix} 7.979829 & 5.6737590 & 1 \\ 5.639219 & 0.6246537 & 1 \\ 0 & 6.3902016 & 1 \\ 0 & 1.2161476 & 1 \\ 0 & 0 & 1 \end{bmatrix}.$$

With $\beta_k \sim \mathcal{N}_{20}\left(\mathbf{0}_3, \gamma_k I_3\right)$, $k = 1, 2$, and $e \sim \mathcal{N}_{20}(\mathbf{0}_{20}, \gamma_3 I_{20})$, and taking $\gamma_3 = 1$, the model can be rewritten as $y = B_o^\top X_1\beta_1 + B_o^\top X_2\beta_2 + B_o^\top e$.

We consider γ_1 and γ_2 taking values in $\{0.1, 0.25, 0.5, 0.75, 1, 2, 5, 10\}$. Thus, for each possible combination of γ_1 and γ_2, the model y is observed 1000 time, and for each observation the sub-diagonalization method is applied and the variance components estimated for each observed y. The Tables 1 and 3 present the average of the estimated values of γ_1 and γ_2, respectively. In order to compare the sub-diagonalization method performance with the REML, for the same 1000 observations of y, the REML method is applied and the results presented in both Tables 2 and 4.

Taking a look at tables, and comparing the averages estimated values from the sub-diagonalization method to the ones of the REML methods (see Tables 1, 2, 3, and 4), the reader may easily concludes that the results provided by the sub-diagonalization method are in general slightly more realistic. In other hand, the averages variability of the sub-diagonalization methods is relatively higher than those of REML method (see Tables 5, 6, 7, and 8); this is because of the correlation between the sub-models. This gap will be fixed in future works.

5 Concluding Remarks

Besides its simple and fast computational implementation once it depends only on the information retained on the eigenvalues of the design matrices and the quadratic errors of the model, *Sub-D* provides centered estimates whether for balanced or unbalanced designs, which is not the case of estimators based on ANOVA methods. As seen at Sect. 4, *Sub-D* provides a slightly more realistic estimates than the REML estimator, but with more variability (when the model is balanced they have a comparable variability). However, since in any computational program (source code) when we are interested in share the code, create package or use it repeatedly, we might consider its efficiency and, for this matter, the code run-time constitutes a good start point. Doing so, to compute the estimates and the corresponding variance for each pair γ_1 and γ_2 taking values in $\{0.25, 0.5, 1, 2, 5, 10\}$, for 1000 observations of the model, we found that the *Sub-D* run-time is about 0.25 s while the REML estimator run-time is about 35.53 s, which means that the code for *Sub-D* is more than 70 times faster than the one for REML. The code was run using R software.

It seems that the problem of the little higher variability in *Sub-D* comparing to REML estimator is due to the correlation between the sub-models (for the case of models with three variance components, for example) y_{ij}, $i = 1, \ldots, h_1$, $j = 1, \ldots, h_{2h_i}$. From (10) we see that the variance components matrix of the model $w_2 = P_2 P_1 y$ is a blockwise matrix whose diagonal matrices are $D_1, \ldots, D_{h_1}$, where $D_i = diag(\lambda_{i1}, \ldots, \lambda_{ih_{2i}})$, corresponding to $cov(y_{ij}, y_{sk})$ for $i = s$, $j = k$, and the off diagonal matrices are the non-null matrices $\gamma_2 A_{2ij} W_{is}^2 A_{2sk}$, corresponding to $cov(y_{ij}, y_{sk})$ for $i \neq s$. This problem will be handled in future work. Confidence region will be obtained and tests of Hypothesis for the variance components will be derived in future works.

Acknowledgements This work was partially supported by the Fundaç ao para a Ciência e a Tecnologia (Portuguese Foundation for Science and Technology) through PEst-OE/MAT/UI0297/2011 (CMA), and by the Fundaç ao Calouste Gulbenkian Through a PhD Grants. It was also partially supported by Universidade de Cabo Verde.

Appendix

Table 1 Sub-diagonalization method - average estimate for γ_1

γ_1/γ_2	0.1	0.25	0.5	0.75	1	2	5	10
0.1	0.0917	0.0984	0.0828	0.1162	0.0833	0.1052	0.1102	0.1053
0.25	0.2716	0.2954	0.2698	0.2538	0.3041	0.2882	0.1993	0.3322
0.5	0.5010	0.5127	0.4929	0.5088	0.5297	0.4613	0.5314	0.5569
0.75	0.7279	0.7683	0.7685	0.7755	0.7693	0.7504	0.6982	0.8215
1	1.0305	1.0293	1.0143	0.9971	1.0309	1.0013	1.0046	1.0809
2	1.9844	2.0004	2.0032	1.9702	2.0827	2.0893	2.0643	2.2640
5	5.1864	5.0386	4.9128	5.0722	5.2111	5.0170	4.8472	5.1269
10	9.6167	10.1588	10.2468	10.1263	9.6940	9.9046	10.0246	9.8474

Table 2 REML method - average estimate for γ_1

γ_1/γ_2	0.1	0.25	0.5	0.75	1	2	5	10
0.1	0.1431	0.1683	0.1779	0.1884	0.1975	0.2154	0.2189	0.2156
0.25	0.2872	0.3157	0.3379	0.3286	0.3416	0.3316	0.3740	0.3480
0.5	0.5191	0.5546	0.5244	0.5637	0.6110	0.5897	0.6469	0.6281
0.75	0.7271	0.7620	0.7587	0.7908	0.8159	0.8245	0.8373	0.8241
1	1.0300	1.0026	1.0245	1.0172	1.0138	1.0726	1.0352	1.0515
2	1.9343	1.9884	1.9565	2.0178	2.1510	2.1482	2.0774	2.2323
5	5.1267	4.9747	4.7743	5.0955	5.1395	4.9907	4.8066	4.8150
10	9.5043	10.0881	10.1912	10.0269	9.4706	9.7784	9.9445	9.6754

Table 3 Sub-diagonalization method - average estimate for γ_2

γ_1/γ_2	0.1	0.25	0.5	0.75	1	2	5	10
0.1	0.1026	0.2643	0.5147	0.7147	1.0286	1.9595	4.9390	9.9718
0.25	0.1051	0.2589	0.4918	0.7827	1.0172	2.0427	4.8713	9.7690
0.5	0.0903	0.2323	0.5043	0.7865	1.0117	1.9496	4.8136	9.8913
0.75	0.0855	0.3068	0.5144	0.7676	1.1207	2.0762	4.7910	9.7847
1	0.0581	0.2746	0.5052	0.7969	1.0035	2.1009	5.0871	10.2702
2	0.0902	0.2966	0.6198	0.7870	0.9909	1.9605	5.217	9.7318
5	0.1759	0.3403	0.5565	0.7276	1.0007	2.036	4.8617	9.7160
10	0.1614	0.2562	0.5649	0.7481	0.9934	2.1402	5.1631	10.1369

Table 4 REML method - average estimate for γ_2

γ_1/γ_2	0.1	0.25	0.5	0.75	1	2	5	10
0.1	0.1539	0.2701	0.5143	0.7095	0.9992	1.9007	4.9153	9.9579
0.25	0.1630	0.2965	0.5165	0.7840	1.0271	2.0990	4.7929	9.5820
0.5	0.1867	0.3061	0.5490	0.7964	1.0400	1.9358	4.7022	9.6481
0.75	0.1976	0.3501	0.5480	0.8079	1.0678	2.1196	4.6759	9.7793
1	0.2008	0.3289	0.5488	0.8134	1.0282	2.0205	5.0126	10.3663
2	0.2186	0.3379	0.5703	0.8469	1.0249	1.9900	5.4291	9.5900
5	0.2198	0.3799	0.5603	0.7773	1.0027	2.0142	4.7727	9.6886
10	0.2284	0.3551	0.5906	0.7792	1.1087	2.0735	4.9235	10.0843

Table 5 Sub-diagonalization method - variation of the estimated γ_1

γ_1/γ_2	0.1	0.25	0.5	0.75	1	2	5	10
0.1	0.1264	0.2253	0.4626	0.8296	1.2005	4.3832	19.6631	83.6993
0.25	0.2637	0.3814	0.6248	1.0775	1.5931	4.7676	20.1332	72.7948
0.5	0.5737	0.7863	1.1830	1.7217	2.3142	4.7103	22.8545	78.2997
0.75	0.9224	1.2110	1.5779	2.0896	3.3078	7.4140	20.7793	77.7225
1	77.7225	1.8328	2.4022	2.9417	3.8380	7.6562	27.1356	101.9337
2	4.8401	5.6613	6.9492	6.8652	8.4356	13.2666	37.4524	107.8436
5	30.5767	31.3904	34.2362	36.0102	36.5273	43.1085	72.8085	157.0055
10	111.1505	117.9503	114.2234	120.8808	124.3445	138.0213	192.7288	288.9592

Table 6 Sub-diagonalization method - variation of the estimated γ_2

γ_1/γ_2	0.1	0.25	0.5	0.75	1	2	5	10
0.1	0.1532	0.2972	0.6524	1.1154	2.0379	6.4364	33.8728	138.7916
0.25	0.2379	0.4537	0.7838	1.3616	2.0686	7.7435	32.4170	112.701
0.5	0.5232	0.7162	1.1545	1.7515	2.7932	6.1609	31.2810	117.2392
0.75	0.7703	1.0841	1.4314	1.9380	3.3226	7.6266	35.7370	139.0834
1	1.1496	1.4291	1.8988	2.6630	3.6221	8.7960	39.6377	159.5489
2	3.8362	4.5207	4.6976	5.5365	6.9396	11.6933	47.5170	140.7587
5	21.0152	22.2408	24.2194	24.0984	29.4643	34.2175	65.9059	176.7041
10	81.3183	82.3035	89.9235	85.9040	85.1849	93.4313	153.1855	265.6179

Table 7 REML method - variation of the estimated γ_1

γ_1/γ_2	0.1	0.25	0.5	0.75	1	2	5	10
0.1	0.07807	0.0880	0.1324	0.1579	0.1801	0.2524	0.2679	0.2052
0.25	0.20365	0.2229	0.2729	0.2676	0.3350	0.3365	0.4485	0.3235
0.5	0.4747	0.6030	0.5822	0.7576	0.8165	0.7607	0.8321	0.9255
0.75	0.8896	0.9458	1.0035	1.1702	1.2667	1.2627	1.2131	1.4153
1	1.4500	1.4368	1.7622	1.7407	1.8813	1.9144	1.8597	1.9659
2	4.6049	4.9522	4.8249	5.6586	6.0638	6.3735	6.0565	7.8698
5	28.4367	29.6686	29.0413	32.1312	29.1439	28.4656	28.1731	29.3058
10	106.6903	108.3732	106.734	105.7222	106.4887	101.2775	111.1112	104.9005

Table 8 REML method - variation of the estimated γ_2

γ_1/γ_2	0.1	0.25	0.5	0.75	1	2	5	10
0.1	0.0833	0.1798	0.5192	0.7836	1.4306	4.8877	27.2749	100.2321
0.25	0.0914	0.2295	0.5842	0.9688	1.5517	6.1586	25.9314	92.9996
0.5	0.1260	0.2744	0.5607	1.2902	1.8142	4.4948	23.3488	94.9688
0.75	0.1534	0.3081	0.6120	1.2712	1.6747	5.9940	26.5791	110.6777
1	0.1732	0.3270	0.6852	1.2331	1.8197	5.2857	29.3231	126.1761
2	0.2289	0.3608	0.7416	1.5226	1.7834	5.7763	31.7812	101.8187
5	0.2399	0.4452	0.8946	1.2738	1.6384	5.2879	26.9691	97.7408
10	0.2280	0.4149	0.7789	1.2234	2.1941	5.7251	31.2616	98.4346

References

1. Anderson, R.L.: Use of variance component analysis in the interpretation of biological experiments. Bull. Int. Stat. Inst. **37**, 71–90 (1960)
2. Anderson, R.L.: Designs and estimators for variance components. Statistical Design and Linear Model, pp. 1–30. North-Holland, Amsterdam (1975)
3. Anderson, R.L.: Recent developments in designs and estimators for variance components. Statistics and Related Topics, pp. 3–22. North-Holland, Amsterdam (1981)
4. Anderson, R.L., Crump, P.P.: Comparisons of designs and estimation procedures for estimating parameters in a two-stages nested process. Tecnometrics **9**, 499–516 (1967)
5. Casella, G., Berger, R.L.: Statistical Inference. Duxbury Pacific Grove (2002)
6. Hartley, H.O., Rao, J.K.: Maximum likelihood estimation for the mixed analysis of variance model. Biometrika **54**, 93–108 (1967)
7. Khuri, A.I.: Design for variance components estimation: past and present. Int. Stat. Rev. **68**, 311–322 (2000)
8. Khuri, A.I., Sahai, H.: Variance components analysis: a selective literature survey. Int. Stat. Rev. **53**, 279–300 (1985)
9. Miller, J.J.: Asymptotic properties and computation of maximum likelihood estimates in the mixed model of the analysis of variance. Technical report 12, Department of Statistics, Stanford University, Stanford, California (1973)
10. Moser, B.: Linear Models: A Mean Model Approach. Elsevier, New York (1996)
11. Rencher, A.C., Schaalje, G.B.: Linear Models in Statisitcs. Wiley, New York (2008)

12. Schott, J.R.: Matrix Analysis for Statistics. Wiley, New York (1997)
13. Searle, S.: Topics in variance component estimation. Biometrics **27**, 1–76 (1971)
14. Searle, S.: An overview of variance component estimation. Metrika **42**, 215–230 (1995)
15. Searle, S., Casella, G., McCulloch, C.: Variance Components. Wiley, New York (2009)

Index

N. Bebiano (ed.), *Applied and Computational Matrix Analysis*,
Springer Proceedings in Mathematics & Statistics 192,
DOI 10.1007/978-3-319-49984-0

T

V

Z

MIX
Papier aus verantwortungsvollen Quellen
Paper from responsible sources
FSC® C105338

If you have any concerns about our products,
you can contact us on
ProductSafety@springernature.com

In case Publisher is established outside the EU,
the EU authorized representative is:
Springer Nature Customer Service Center GmbH
Europaplatz 3, 69115 Heidelberg, Germany

Printed by Libri Plureos GmbH
in Hamburg, Germany